# HEAT TRANSFER IN GAS–COOLED ANNULAR CHANNELS

# Experimental and Applied Heat Transfer Guide Books

## A. Žukauskas, *Editor*

*A. Žukauskas* and *J. Žiugžda*, Heat Transfer of a Cylinder in Crossflow
*J. Vilemas, B. Česna*, and *V. Survila*, Heat Transfer in Gas-Cooled Annular Channels
*M. Tamonis*, Radiation and Combined Heat Transfer in Channels

IN PREPARATION

*J. Stasiulevičius* and *A. Skrinska*, Heat Transfer of Bundles of Finned Tubes in Crossflow
*A. Žukauskas* and *A. Slančiauskas*, Heat Transfer in Turbulent Fluid Flows
*A. Žukauskas, V. Katinas*, and *R. Ulinskas*, Vibration and Fluid Dynamics of Tube Bundles in Crossflow
*A. Žukauskas* and *R. Ulinskas*, Heat Transfer of Tube Bundles in Crossflow

# HEAT TRANSFER IN GAS-COOLED ANNULAR CHANNELS

J. Vilemas
B. Česna
V. Survila

*Institute of Physical and Technical Problems of Energetics*
*Kaunas, Lithuanian SSR*

Edited by

## A. Žukauskas
*Academy of Sciences of the Lithuanian SSR, Vilnius*

English-edition editor

## J. Karni
*State University of New York*
*at Stony Brook, N.Y.*

**◉ HEMISPHERE PUBLISHING CORPORATION**
A subsidiary of Harper & Row, Publishers, Inc.
Washington    New York    London

**DISTRIBUTION OUTSIDE NORTH AMERICA**
**SPRINGER-VERLAG**
Berlin    Heidelberg    New York    London    Paris    Tokyo

**HEAT TRANSFER IN GAS–COOLED ANNULAR CHANNELS**

English translation copyright © 1987 by Hemisphere Publishing Corporation. All rights reserved. Printed in the United States of America. Except as permitted under the United States Copyright Act of 1976, no part of this publication may be reproduced or distributed in any form or by any means, or stored in a data base or retrieval system, without the prior written permission of the publisher.

1 2 3 4 5 6 7 8 9 0   BCBC   8 9 8 7 6

This book was set in Press Roman by The Sheridan Press

BookCrafters, Inc. was printer and binder.
Originally published by Mokslas, Vilnius, as Teplootdacha v gazookhlazhdaemykh kol'tsevykh kanalakh in the series Teplofizika, vol. 8

**Library of Congress Cataloging in Publication Data**

Vilemas, Jurgis, date
  Heat transfer in gas-cooled annular channels.

  (Experimental and applied heat transfer guide books)
  Translation of: Teplootdacha v gazookhlazhdaemykh kol'tsevykh kanalakh.
  Bibliography: p.
  Includes index.
  1. Heat—Transmission.  2. Gas flow.  3. Pipe—Fluid dynamics.  I. Chesna, B. (Benediktas)  II. Survila, V. (Viktoras)  III. Zhukauskas, A., date.
TJ260.V5313   1987     621.402′2      86-31840
ISBN 0-89116-364-6   Hemisphere Publishing Corporation

**DISTRIBUTION OUTSIDE NORTH AMERICA:**
ISBN 3-540-17553-9  Springer-Verlag  Berlin

# CONTENTS

|  |  |
|---|---|
| Preface | ix |
| Nomenclature | xi |

**1 INTRODUCTION** — 1

**2 EXPERIMENTAL TECHNIQUES** — 15
2.1 Experimental Units — 16
2.2 Test Sections — 19
2.3 Determination of Thermal and Hydrodynamic Parameters in the Boundary Layer — 32
2.4 Technique for Processing Data on Heat Transfer and Hydraulic Drag — 43
2.5 Calibration and Running of Experiments — 51

**3 VELOCITY AND TEMPERATURE DISTRIBUTIONS IN THE BOUNDARY LAYER OF A CYLINDER IN AXIAL FLOW** — 55
3.1 Velocity Profiles without Heat Transfer — 55
3.2 Distortion of Velocity and Temperature Fields in the Case of Variable Physical Properties of the Coolant — 65
3.3 Distortion of the Velocity Profile Due to Variability of Physical Properties of a Flow with Surface-Type Turbulence-Generating Mechanism — 73

## 4 SKIN FRICTION OF A CYLINDER IN AXIAL FLOW 77

4.1 Skin Friction without Heat Transfer 79
4.2 Effect of Temperature Factor on Friction 83
4.3 Effect of the Curvature of the Cylinder on Friction 87
4.4 Mean Skin Friction Coefficient 90

## 5 HEAT TRANSFER IN THE ENTRANCE REGION OF AN ANNULUS 91

5.1 Distribution of Heat Transfer Coefficients Along the Cylinder 91
5.2 Local Heat Transfer from a Cylinder 94
5.3 Effect of Temperature Factor on Heat Transfer from a Cylinder 99
5.4 Effect of the Curvature of the Cylinder Surface on Its Heat Transfer Characteristics 102
5.5 Heat Transfer with Enhanced Turbulence Intensity in the Boundary Layer and Free Stream 106

## 6 HEAT TRANSFER IN ANNULI UNDER FULLY DEVELOPED FLOW CONDITIONS 109

6.1 Local Heat Transfer from the Inner Tube of an Annulus Heated from One Side 110
6.2 Local Heat Transfer in an Externally Heated Annulus 119
6.3 Local Heat Transfer in an Annulus Heated from Both Sides 122
6.4 Correlation of Results on Heat Transfer in Annuli 129
6.5 Heat Transfer in an Annulus with an Obstacle 134
6.6 Hydraulic Drag at Large Temperature Differences 139

## 7 HEAT TRANSFER AND HYDRAULIC DRAG IN AN ANNULUS WITH A HELICALLY TWISTED INNER TUBE 141

7.1 Heat Transfer from a Helically Shaped Inner Tube of an Annulus 141
7.2 Hydraulic Drag of an Annulus with a Shaped, Helically Twisted Inner Tube 149

## 8 CONCLUSIONS AND PRACTICAL RECOMMENDATIONS 151

8.1 Features of the Boundary Layer on a Cylinder in Axial Flow 151
8.2 Effect of Physical Properties Variation on Velocity and Temperature Profiles 155
8.3 Drag and Heat Transfer of a Cylinder in Axial Flow 155
8.4 Heat Transfer and Hydraulic Drag in Hydrodynamically Fully Developed Turbulent Flow in Annuli 158
8.5 Conclusion 161

Appendixes

| | | |
|---|---|---|
| 1 | Experimental Data on Isothermal Velocity Profiles in the Boundary Layer of a Cylinder in Axial Flow | 163 |
| 2 | Experimental Data on Distortion of Velocity and Temperature Profiles in the Boundary Layer Due to Physical Properties Variations of the Coolant | 169 |
| 3 | Experimental Data on Friction of a Cylinder in Axial Air Flow | 173 |
| 4 | Experimental Data on Local Heat Transfer from a Cylinder with $d = 15.5$ and Blunt Leading Edge | 179 |
| 5 | Experimental Data on Local Heat Transfer from a Cylinder ($d = 4.02$ mm) with Significant Surface Curvature | 191 |
| 6 | Experimental Data on Heat Transfer in an Internally Heated Annulus | 197 |
| 7 | Experimental Data on Heat Transfer in an Externally Heated Annulus | 205 |
| 8 | Experimental Data on Heat Transfer in Annuli Heated from Both Sides | 213 |

References   215

Index   221

# PREFACE

This volume in the Experimental and Applied Heat Transfer Guide Books series is concerned with studies of turbulent heat transfer at large temperature differences, performed at the Institute of Physical and Technical Problems of Energetics of the Academy of Sciences of the Lithuanian SSR over a number of years.

The book presents data on local heat transfer and drag in gas-cooled annuli over wide ranges of Reynolds number and temperature. A great deal of attention is devoted to the study of velocity and temperature fields in the entrance region of the annulus. The effects of physical properties variation and of the surface curvature on heat transfer and drag are determined. The effect of freestream turbulence on heat transfer and drag is also examined. The results obtained over the entrance length of the annulus can also be used for calculating the heat transfer and drag characteristics of a single cylinder or plate in axial flow.

A large volume of data is presented on heat transfer in fully developed flow. The dependence of heat transfer on the ratio of diameters of the cylinders forming the annulus and on the ratio of wall to freestream temperature (temperature factor) in one- as well as in two-sided heating is investigated.

The studies were performed with annuli made up of concentric tubes, as well as with annular passages whose inner wall consisted of a specially shaped helical pipe.

A detailed description of experimental equipment and techniques is given. Some of the most characteristic experimental data are tabulated. The results are correlated in nondimensional form. Recommendations for calculating heat transfer and frictional drag at high temperature differences are given.

The authors wish to express their heartfelt thanks to Academician A. A. Žukauskas for his invaluable assistance in preparing the manuscript of this book, to M. A. Nemira, who actively participated in experiments with heat transfer in fully developed flow, and to all others who assisted in preparing the book for print.

*J. Vilemas*
*B. Česna*
*V. Survila*

# NOMENCLATURE

| | |
|---|---|
| $A, B$ | coefficients in Eq. (2.30) |
| $a$ | thermal diffusivity, m²/s |
| $c_f = \dfrac{2\tau_w}{\rho u_\infty^2}$ | skin friction coefficient |
| $c_p$ | specific heat, J/kg · K |
| $d$ | hydraulic diameter of annulus, diameter of cylinder, m |
| $d_1, d_2$ | diameters of the inner and outer cylinders forming the annulus, respectively, m |
| $F$ | heated surface, flow-passage surface, m² |
| $G$ | mass flow rate, kg/s |
| $H = \dfrac{\delta^*}{\delta^{**}}$ | shape factor |
| $\Delta h$ | dynamic pressure, Pa |
| $I$ | current, A |
| $K = \dfrac{\nu}{u_\infty^+} (du/dx)$ | acceleration parameter |
| $K' = \dfrac{4 q_w}{\overline{\rho u}\, c_p\, T_f} \dfrac{1}{\text{Re}}$ | acceleration parameter in heated flows |
| $K_f = \dfrac{Q^+}{\text{Nu}_{\Psi=1}}$ | heating-rate parameter |
| $l$ | mixing length, m |
| $L = L_0 + x$ | distance from inlet edge, m |
| $L_0$ | nonheated length of cylinder, m |
| $n_\rho, n_\lambda, n_\mu, n_c$ | power exponents in expressions for the temperature dependence of density, thermal conductivity, viscosity, and specific heat, as in Eq. (1.31) |
| $P$ | pressure, Pa |
| $\Delta p$ | pressure drop, Pa |

## xii NOMENCLATURE

| | |
|---|---|
| $Q$ | heat transfer rate, W |
| $Q^+ = \dfrac{qd}{\lambda T}$ | dimensionless heat flux parameter |
| $q$ | heat flux, W/m² |
| $q^+ = \dfrac{q}{\overline{\rho u}\, c_p\, T_f}$ | heat flux parameter |
| $r_0$ | radius of cylinder, m |
| $T$ | temperature, K |
| $Tu = \dfrac{\sqrt{\overline{u'^2}}}{u_\infty}$ | turbulence intensity |
| $u$ | axial velocity, m/s |
| $u_* = \sqrt{\tau_w/\rho}$ | friction velocity, m/s |
| $u^+ = \dfrac{u}{u_*}$ | nondimensional velocity |
| $U$ | potential drop along heated length, V |
| $x, y$ | Cartesian coordinates |
| $x$ | distance from start of heating, m |
| $x_0$ | virtual origin of turbulent boundary layer, m |
| $y^+ = \dfrac{u_* y}{\nu}$ | nondimensional coordinate |
| $\alpha$ | heat transfer coefficient, W/m² · K |
| $\delta$ | boundary layer thickness, m |
| $\delta^*$ | boundary layer displacement thickness, m |
| $\delta^{**}$ | boundary layer momentum thickness, m |
| $\epsilon$ | emissivity |
| $\xi$ | frictional drag coefficient |
| $\vartheta = T_w - T$ | temperature difference, K |
| $\vartheta_* = \dfrac{q_w}{\rho c_p u_*}$ | characteristic temperature, K |
| $\vartheta^+ = \dfrac{\vartheta}{\vartheta_*}$ | nondimensional temperature |
| $\kappa$ | universal constant |
| $\Psi = \dfrac{T_w}{T_f}$ | temperature factor |
| $\lambda$ | thermal conductivity, W/m · K |
| $\mu$ | dynamic viscosity, N · s/m² |
| $\nu$ | kinematic viscosity, m²/s |
| $\rho$ | density, kg/m³ |
| $\tau$ | shear stress, N/m² |
| $\omega_1, \omega_2$ | coefficients defined in Eqs. (6.21) and (6.22) |
| $\Omega = 1 + \dfrac{\delta}{r_0}$ | transverse curvature parameter |
| Nu | Nusselt number |
| Pr | Prandtl number |
| $\text{Pr}_t$ | turbulent Prandtl number |

| | |
|---|---|
| Re | Reynolds number |
| St | Stanton number |

**Subscripts**

| | |
|---|---|
| 1 | refers to inner wall of an annulus |
| 2 | refers to outer wall of an annulus |
| dc | describing circumference |
| $E$ | entrance |
| $f$ or $\infty$ | refers to freestream conditions |
| $fp$ | refers to a flat plate |
| $q$ | refers to heat transfer conditions |
| sh | shield |
| $w$ | refers to wall conditions |

**Superscripts**

| | |
|---|---|
| $(\bar{\phantom{x}})$ | average |
| ad | adiabatic |

CHAPTER
# ONE

## INTRODUCTION

Current developments in thermal energy, nuclear and missile power, chemical industries and other areas require further enhancement of heat transfer rates, design of efficient shapes of heated surfaces, and development of methods of calculation for them. This applies in particular to systems with high heat transfer rates employing gases as coolants. To ensure sufficiently high heat flux, one must design for a large temperature difference between the wall and the coolant, which inevitably leads to the problem of variability of physical properties. Gases have much lower heat transfer coefficients than liquids, and hence small errors in estimating these coefficients result in large differences between the actual and calculated wall temperatures. This may result in failure of heated surfaces, which, as a rule, are operated at the very limit of their high-temperature strength. This imparts great importance to accuracy in calculating heat transfer when working with gases as coolants.

Turbulent flow is the predominant flow mode both in nuclear reactors and in most other heat exchange devices. The fact that the governing time-averaged equations of turbulent motion do not form a complete set makes it necessary to use empirical transport parameters, which have not been sufficiently explored in the case of variable physical properties. Additional difficulties arise in solving the resultant differential equations, since these become nonlinear in the case of variable physical properties. Despite the substantial achievements in the analytic approach to the solution of these problems, experimental results still play the predominant role in working out recommendations for engineering calculation of heat transfer in turbulent flow of coolants with variable physical properties.

The incomplete set of equations for turbulent flow is usually solved with the aid of semiempirical theories, based on various models of turbulent transport, or by experimental determination of terms accounting for this transport. A detailed analysis of these problems is presented in the books by Monin and Yaglom [1], Hinze [2], Schlichting [3], and Ievlev [4].

Fuel-element bundles of nuclear reactors and the flow paths of many other heat exchanging devices usually consist of a system of longitudinal ducts having various cross-section shapes. A place of particular importance among this entire variety of cross sections is occupied by the annulus, which, in limiting cases, can be treated as a circular pipe or a flat slot. Moreover, as shown by Hall [5], Kjellström and Hedberg [6], Maubach and Rehme [7], and Dalle Donne and Meerwald [8], the annular geometry can be used as a basis in determining the heat transfer and flow behavior of bundles of cylindrical rods in axial flow.

Unlike the case of circular tubes, the study of heat transfer and flow patterns in annuli is complicated by two factors. First, annuli are not in general geometrically similar, and hence, in addition to the hydraulic diameter, one must use another controlling geometric parameter, which is usually the diameter ratio of the inner and outer tubes $(d_1/d_2)$. Second, the boundary conditions at the inner and outer surfaces may be different. Processes occurring at one surface are interrelated with those occurring at the second.

Extensive experimental studies of heat transfer to a turbulent flow in an annulus have been conducted, particularly in the case of one-sided heating. Surveys of earlier work, performed by Quirrenbach [9] and by Doroshchuk and Frid [10], showed that the results obtained by different investigators are highly contradictory. Given all the circumstances, they should be regarded as excessively approximate and incorrect.

Studies by Puchkov and Vinogradov [11, 12] were performed over a rather wide range of diameter ratio ($0.121 < d_1/d_2 < 0.844$) and Reynolds number ($10^4 < \mathrm{Re} < 10^5$). However, due to methodical errors, the heat transfer rates obtained by them at small $d_1/d_2$ are on the high side.

One of the more thorough studies of heat transfer in annuli was performed by Kays and Leung [13], who showed for the first time that the Nusselt numbers can be calculated in the case of arbitrary heat fluxes at the walls and for any axial distribution of these heat fluxes. This is done by superposing known values of Nu for one-sided uniform heating. When the inner and outer surfaces are heated by an axially constant heat flux, their Nusselt numbers, defined as $\mathrm{Nu}_{12} = \alpha_1 d/\lambda$ and $\mathrm{Nu}_{21} = \alpha_2 d/\lambda$ for the inner and outer surfaces, respectively, can be expressed by the simple formulas

$$\mathrm{Nu}_{12} = \frac{\mathrm{Nu}_1}{1 + C_1 \dfrac{q_2}{q_1}} \tag{1.1}$$

$$\mathrm{Nu}_{21} = \frac{\mathrm{Nu}_2}{1 + C_2 \dfrac{q_1}{q_2}} \tag{1.2}$$

where $\mathrm{Nu}_1$ and $\mathrm{Nu}_2$ are the Nusselt numbers for heating only the inner or outer wall, respectively (the opposite wall being adiabatic). $C_1$ and $C_2$ are the so-called influence coefficients, which are functions of certain flow regimes and geometric parameters.

Numerical integration of the energy equation using experimental profiles of velocity and eddy diffusivity coefficients was employed by Kays and Leung [13] to obtain solutions for $Nu_1$, $Nu_2$, $C_1$, and $C_2$ in the fully developed heat transfer region for a wide range of controlling parameters ($10^4 <$ Re $< 10^6$, $0 <$ Pr $\leq 10^3$, $0.1 \leq d_1/d_2 \leq 1$). The results of calculations are tabulated, which makes their practical use difficult.

The reliability of their analytic calculations for air in one-sided heating, as well as the validity of Eqs. (1.1) and (1.2), was checked by Kays and Leung experimentally. The agreement between analytic and experimental values of Nu at Re $> 3 \times 10^4$ is quite satisfactory; however, at Re $< 3 \times 10^4$, the analytic results are as much as 10–15% higher than experimental values. In addition, at high Pr the results of Kays and Leung for the limiting case of a circular tube are significantly lower than experimental data.

In their study of steady-state heat transfer in the annulus, Petukhov and Royzen [14] obtained, for one-sided heating, Lyon-type integrals for the Nusselt numbers. They used adiabatic temperatures of the nonheated walls, which, in nondimensional form, are

$$\Theta_1 = \frac{(T_{w1}^{ad} - T_f)\lambda}{q_2 d}, \quad \Theta_2 = \frac{(T_{w2}^{ad} - T_f)\lambda}{q_1 d} \qquad (1.3)$$

Using these expressions, and the principle of superposition of wall temperature solutions, Petukhov and Royzen expressed Nu for two-sided heating in terms of Nu values for one-sided heating:

$$Nu_{12} = \frac{Nu_1}{1 + \frac{q_2}{q_1} \Theta_1 Nu_1} \qquad (1.4)$$

$$Nu_{21} = \frac{Nu_2}{1 + \frac{q_1}{q_2} \Theta_2 Nu_2} \qquad (1.5)$$

The nondimensional temperatures $\Theta_1$ and $\Theta_2$ are related by

$$\Theta_2 = \Theta_1 \frac{d_1}{d_2} \qquad (1.6)$$

Equations (1.4) and (1.5) are analogous to Eqs. (1.1) and (1.2) and are valid for both laminar and turbulent flows, as was confirmed experimentally by Petukhov and Royzen [15], whose study was performed with air at $0.0718 \leq d_1/d_2 \leq 0.692$. This study remains to this day a very significant experimental milestone in investigating heat transfer in annuli because of the range of flow variables and geometric parameters and the accuracy and reliability of results. Petukhov and Royzen determined, together with Nu, the adiabatic temperatures of the nonheated walls, measurement of which involves great difficulties and thus is not usually performed.

The practical application of Eqs. (1.4) and (1.5) is limited by the lack of reliable data on adiabatic temperatures for most liquids and gases, with the exception of air.

Petukhov and Royzen [16] correlated the results of their experiments at Pr = 0.7 [15] in a form relating the ratio of fully developed values of Nu for both surfaces of the annulus to Nu of a circular pipe.

Subsequently, the same investigators [17] suggested, on the basis of analytic results by Kays and Leung [13], more general formulas for Nu and adiabatic temperatures at different Pr:

$$\frac{Nu_{1\infty}}{Nu_{t\infty}} = [1 - \varphi(Pr)] \left(\frac{d_2}{d_1}\right)^{n(Pr)} \cdot \varepsilon \qquad (1.7)$$

$$\frac{Nu_{2\infty}}{Nu_{t\infty}} = 1 - \varphi(Pr) \left(\frac{d_1}{d_2}\right)^{0.6} \qquad (1.8)$$

$$\Theta_{1\infty} = 22 \left[0.27 \left(\frac{d_1}{d_2}\right)^2 - 1\right] Re^{-0.87} Pr^{-0.18}, \quad \Theta_{2\infty} = \frac{d_1}{d_2} \Theta_{1\infty} \qquad (1.9)$$

where

$$\varphi(Pr) = \frac{0.45}{2.4 + Pr}, \quad n(Pr) = 0.16 \, Pr^{-0.15}$$

$$\varepsilon = 1 + 7.5 \left(\frac{d_2/d_1 - 5}{Re}\right)^{0.6} \text{ at } \frac{d_1}{d_2} < 0.2$$

$$\varepsilon = 1 \text{ at } \frac{d_1}{d_2} \geq 0.2$$

Also in [17], the authors suggested for the first time expressions for incorporating the change in heat transfer rate along the channel under conditions of hydrodynamic developing flow at the start of the heated length:

$$\frac{Nu_1}{Nu_{1\infty}} = 0.86 + 0.8 \left(\frac{d}{x}\right)^{0.4} \left(\frac{d_1}{d_2}\right)^{0.2} \qquad (1.10)$$

$$\frac{Nu_2}{Nu_{2\infty}} = 0.86 + 0.54 \left(\frac{d}{x}\right)^{0.4} \left[1 + 0.48 \left(\frac{d_1}{d_2}\right)^{0.37}\right] \qquad (1.11)$$

which are applicable from $x/d = 1$ to values of $x/d$ corresponding to $Nu/Nu_\infty = 1$.

The expressions suggested in [17] require experimental validation at high Pr, since, as previously mentioned, the data underlying their derivation are of doubtful validity.

From the time of appearance of [13–15], many investigations were performed regarding velocity fields and turbulent heat transfer coefficients in annuli. This made it possible for Bobkov et al. [18] to perform a more rigorous calculation of temperature fields and thus also of Nu in turbulent flow of coolants in annuli at $0.0625 \leq d_1/d_2 \leq 1$, $10^4 < \text{Re} < 10^5$, and Pr from 0 to 10. Integrating the equation of energy numerically, they performed calculations of the hydrodynamic and thermal entrance lengths for one-sided axially uniform heating. Using analytic results and the most reliable experimental data, they obtained in the fully developed region the expression

$$\text{Nu} = \text{Nu}_0 + \beta \text{Re}^{0.87} \text{Pr}^n, \tag{1.12}$$

where Nu is the heated wall Nusselt number. For the inner wall,

$$\text{Nu}_0 = \text{Nu}_{01} = \left(6.4 - \frac{3}{\lg \text{Re}}\right)\left(\frac{d_1}{d_2}\right)^{-0.24}$$

$$\beta = \beta_1 = 0.008\left[1 + 0.5 \exp\left(-4\frac{d_1}{d_2}\right)\right] \tag{1.13}$$

and for the outer wall,

$$\text{Nu}_0 = \text{Nu}_{02} = \left(6.4 - \frac{3}{\lg \text{Re}}\right)\left[1 + 0.2 \exp\left(-4.5\frac{d_1}{d_2}\right)\right]$$

$$\beta = \beta_2 = 0.008\left[1 + 0.2 \exp\left(-4\frac{d_1}{d_2}\right)\right] \tag{1.14}$$

The exponent $n$ of Pr remains almost constant with changes in Re and $d_1/d_2$; it is independent of the direction of the heat flux, and, as a function of Pr, it is given by the expression

$$n = 0.4 + \frac{0{,}45}{1 + 2\,\text{Pr}} \tag{1.15}$$

The technique for determining heat transfer parameters for two-sided heating is analogous to that used by Petukhov and Royzen [13], whereas the influence coefficients in Eqs. (1.1) and (1.2) are expressed by Bobkov et al. [18] as

$$C_1 = \frac{-0.45\left(\frac{d_1}{d_2}\right)^{-0.2}\text{Nu}_{1\infty}}{(1 + 2\,\text{Pr}^{0.8})\,\text{Nu}_{2\infty}} \tag{1.16}$$

$$C_2 = \frac{-0.45\left(\frac{d_1}{d_2}\right)^{0.8}}{1 + 2\,\text{Pr}^{0.8}} \tag{1.17}$$

Among analytic studies of heat transfer in turbulent flows, studies by Buleyev et al. [19] and Dwyer and Tu [20] deserve special mention. Rather interesting work was performed by Petrikevich [21] and by Wilson and Medwell [22, 23], but these are basically methodological.

All the above results are applicable only in the case of constant physical properties of the coolant, that is, at small temperature differences. In addition, all of them, with the exception of work by Petukhov and Royzen, pertain only to fully developed heat transfer, far from the inlet.

The effect of variable physical properties of the coolant on heat transfer in annuli has been rather poorly explored. The first studies in this direction were experimental investigations by Dalle Donne and Meerwald [8, 24] and Presser et al. [25], performed over a rather narrow range of $d_1/d_2$ and with only the inner tube heated. The most interesting of the three is actually [8], where rather exact experimental results were obtained at $d_1/d_2$ of 0.503 and 0.725, $8 \times 10^3 <$ Re $< 2 \times 10^5$, and $1.3 < (T_w/T_f) < 2.6$. These experiments showed that under the above conditions the effect of the temperature factor $\Psi = T_w/T_f$ varies along the channel and does not attain the asymptotic value, even at $x/d = 90$. The experimental results are represented by the expression

$$\text{Nu}_1 = 0.018 \left(\frac{d_1}{d_2}\right)^{-0.16} \text{Re}^{0.8} \text{Pr}^{0.4} \Psi^{-(0.25 + 0.0018\, x/d)} \tag{1.18}$$

Note that the exponent of $\Psi$ here is simpler than in the expression for round tubes.

It is additionally suggested in [8] that the variability of physical properties be corrected for by the ratio of wall temperature $T_w$ in the point under study to the flow temperature entering the annulus $T_E$. Thus the equation

$$\text{Nu}_1 = 0.018 \left(\frac{d_1}{d_2}\right)^{-0.16} \text{Re}^{0.8} \text{Pr}^{0.4} \left(\frac{T_w}{T_E}\right)^{-0.2} \tag{1.19}$$

in which the exponent of the temperature factor is independent of $x/d$, is recommended by [8].

There is some doubt whether the factor $(T_w/T_E)^{-0.2}$ in Eq. (1.19) is capable of correcting for variability of physical properties. Equations (1.18) and (1.19) differ only by terms $(T_w/T_f)^n$ and $(T_w/T_E)^{-0.2}$. If the flow temperature at the inlet and the wall temperature in the section under study are held constant, the value of $T_w/T_E$ does not change with Re, whereas that of $T_w/T_f$ increases with increasing Re and decreases with decreasing Re due to changes in the flow temperature. Under identical operating conditions the values of the coefficients $(T_w/T_E)^{-0.2}$ and $(T_w/T_f)^n$ should be identical, which is possible only when one of the power exponents of $T_w/T_f$ or $T_w/T_E$ changes with Re. Analysis of results obtained for circular tubes, as well as of studies by Nemira and Vilemas [26, 27], who investigated fully developed local heat transfer in air-cooled annuli under various conditions, do not indicate any perceptible dependence of the power exponent of $T_w/T_f$ on Re. In that case the power exponent of $T_w/T_E$ should have different values at different Re, and hence the use of Eq. (1.19) over a wide range of Re can result in significant error.

The analytic study by Galin and Yesin [28] is very interesting. The authors, using various models of turbulent transport, performed a detailed analysis of heat transfer in annuli when the coolant exhibits variable physical properties. Their calculations of heat transfer in the (thermally) fully developed region were performed by an integral method developed by Petukhov with Popov [29], based on the idea of quasi stabilization of the flow. They obtained analytic expressions for constant as well as variable physical properties that, in the limiting case of a circular pipe, become identical to the familiar Petukhov-Popov integral equation [29]. It is suggested that the variability of physical properties be estimated with the Petukhov formula for a circular tube,

$$\frac{\mathrm{Nu}}{\mathrm{Nu}_{\Psi=1}} = \Psi^{a \lg \Psi + 0.36} \tag{1.20}$$

where $a = 0$ for cooling and $a = 0.3$ for heating of the gas.

To be able to calculate the local heat transfer at the start of the heated length, Galin and Yesin [28] used the boundary layer approximations, to solve a complete set of differential equations of motion and energy using the velocity fields and features of turbulent transport obtained experimentally by other investigators. The calculations were performed for annuli with $d_1/d_2 = 0.375, 0.503,$ and $0.725$ with only the inner wall heated. They suggested the following equation, incorporating the effect of the temperature factor on heat transfer, at $\Psi \leq 3$:

$$\frac{\mathrm{Nu}_1}{\mathrm{Nu}_{1,\,\Psi=1}} = \Psi^n, \quad \text{where} \quad n = \frac{-0.448\,\dfrac{x}{d}}{12.4 + \dfrac{x}{d}} \tag{1.21}$$

They suggested that the Nusselt number correction for the thermal entrance length be found from Eq. (1.10) and the adiabatic temperatures be approximated by the expression

$$\Theta_i = \Theta_{i\infty}\left[1 - \exp\frac{\left(\dfrac{x}{d}\right)^{3/2}}{400 + 0.01\,\mathrm{Re}}\right] \tag{1.22}$$

Equations (1.20)–(1.22) require a further experimental check since at the time when they were obtained no sufficient body of experimental data was available.

In addition, the calculations by Galin and Yesin [28] were in all cases performed only for a fully developed velocity profile, that is, when the start of the heating length is sufficiently far from the inlet.

The solution of heat transfer problems in annuli with turbulent flow of gases and variable physical properties is closely related to the overall advances in the field of heat transfer at high temperature differences and, in the first place, heat transfer in tubes. Hence, to obtain a clearer idea of the effect of large temperature differences on heat transfer in gas-cooled channels, it is useful to familiarize oneself with the principal studies performed with circular tubes.

A large number of experimental studies have been performed in which sufficiently accurate and general empirical formulas were obtained for calculating heat transfer in pipes under large heat fluxes. A large body of work on measuring local heat transfer from heating air, argon, and carbon dioxide in a circular pipe was performed at the All-Union Heat Engineering Institute [31–33].

Lel'chuk and Dyadyakin [31] were the first to show that when the variability of physical properties is incorporated by using the expression

$$\frac{\text{Nu}}{\text{Nu}_{\Psi=1}} = \Psi^n$$

the power exponent $n$ is variable along the tube. The experimental results obtained at the Institute were correlated by them in the form

$$\text{Nu} = 0.023\ \text{Re}^{0.8}\ \text{Pr}^{0.4}\ \Psi^{-(c+0.00272\, x/d)} \quad (1.23)$$

where $c = 0.47$ for air and 0.45 for argon. By introducing an additional factor accounting for the temperature level and the compressibility, it became possible to obtain a constant power exponent of $\Psi$ at $x/d > 50$. In this case it was found that

$$\text{Nu} = 0.024\ \text{Re}^{0.8}\ \text{Pr}^{0.4}\ \Psi^{-0.5} \left(\frac{T}{T_{\text{cr}}}\right)^{-0.15} (1 - \Lambda)^{0.5} \quad (1.24)$$

where $T_{\text{cr}}$ is the thermodynamic critical temperature, $\Lambda = \bar{u}_x/\sqrt{2i^0}$ is the velocity coefficient, and $i^0$ is the total stagnation enthalpy.

Dalle Donne and Browdich [34], who performed their experiments with an air-cooled tube lacking a nonheated hydrodynamic entrance region, at $\Psi$ from 1.4 to 2.5, Re from $10^4$ to $1.3 \times 10^5$, and $x/d$ from 26 to 166, suggested the following expression:

$$\text{Nu} = 0.024\ \text{Re}^{0.8}\ \text{Pr}^{0.4}\ \Psi^{-(0.29+0.0054\, x/d)} \quad (1.25)$$

Analogous results, but with a somewhat different power exponent of $\Psi$, were obtained by Taylor [35] and by Perkins and Vorsoe-Schmidt [36].

A large series of studies of heat transfer to pipe flows of gases [37–39] and on developing a technique for correlation of experimental data [40–42] was performed by Petukhov et al. They showed that at a given value of $x/d$ the curve of

$$\frac{\text{Nu}}{\text{Nu}_{\Psi=1}} = f\left(\Psi, \frac{x}{d}\right) \quad (1.26)$$

is not a straight line in logarithmic coordinates. Irrespective of the conditions at the inlet of the heated length, the effect of variable physical properties was the same. In addition, as opposed to the paper by Lel'chuk and Elfimov [32], the inlet temperature was not found to affect the heat transfer, that is, the heat transfer rate was not found to depend on the absolute temperature level.

The results obtained by Lel'chuk et al. [31, 32] and by Petukhov et al. [37, 38] are correlated by Petukhov et al. [40] by the formula

$$\frac{\mathrm{Nu}}{\mathrm{Nu}_{\Psi=1}} = \Psi^{0.53\,n_\rho + \frac{1}{3}n_\lambda + \frac{1}{4}n_c - \Phi\left(\frac{x}{d}\right)n_\mu \lg \Psi} \tag{1.27}$$

in which, using power exponents from the relations

$$\frac{\rho}{\rho_1} = \left(\frac{T}{T_1}\right)^{n_\rho},\quad \frac{\lambda}{\lambda_1} = \left(\frac{T}{T_1}\right)^{n_\lambda},\quad \frac{\mu}{\mu_1} = \left(\frac{T}{T_1}\right)^{n_\mu},\quad \frac{c_p}{c_{p_1}} = \left(\frac{T}{T_1}\right)^{n_c} \tag{1.28}$$

it is possible to make allowance for individual properties of gases. The function $\Phi(x/d)$ is tabulated in the paper by Petukhov et al. [40]. Equation (1.27) is recommended at $\mathrm{Re} > 7 \times 10^3$, $q_w = $ constant, and $\Psi$ between 0.5 and 4. The standard deviation of the experimental results is 3.8%. The same authors suggest, in [41], the following expression for calculating heat transfer at a constant property gas flow:

$$\mathrm{Nu}_{\Psi=1} = 0.0225\,\mathrm{Re}^{0.8}\,\mathrm{Pr}^{0.6}\,\epsilon_x, \tag{1.29}$$

where $\epsilon_x$ is a function making allowance for Nu variation in the entrance region.

As seen from the above studies, the majority of investigators make allowance for the variability of physical properties by means of the temperature factor.

In some studies an attempt was made to allow for variability of physical properties by a specially selected boundary layer temperature or temperature of the wall. This temperature is used for determining the physical properties in dimensionless numbers in the familiar relation

$$\mathrm{Nu} = c\,\mathrm{Re}^{0.8}\,\mathrm{Pr}^{0.4}\,\epsilon, \tag{1.30}$$

However, this did not yield satisfactory results.

It is also inconvenient to correct for variable physical properties using the temperature factor [an equation such as Eq. (1.26)], since $T_w$ is usually the sought quantity, and in this case it must be found by a trial and error method.

It is better to perform calculations by a formula suggested by Kurganov and Petukhov in [41]:

$$\frac{\mathrm{Nu}}{\mathrm{Nu}_{\Psi=1}} = \exp\left\{-K_f\left[a\varphi\left(\frac{x}{d}\right) + n_\mu \Phi_1\left(\frac{x}{d}\right)K_f\right]\right\} \tag{1.31}$$

where

$$a = -0.53\,n_\rho - 1/3\,n_\lambda - 1/4\,n_c$$

$$\varphi\left(\frac{x}{d}\right) = 1 - \exp(-10\,\tilde{x})$$

$$\Phi_1\left(\frac{x}{d}\right) = 1.25\,\frac{\tilde{x}^2}{1+\tilde{x}^2},\quad \tilde{x} = \frac{x}{100\,d}$$

$$K_f = \frac{qd}{\lambda T_f \mathrm{Nu}_{\Psi=1}}$$

and $\mathrm{Nu}_{\Psi=1}$ is obtained from Eq. (1.29).

Equation (1.31), written in the form

$$\Psi = 1 + K_f \exp\left\{K_f\left[a\varphi\left(\frac{x}{d}\right) + n_\mu\,\Phi_1\left(\frac{x}{d}\right)K_f\right]\right\} \qquad (1.32)$$

makes it possible to immediately calculate the local wall temperature from specified values of $q$, $d$, $\rho$, $u$, and $T_f$. The calculations can be performed using the approximate values of $a$ and $n_\mu$ listed in Table 1.1.

Equations (1.31) and (1.32) correlate the majority of presently known experimental data, obtained at $q_w = $ constant with a standard deviation of 7% for all gases at $(x/d) > 1-5$. They are the most accurate of all the currently available design equations.

Kurganov and Petukhov [42] showed that, using superposition, it is possible to extend Eq. (1.31) with sufficient accuracy also to the case of arbitrary variation in $q_w$ along the tube.

This means that currently the effect of variability of physical properties on heat transfer in gas-cooled circular tubes has been investigated rather extensively and that quite exact general design equations are available.

The situation concerning analytic estimation of the effect of physical properties variation on heat transfer is different. A rigorous solution of the problem of convective heat transfer in turbulent flow of gases with variable thermophysical properties is currently impossible. The main reason for this lies in incompleteness of the understanding of the internal mechanism of processes associated with turbulent transport during a significant change in thermophysical properties in the boundary layer. The semiempirical relationships governing turbulent transport, obtained in isothermal flows, are unsuitable at variable properties, whereas sufficiently accurate data for nonisothermal flow are highly limited.

Nonetheless, attempts have been made to investigate this question analytically with many simplifications and various assumptions. Deissler [43–45], Kutateladze and Leont'yev [46], Petukhov and Popov [29], Bankston and McEligot [47], and McEligot et al. [48] used various flow models and, accordingly, determined the eddy viscosity in different ways. They obtained results that, in the case of heated gas in a fully developed flow, are in satisfactory agreement with experimental data, while differing greatly from them in the case of cooling.

**TABLE 1.1** Values of $a$ and $n_\mu$

| Gas | $a$ | $n_\mu$ | Gas | $a$ | $n_\mu$ |
|---|---|---|---|---|---|
| Monatomic gases | 0.30 | 0.67 | $H_2O$ (375—1200 K) | 0.013 | 1.18 |
| Diatomic gases | 0.26 | 0.70 | $NH_3$ | −0.04 | 0.92 |
| $CO_2$ | 0.09 | 0.77 | $CH_4$ (300—1200 K) | −0.097 | 0.71 |

Whereas rather reliable design correlation equations, satisfying practical needs, are available at present for estimating the effect of physical properties variation on heat transfer in a fully developed duct flow (primarily in circular tubes), the situation is significantly poorer in the entry length of ducts, including annuli. However, in many cases engineering devices operate under conditions where duct flows are not hydrodynamically fully developed over significant parts of their length. The calculation of heat transfer and drag in this case requires, in the first place, information on the behavior of the flow in the entrance region.

The formation of boundary layers at the inner and outer surfaces of annuli occurs in a very different manner, particularly when the difference between the diameters of the outer and inner cylinders is large. In this case the effect of transverse surface curvature of the inner cylinder becomes significant. Within the entrance region of an annulus, the inner tube can be treated as a circular cylinder in a uniform axial flow. When the cylinder diameter significantly exceeds the thickness of the boundary layer, its heat transfer and drag do not differ significantly from these quantities for a flat plate. In this case they can be calculated by one of the techniques suggested by Patankar and Spalding [49], Popov [50], Van Driest [51], and Žukauskas and Šlančiauskas [52]. In other cases, the effect of surface curvature must be taken into account. However, this problem has not been sufficiently explored.

The first approximate study, performed by Jacob and Dow [53], showed that the heat transfer coefficient is larger on a cylinder than on a plate.

Eckert [54] assumed that the velocity profiles on a cylinder and a plate are the same and that the relationship between the friction coefficient and the thickness of the boundary layer for a cylinder remains the same as for a plate. He obtained the following relationship between the boundary layer thicknesses on a cylinder and a plate (at the same $Re_x$):

$$\frac{\delta}{\delta_{fp}} = \frac{1}{1 + \left(\frac{\delta}{3r_0}\right)^{0.8}} \quad (1.33)$$

He recommends that the skin friction coefficient under these conditions be calculated from the expression

$$c_f = c_{f\,fp}\left(1 + \frac{\delta}{3r_0}\right)^{0.2} \quad (1.34)$$

The most detailed analytic study of heat transfer and fluid mechanics for a cylinder in axial flow with a turbulent boundary layer and constant physical properties was performed by Sparrow et al. [55]. They calculated the velocity and temperature fields and the distribution of shear stress in the boundary layer using turbulent transport coefficients suggested by Deissler [44] for circular pipes. They additionally assumed that

$$\frac{q}{q_w} = \frac{\tau}{\tau_w} = \frac{r_0}{r_0 + y} \qquad (1.35)$$

Calculations showed that the velocity and temperature curves become increasingly steeper with reduction in the radius of the cylinder, the boundary layer thickness decreases, and hence the skin friction and heat transfer coefficients increase.

An analogous result was obtained by Ginevskiy and Solodkin [56], who calculated the skin friction of a cylinder in axial flow.

The skin friction coefficients measured by Hughes [57] and Kemph [58] are in satisfactory agreement with those calculated by Eckert et al. [54, 55], and Ginevskiy and Solodkin [56].

It follows from the above brief survey that most studies of heat transfer in annuli have been performed with a constant-property flow. There are only a small number of studies of heat transfer with variable physical properties at the entrance as well as in the fully developed flow region, and they do not give sufficient knowledge on the manner in which heat transfer occurs under these conditions. The effect of conditions at the duct inlet on heat transfer and drag at the entrance region of an annulus has not been explored at all. There are no data regarding the effect of inlet conditions on velocity and temperature profiles in annuli for constant as well as variable physical properties of the coolant, which significantly retards the development of analytic methods of calculation.

All the above suggests the conclusion that heat transfer in annuli with turbulent flow of coolants with variable physical properties has been little explored. Such a situation does not satisfy growing requirements for accuracy of engineering calculations and has a negative effect on the development of analytic methods. In addition, this complicates obtaining sufficiently accurate relationships for a constant-property flow, since correlating the large volume of published data, obtained at temperature differences, significantly different from zero, to constant physical properties is possible only if the extent of the effect of variable properties on heat transfer and friction is known. An excessive reduction of the temperature factor during experiments resulted in a large increase in errors.

Our studies were intended for investigating heat transfer and drag in gas-cooled annuli under variable physical conditions. The experiments were performed over a wide range of flow and geometric parameters. The effect of freestream turbulence on heat transfer and friction in the entrance region of an annulus has been investigated separately. Measurements of velocity and temperature profiles in the boundary layer and the degree of their distortion as a result of the nonisothermicity made possible a deeper understanding of the mechanism of heat transfer under these conditions.

The large body of experimental data that was accumulated made it possible to obtain an accurate quantitative estimate of the effect of physical properties variation on heat transfer, to find efficient forms of correlating experimental data, and to derive general design equations that can be recommended for practical use.

CHAPTER
# TWO

## EXPERIMENTAL TECHNIQUES

Heat transfer in gas-cooled annuli at high heat fluxes involves large temperature gradients that limit to a significant degree the freedom of an investigator to select methods of solution to the problems under study.

At the same time, the significant difference in the flow at the entrance (prior to merging of boundary layers) and fully developed parts of the annulus not only limits the range of problems that can be investigated in a given case but also the kinds of experimental techniques that can be used. In our studies of flow in the annulus entrance region, the main attention was devoted to the study of transfer of heat and momentum in the turbulent boundary layer forming on the inner tube. We investigated the relationship governing velocity and temperature distributions in the boundary layer, skin friction, local heat transfer, and the dependence of these quantities on the turbulence intensity of the free stream, as well as on local disturbances in the boundary layer. In all this, paramount attention was paid to the study of the effect of large heat fluxes on principal parameters of the boundary layer.

In the fully developed flow region we investigated primarily the characteristics of heat transfer and its dependence on flow parameters and geometry for one- as well as for two-sided heating of the annulus surface. The effect of a sharp disturbance of a stable flow (by placing a transverse step) on the degree of influence of variable physical properties on heat transfer was investigated separately. Extensive studies were performed in annuli with a helically shaped inner tube. Hydraulic drag coefficients were also measured and analyzed.

Such a wide range of problems required the use of various experimental techniques and design of various pieces of experimental equipment and test sections. To attain maximum accuracy in all the measurements, we used automated systems for collecting and recording experimental data.

The experiments were performed over a wide range of flow parameters: $Re_x$ was varied from $10^5$ to $5 \times 10^7$ and $T_w/T_f$ from 1 to 4 in the entrance region, and Re was varied from $5 \times 10^3$ to $5 \times 10^5$ and $T_w/T_f$ from 1 to 3 in the fully developed region. The coolant used was air, and its physical properties were determined according to data of Vasserman et al. [59] and Vargaftik [60].

## 2.1 EXPERIMENTAL UNITS

The boundary layer structure can be investigated only if its dimensions are significantly larger than the measuring device. Hence, to investigate the relationships governing the formation of the boundary layer at the entrance region of an annulus, it was necessary to construct a rather large test section.

High Re can be attained either by increasing the reference dimension and flow velocity or by reducing the kinematic viscosity of the gas flow by raising its pressure. Increasing the reference dimension is limited by technical factors, whereas increasing the gas velocities is controlled by compressibility effects. Hence reduction of the kinematic viscosity of the gas is the simplest method of attaining high Re in incompressible flows. For the test-section dimensions we selected, the gas mass flow rate was as high as 10 kg/s, and hence experiments in the entrance part of annuli were performed in a closed-cycle wind tunnel (Fig. 2.1) in which the pressure could be maintained between 0.1 and 2.5 MPa.

Air sucked in by a blower (10) from the drum (9) moved through moisture and oil trap (5) and entrance length (3) to test section (1), from which it returned to the drum (9) by way of connecting tubes ($d = 200$ mm).

The air circulation was produced by an aircraft blower that was driven by a brush-type 130-kW ac motor. Its rpm was adjustable from 125 to 730. Meshes and grids with different-sized cells were placed in the stabilizing section in order to make the velocity field uniform and damp out the turbulence.

The required air pressure in the cycle was produced by a compressor unit. The compressor pumped air from the room through a moisture and oil trap to a receiver. From the latter the air was fed through a pressure-reduction valve to a drum, preceded by a moisture and oil trap. The air pressure in the receiver was maintained automatically within specified limits.

The value of Re could be varied in three ways: (1) changing the motor rpm, (2) reducing the intake using a valve (8), (3) or changing the air pressure in the loop. The last method was the one used predominantly.

The temperature of air circulating within the cycle could be adjusted by a water cooling loop. The cooling water was supplied by a pump (15) from a tank (16) and, passing through water jacket (6), fitted around the connecting tubes and the moisture and oil trap (5), and then returned to the tank. The latter was

EXPERIMENTAL TECHNIQUES **17**

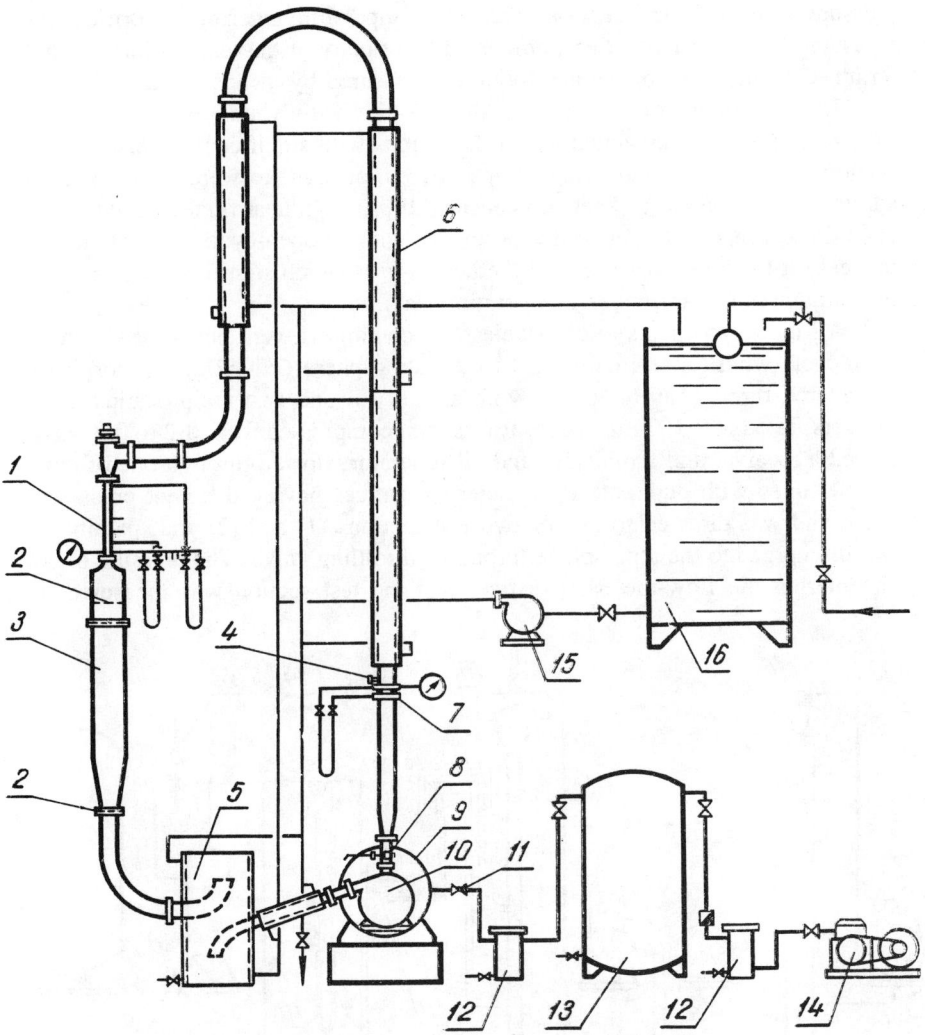

**FIG. 2.1** Schematic of experimental apparatus. (1) Test section, (2) meshes, (3) stabilization section, (4) resistance-thermometer well, (5) moisture and oil trap, (6) cooling jacket, (7) metering orifice, (8) control valve, (9) air-blower drum, (10) air blower, (11) pressure reduction valve, (12) moisture and oil trap, (13) receiver, (14) compressor, (15) water pump, (16) water tank.

refilled with tap water by way of an automatic float-type valve and the heated water was drained.

The flow rate of gas in the loop was measured by standard metering orifices, produced and installed in full conformance with the standard listed by Preobrazhenskiy [61].

The pressure drop across the orifice was measured by a water-type differential manometer. Four such manometers were used for measuring the static pressure drop in the test section. The flow temperature entering the orifice was measured by a resistance thermometer. The velocity in the test section was determined on the basis of the air flow rate, measured by means of the orifice.

The heat transfer and hydraulic drag in the hydrodynamic fully developed flow region were investigated with test sections with significantly smaller cross sections. However, the pressure drops in them at high Re were significant. In addition, the cooling gas heats up considerably at high heat fluxes. Under these conditions, it is best to perform experiments with an open-cycle wind tunnel. In that case it is unnecessary to cool the heated gas, which significantly simplifies the equipment and improves its operating stability.

All the experiments under stable flow conditions were performed with an open-cycle wind tunnel, exhausted to the atmosphere (Fig. 2.2). Atmospheric air entered through an air intake with a filter (6) and two reciprocating compressors, rated at 0.107 kg/s each. It was then compressed from 0.2 to 0.8 MPa, passed a receiver and a moisture and oil separator, flowed through one of three parallel lines with duplicate flow-metering orifices having different cross sections, and was directed to one of two test sections (1) and (2) and, passing it, was discharged to the atmosphere through a throttling valve. The pressure of the air entering the flow-metering orifices and the test section was measured by

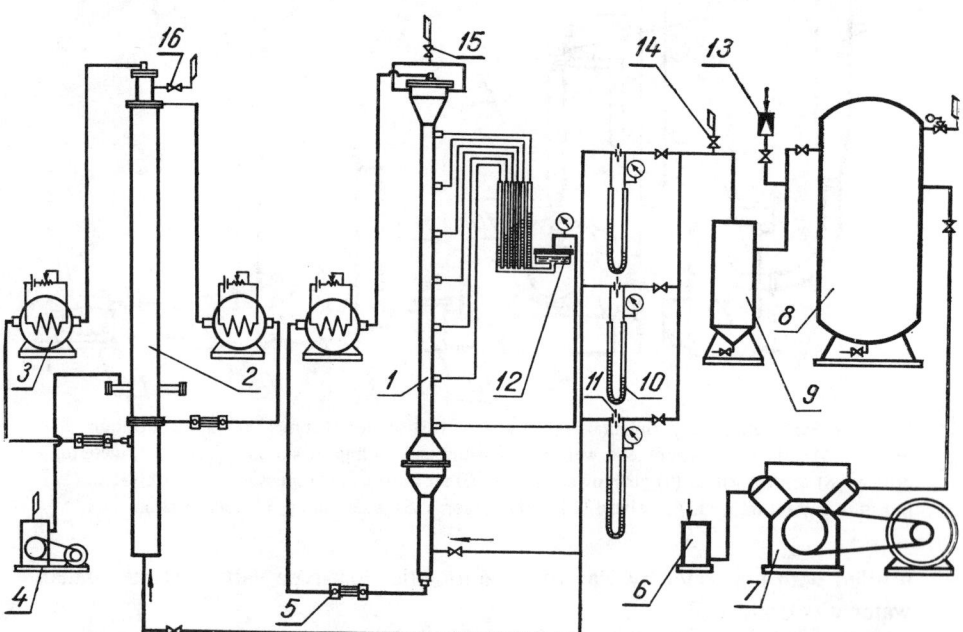

**FIG. 2.2** Schematic of test equipment for investigating heat transfer and flow behavior in an annulus. (1) Test section 1, (2) test section 2, (3) generator, (4) vacuum pump, (5) shunt, (6) air intake, (7) compressor, (8) receiver, (9) moisture and oil trap, (10) differential manometer, (11) flow-metering orifice, (12) gang differential manometer, (13) pressure-reduction valve, (14) bypass valve, (15) and (16) discharge valves.

standard pressure gauges and tensometric pressure transducers (0.25 precision). The pressure drop across the orifices were measured by well-type water-filled differential manometers (10) or by inductive pressure-drop sensors. The throttling orifices were fabricated and installed in accordance with the specifications listed in the book by Makarov and Sherman [62]. The temperature entering the flow-metering orifices was measured by chromel-alumel thermocouples. The pressure and flow rate were controlled by the throttling and bypass valves. A part of the experiments was performed with air supplied from a centralized compressor station. In this case the pressure was controlled by pressure reducer (13), and the flow rate by throttling valves (15) or (16).

The heating elements in the experiments were, in most cases, heated by direct current, supplied from independent sources. Some of the experiments with heat transfer at the entrance region were performed with ac heating. The current was generated by two electromechanical converters (maximum current of one was 2500 A at a voltage of 36 V, and the second was rated at 1000 A at 60 V). The current was controlled by changing the voltage supplied to the excitation winding of the converter. To ensure high stability of the generated voltage, and thereby also the stability of heat release, we designed a special control device with feedback, placed on the generator excitation circuit. The feedback monitored the pressure drop along the test section. This device made it possible to eliminate the voltage drift, thus the voltage fluctuation at the generator terminals did not exceed 0.5%. In the absence of voltage regulation in the excitation circuit, when the excitation winding was supplied from a separate self-contained stabilizer, the voltage fluctuations were as high as 5%. The current in the circuits was determined from the voltage drop using 0.2 and 0.5% precision class shunts.

## 2.2 TEST SECTIONS

The experiments were performed with several test sections. In all cases these were annuli whose geometries and other parameters were controlled by the specific intents of the experiments. The ratio of diameters was varied by replacing the inner cylinders.

*The investigations in the entrance region of the annulus* were performed with a test section (Fig. 2.3) where, when the outer cylinder was of sufficiently large diameter, the ratio of diameters of the inner to the outer cylinders was very small. This was done to obtain an extended annulus length over which velocity boundary layers formed without merging. In addition, the effect of transverse curvature of the surface on the boundary layer and on transport parameters can be obtained only with a rather small inner cylinder. All the experiments were performed on inner cylinders that could be changed between tests.

The 100-mm-diameter and 1000-mm-long outer tube was made of stainless steel. Its inner surface was carefully polished. Ten holes, with axes directed

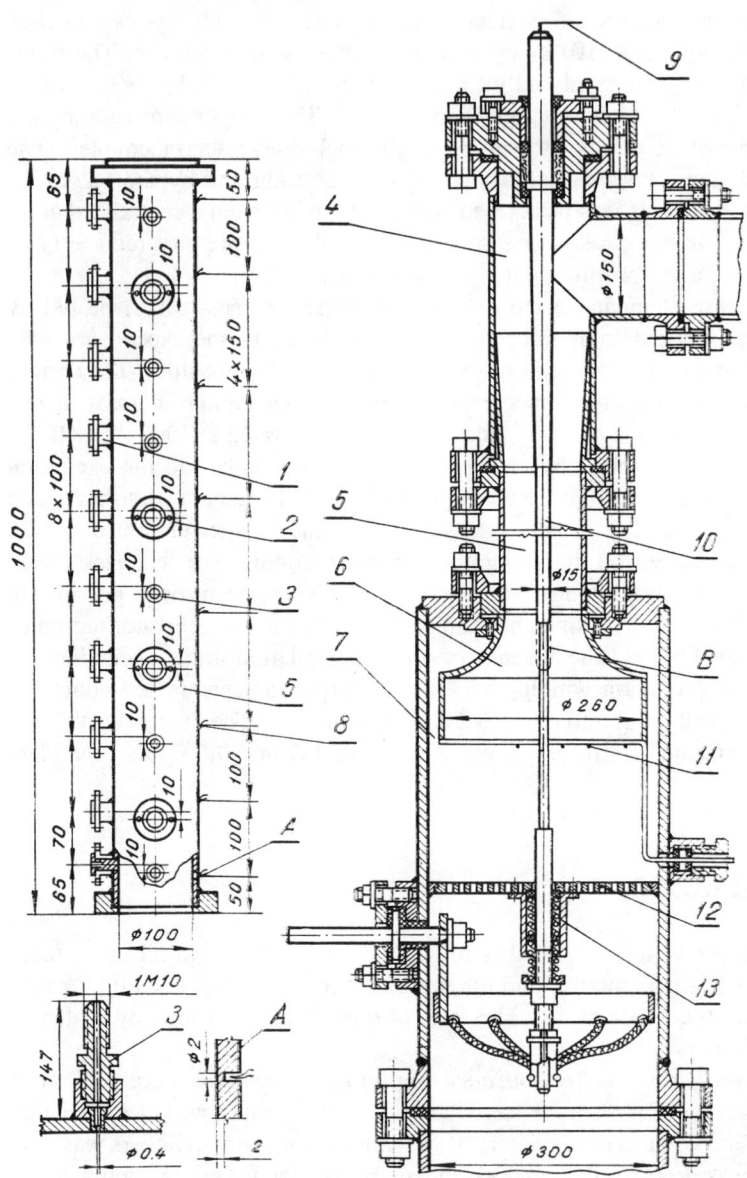

**FIG. 2.3** Test section. (1) Stubs for inserting the velocity probe, (2) stubs for inserting the temperature probe, (3) stubs for measuring the static pressure, (4) discharge port, (5) test section, (6) smooth inlet, (7) expansion chamber, (8) duct thermocouples, (9) lead-out of cylinder's thermocouples, (10) cylinder (inner tube) under study, (11) thermocouples for measuring the freestream temperature, (12) grid, (13) cylinder stretching spring. B is the location of surface-type turbulence promoters. Dimensions are in millimeters.

toward the center of the channel, were drilled on each of two axially perpendicular lines, situated at 90° from one another. The hole diameter in one of the rows was 14 mm, whereas that of 4 holes in the other row was 12 mm, with the remaining 6 holes being 4 mm in diameter. The 14 and 12 mm diameter holes were covered by plugs, fitted flush with the inner surface of the tube. Upon removing the plugs, these holes could be used for placing velocity or temperature probes, or for measuring the turbulence intensity. The remaining 6 holes were closed by special plugs, in which static-pressure taps were drilled. These plugs were placed precisely perpendicular to the flow and were precisely fitted to the inner surface of the channel.

A pressure-tap hole disturbs the flow near it, which to some extent distorts the magnitude of the measured pressure, introducing an error in the experiment results. As established by Povkh [63], Shaw [64], and Ray [65], the magnitude of this error decreases with hole diameter. Povkh, Shaw, and Ray investigated the effect of the diameter and depth of the hole on the magnitude of the error. These studies give an idea about the approximate error magnitude and the reason for its existence. According to all these recommendations, the diameter of holes of the static-pressure taps was selected at 0.4 mm, and the depth of drill hole with this diameter was also 0.4 mm. The remaining part of the hole was 4 mm in diameter.

The inlet to the test section was preceded by a convergent length, in which the air was compressed ninefold, which, combined with suitable selection of equalizing meshes (Fig. 2.1), ensured uniform velocity distribution across the test section.

Four copper-constantan thermocouples for measuring the freestream temperature were placed in the wide part of the convergent segment. These were prepared from 0.15-mm-diameter wires, insulated by lacquer and silk, and installed in horizontal stainless steel tubes ($d = 2$ mm), which also served as a shield for protection from radiation by the cylinder. The thermocouple leads were led in the direction of the incoming flow through holes in the walls of the tubes.

A check of velocity fields in several locations along the test section showed that these fields are axysymmetrical over all the cross sections of the annulus. The symmetry of velocity fields was checked additionally on the basis of heat transfer from the tube under study by rotating the latter (and the thermocouples that were welded to its wall) about its axis. No changes in thermocouple readings were observed when this rotation was performed under steady flow conditions.

The cooling flow moved upward in the test section. The exit from the test section was one sided (asymmetric); thus, calculations for determining the heat transfer and drag coefficients were performed only up to a distance of 0.86 m from the inlet. The rear part of the cylinder, which could be affected by the discharge of air, was not included in the calculations.

The inner cylinder under study (Fig. 2.3) was placed concentrically in the outer tube and was electrically insulated from it. It was fastened rigidly at its

## 22 HEAT TRANSFER OF GAS COOLANTS

upper end by tightening of the upper part of the test section, while its lower part was stretched by a spring, ensuring movement of the lower inlet with elongation of the cylinder as it was heated. Current busbars, placed in the wide part of the test section, were constructed in such a manner that they would create only small blockage to the flow of air and would not significantly distort the velocity field entering the duct.

The experiments in the test section described here were performed with three 15.5-mm-diameter, one 10.5-mm-diameter, and one 4-mm-diameter inner cylinder. The 1000-mm heated lengths of the cylinders were constructed of thin-walled, polished stainless steel tubes. The design of two such cylinders (15.5 mm diameter) is shown in Fig. 2.4. A cylindrical copper busbar is welded to the leading end of the heated tube; the diameter of the busbar near the tube is 3.5 mm less than the diameter of the tube proper. The point of junction between the tube and busbar formed a smooth transition from one diameter to the other.

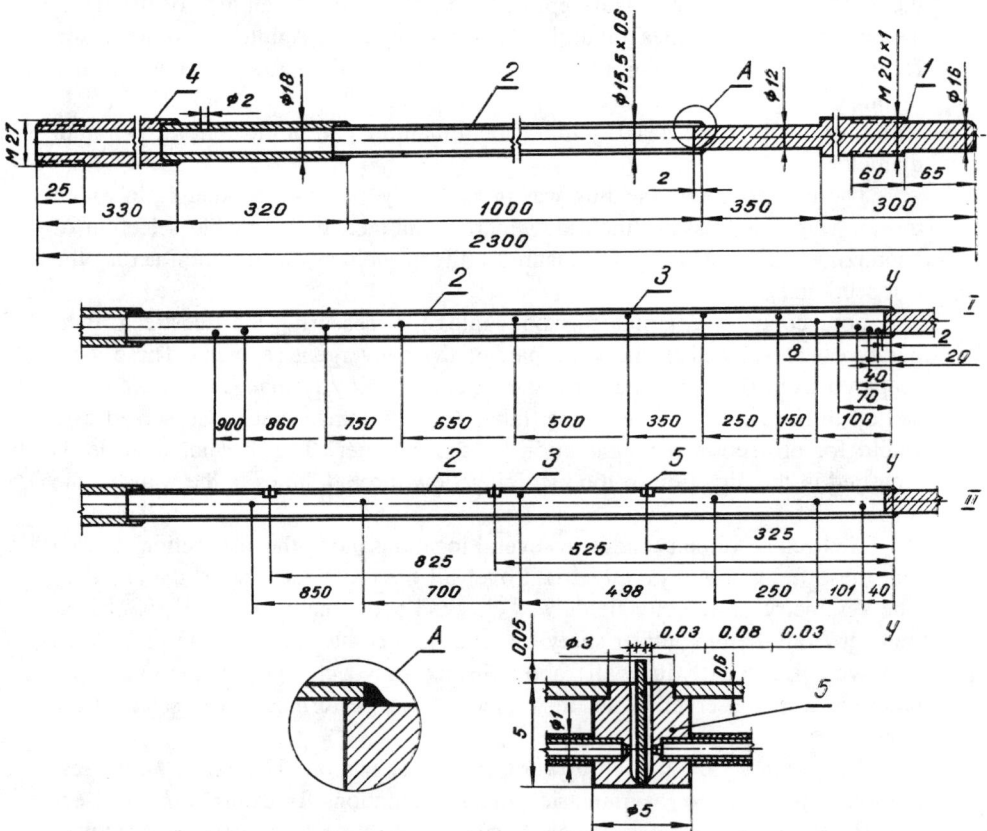

**FIG. 2.4** Design of inner tubes 1 and 2. (1) Lower inlet, (2) cylinder under study, (3) thermocouples, (4) upper inlet, (5) sensor of cylinder 2 for measuring the skin friction by the protruding plank method. Dimensions are in millimeters.

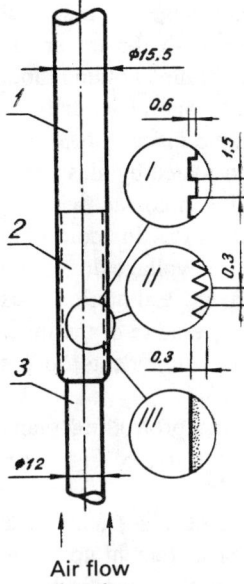

Air flow direction

**FIG. 2.5** Design of surface-type turbulence promoter. (1) Experimental cylinder, (2) turbulence promoter (I, type 1; II, type 2; III, type 3), (3) copper busbar.

The wall temperature was measured by welding 14 thermocouples to the inner surface of the 15.5-mm tube and 10 thermocouples in the case of the 10-mm tube. The thermocouples were made of 0.3 mm chromel and alumel wires, coated by heat-resistant insulation, and arranged helically, one thermocouple in each cross section. The thermocouple wires were led outside through a longitudinal hole in the upper busbar with a tightening device, placed on its outer end.

To prevent flattening at high temperatures, the 15.5- and 10-mm-diameter cylinders were relieved during the experiments. For this their inner space was connected, through a special hole, to a circulating air system, and the pressure was equalized. The air inside the inner cylinder did not flow.

One of the 15.5-mm-diameter tubes was used for investigating the effect of a surface-type turbulence promoter on heat transfer and drag. For this purpose either a rectangular or a triangular thread was cut on part of the forward busbar directly adjoining the heated tube, or a rough emery cloth was pasted on it (Fig. 2.5). (Henceforth these types of roughness will be referred to as turbulence promoters 1, 2, and 3, respectively.) The diameter of the turbulence generating segment was equal to that of the tube. The transition from the smaller dimension of the lower busbar ($d = 12$ mm) to the larger dimension ($d = 15.5$ mm) of the turbulence promoter was smooth.

Since the number of thermocouples that can be welded to a small-diameter tube is insufficient, the wall temperature of the cylinder with $d = 4$ mm was measured by means of a single traversing thermocouple, fastened to a special device with a micrometer-screw mechanism. The device with the traversing thermocouple (Fig. 2.6) consisted of a 3-mm-diameter thin-walled holding tube, with the hot junction bead of the thermocouple fastened to one of its ends by means of flexible springs. The chromel-alumel thermocouple made of 0.3-mm-diameter wires and the tube in which it was placed were insulated by glass fiber saturated with heat-resistant lacquer. The holding tube (3) is in contact with the inner surface of the cylinder about the thermocouple junction (5). In addition to measuring the wall temperature, this device also measures the voltage drop from the start of the heated length to the point under measurement. Calibration tests showed that the thermocouple does indeed measure the temperature of the inner surface of the tube with sufficient accuracy, and that its components do not introduce detectable errors to the results.

Three sensors for measuring the surface drag by the protruding plank method were placed on one of the cylinders with $d = 15.5$ mm (see Fig. 2.4). These experiments were performed without heating.

The sensor itself consists of a thin plate, slightly protruding above the flow-washed surface, with previsions for static-pressure measurement upstream and downstream of the plate. The latter, which was 0.08 mm thick was installed into a split metal housing with 0.03-mm-deep and 1-mm-wide grooves in the plane of the slit (Fig. 2.4). A part of the plate protrudes 0.05 mm above the back surface of the housing, which has the same curvature as the test cylinder.

*Studies with hydrodynamicfully developed flow* were performed using one of two test sections. The first of these (Fig. 2.7) could be used only for experiments with heating of the inner tube. The same section was used for measuring the hydraulic drag coefficient with and without heating. The annulus was formed by a stainless steel outer tube having outer diameter of 30 mm and 1 mm

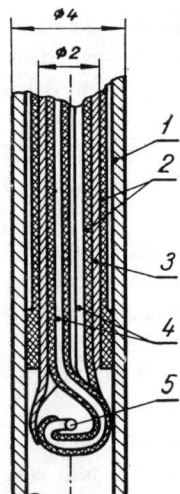

FIG. 2.6 Design of traversing thermocouple (for cylinder with $d = 4$ mm). (1) Test cylinder, (2) insulation, (3) holding tube, (4) thermocouple wires, (5) thermocouple junction.

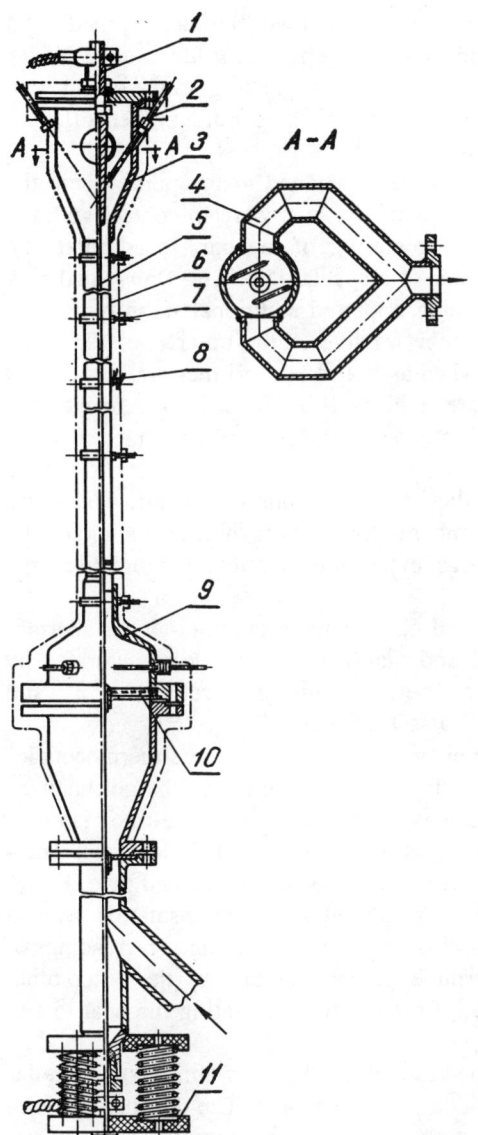

**FIG. 2.7** Test section 1. (1) Upper busbar and thermocouple-wire lead-out point, (2) thermocouple, (3) expansion chamber, (4) discharge duct, (5) inner tube, (6) outer tube, (7) pressure-tap stub, (8) thermocouples, (9) inlet chamber, (10) flow-equalizing mesh and centering grid, (11) thermal-expansion compensation unit.

wall thickness and with a circular or helically shaped tube (5), placed coaxially in it. The inner tube (5) also served as a calorimeter. The inner tube was silver soldered at one of its ends to a hollow copper busbar (1), which was tightly fastened to the cover of the discharge chamber (3). This held the upper part of the inner tube in the center of the outer tube. The lower busbar was made of a circular copper bar with $d = 10.5$ mm. Its upper end was soldered to the inner tube, whose lower end was led out, through a Teflon packing, to the outside and was fastened to the set of springs (11) that stretches the bar and prevents flexing of the inner tube due to thermal expansion.

Three 1.0-mm-diameter radial pins were fastened to the point where the copper bar was connected to the inner tube. The pins were in contact with the walls of the outer tube, thus ensuring centering of the bar. In addition, the copper busbar passed through the center of a special disk with flow-equalizing grids, placed in the inlet chamber, which imparted additional stability and ensured reliable centering of the inner tube relative to the outer one. Since the length of the latter was 1800 mm, whereas that of the former was only 1320 mm, the annulus made up by the outer tube and the lower busbar formed the hydrodynamic entrance region, which was 480 mm long and amounted to about 30 equivalent diameters of the annulus.

The coolant was supplied from the bottom. It consisted of air, which was fed from the flow-equalizing inlet chamber to the annulus through a smooth inlet and was removed through a discharge expansion chamber having discharge ducts on both of its sides.

The temperature was sensed by 0.3-mm-diameter chromel-alumel thermocouples, with wires coated by glass and quartz fiber insulation, saturated by heat-resistant organic-silicate lacquer. The maximum temperature that it could withstand for a prolonged time period was 1050 K.

The temperature of the flowing air was measured by three thermocouples placed in the inlet chamber and by four thermocouples in the discharge chamber. To reduce heat losses, the outer tube was insulated by an 8-mm-thick layer of asbestos rope. This insulation was wrapped with cotton and Teflon tape. Thermocouples were welded in six locations along the outer, nonheated tube. Forty-millimeter-diameter thin copper rings were placed under the insulation layer at these locations, and a thermocouple was connected to each ring. The readings of these thermocouples served to determine the temperature of the outer tube, needed for calculating the radiant heat flux and for calculating the heat lost to the surroundings.

The static pressure was determined by drilling three 0.5-mm-diameter holes in the inlet stub and in six locations along the outer tube. The distance between the pressure taps was 300 mm. The static pressure drop along the annulus was measured by a set of water-filled differential manometers or by alcohol-filled, well-type micromanometers.

This test section served for measurements with a single 10.5-mm-diameter (wall thickness 0.5 mm) circular tube and 9 helically shaped tubes.

The geometry of the annulus is described by Table 2.1, whereas the cross-sectional shapes of the tubes and their photographs are shown in Figs. 2.8 through 2.10. The helically shaped tubes were fabricated from a circular polished stainless steel tube with inner diameter of 10 mm and wall thickness of 0.5

**TABLE 2.1** Geometric Parameters of the Annulus and of Cross Sections of the Helically Shaped Inner Tubes

| No. | $s/d_w$ | $F$, mm$^2$ | $d_2$, mm | $d_w$, mm | $b$, mm | $R_1$, mm | $R_2$, mm | $D_2$, mm |
|---|---|---|---|---|---|---|---|---|
| | | | Three-lobe tubes | | | | | |
| 1 | 13.3 | 568.8 | 19.07 | 10.52 | — | 2 | 2.2 | 4.76 |
| 2 | 19.7 | 557.7 | 18.69 | 10.66 | — | 2.2 | 2.6 | 5.84 |
| 3 | 42.2 | 560   | 18.77 | 10.66 | — | 2.2 | 2.6 | 5.68 |
| 4 | ∞    | 558.4 | 18.72 | 10.78 | — | 2 | 2 | 5.84 |
| | | | Oval (two-lobe) tubes | | | | | |
| 1 | 6.16 | 551.6 | 18.48 | 12.25 | 6.2 | | | |
| 2 | 8.26 | 545.4 | 18.27 | 12.1 | 6.9 | | | |
| 3 | 11.8 | 548.2 | 18.38 | 12.27 | 6.3 | | | |
| 4 | 24.4 | 549.9 | 18.42 | 12.3 | 6.2 | | | |
| 5 | 49   | 549.8 | 18.41 | 12.25 | 6.2 | | | |

mm. The shaping and twisting were performed by pulling the tube through a rotating mandrel of the given shape.

To measure the temperature and pressure drop, two 0.2- to 0.3-mm-diameter chromel-alumel thermocouples were soldered in 5 locations along the inner

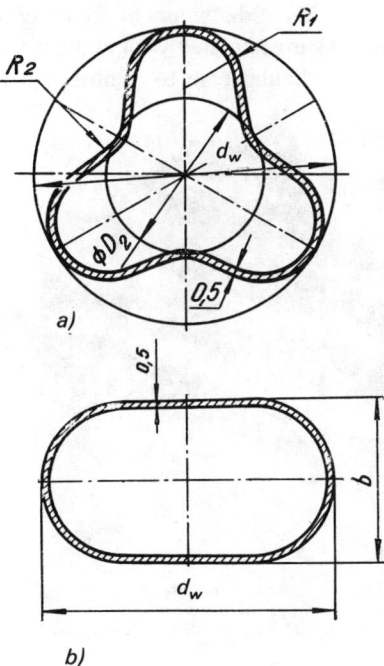

**FIG. 2.8** Geometry of the cross section of helically shaped tubes. (a) Three-lobe tube; (b) two-lobe tube.

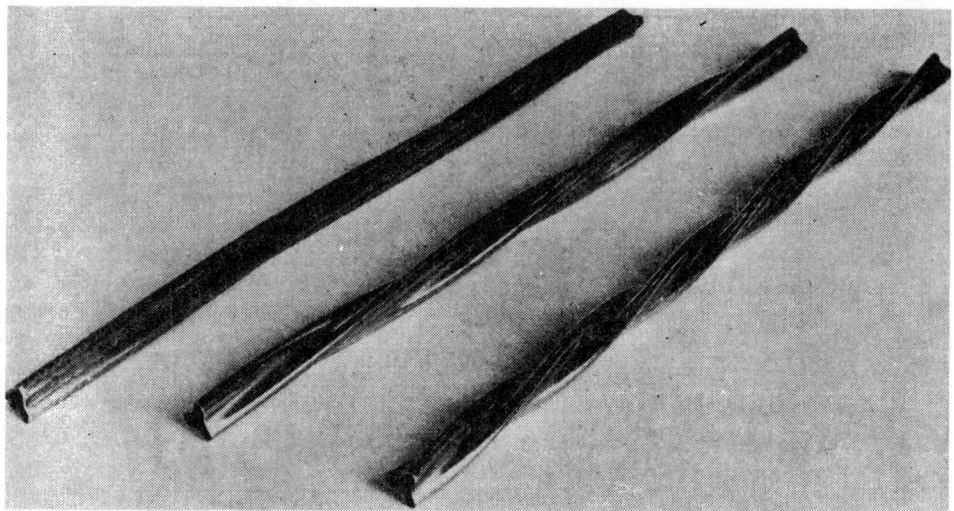

**FIG. 2.9** Helical tubes with three-lobe cross section having $s/d_{dc}$ of 42.2, 19.7, and 13.2.

surface of each helically shaped tube. One thermocouple was welded in a point of minimum distance from the tube axis, and the other in a point of maximum distance from this axis. In the case of a circular inner tube, the thermocouples were placed on opposite sides of the inner surface. The thermocouple wires were led out through the cavity in the upper busbar tip.

Heat transfer from annuli having different ratios of diameters of the outer to the inner tube (heated from one or both sides) was investigated with another test section (Fig. 2.11), which was designed with particular care to minimize heat

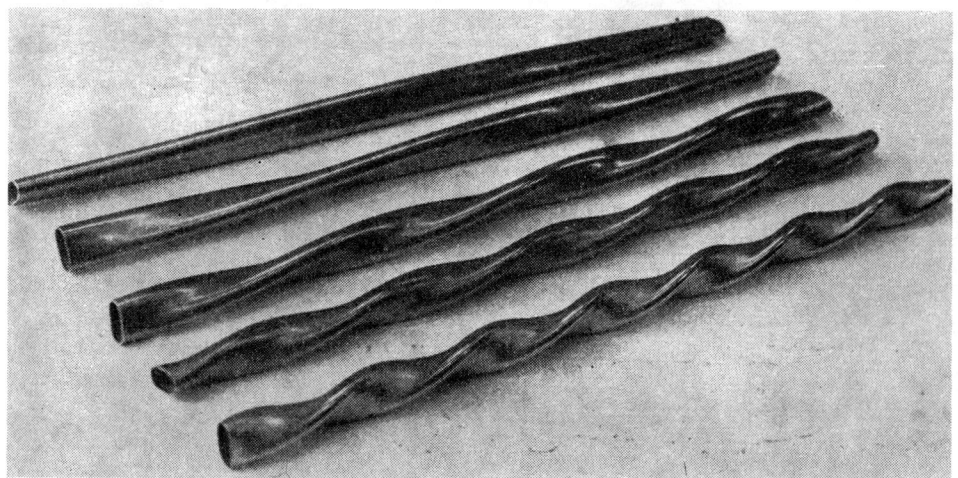

**FIG. 2.10** Helical tubes with oval (two-lobe) cross sections with $s/d_{dc}$ of 49.0, 24.4, 11.8, 8.26, and 6.16.

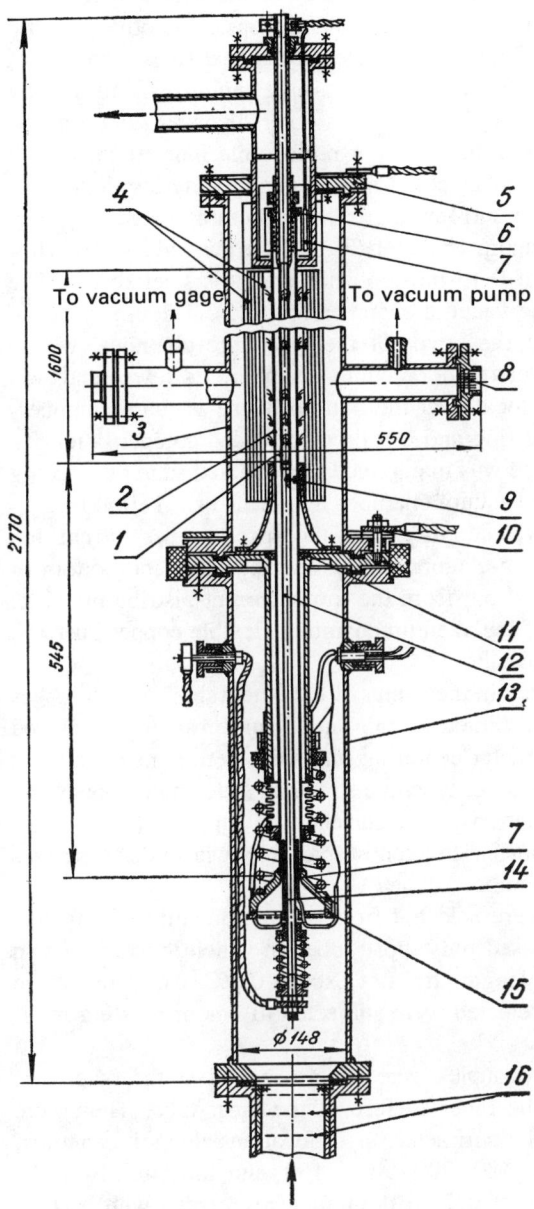

**FIG. 2.11** Test section 2. (1) Inner tube, (2) outer tube, (3) vacuum chamber, (4) reflecting screens, (5) vacuum chamber cover, (6) bearing cone made of devitrified glass, (7) thermocouples, (8) airtight coupling, (9) centering pins, (10) lower current-distributing ring, (11) Teflon centering ring, (12) lower busbar, (13) inlet chamber, (14) static-pressure probe, (15) centering grid, (16) equalizing meshes.

losses to the surroundings. To increase the accuracy of the heat transfer measurements it was necessary to measure the pressure drops, since the pressure taps could have introduced significant disturbances in the heat-release pattern of the electrically heated annulus walls. The 2145-mm-long annulus was formed by a polished stainless steel outer tube, 29.75 mm in outer diameter, and a precision made wall thickness of 0.8 mm, and by a replaceable inner tube, placed coaxially in the outer tube. Both tubes could serve as calorimetering devices.

To reduce heat losses to the surroundings, the 1600-mm heated length of the annulus was placed in a vacuum chamber (3) and surrounded by six cylindrical reflecting screens made of 0.15-mm polished stainless steel sheets. The screens were suspended from the vacuum chamber cover (5). The upper end of the outer tube was soldered to the cover of the discharge chamber, whose housing was welded to the cover of the vacuum chamber. Its lower end was soldered to the elongation stub located at the bottom of the vacuum chamber, which by means of a connecting ring and a bellows provided compensation for thermal expansion. The outer tube was maintained in a stressed state by a spring and was held coaxially with the vacuum chamber by means of a Teflon ring.

Current to the outer tube was supplied from the upper copper current-distributing ring, through the vacuum-chamber cover and the walls and bottom of the discharge chamber. It was led out from the lower current-distributing ring, along bronze pins, the bottom of the vacuum chamber, flexible copper busbars, and a copper ring soldered to the tube.

The inner tube was suspended in the center of the outer tube. Centering was provided in three sections: at the inlet, at the outlet, and prior to the heated length, where three 0.8-mm-diameter centering pins with Teflon insulating tips were placed. The spring held the inner tube in a stretched state and compensated for longitudinal thermal deformations. The current was supplied to the inner tube through the hollow upper busbar and central bar (12), whose diameter was always equal to that of the inner tube.

The blanks for the tubes were selected from a large quantity of polished tubes of the required size. We used only those tubes in which the variation in electrical resistance at 100-mm lengths did not exceed 0.5%. To stabilize the emissivity, the tubes that were selected were subjected to 2-hour roasting in air at 1090 K.

Two chromel-alumel thermocouples were contact-welded at the outer surface of the outer tube and on the inner surface of the inner tube. They were positioned on opposite sides of the perimeter, in 10 locations along the annulus, situated at 30, 55, 95, 160, 280, 480, 700, 960, 1180, and 1400 mm from the start of heating. They were made of 0.2- to 0.3-mm wires, coated with heat-resistant insulation. The thermocouples were used for measuring the temperature as well as for determining the voltage drop along the tube. The locations of thermocouples on both tubes were identical only when either both tubes were not heated or both of them were heated identically. In the remaining cases they were shifted relative to one another due to thermal expansion. A correction for this was made in the test data.

The thermocouple wires were led out from the inner tube through the upper, hollow busbar, whereas those from the outer tube were connected by means of the airtight plug (8).

The air temperature was measured by thermocouples in several locations of the discharge chamber and at the start of the hydrodynamic entrance region. The heat losses to the surroundings were measured by placing eight thermocouples along the first reflecting screen.

Several of the thermocouples from each batch of wires were calibrated over the temperature range of 290 to 900 K. The emf values of thermocouples made from a given batch of wires were virtually identical.

The experiments were performed with 4 smooth circular inner tubes having $d_1$ of 16.47, 10.47, 5.78, and 3.05 mm ($d_1/d_2$ = 0.585, 0.372, 0.205, and 0.108), with respective wall thicknesses of 0.38, 0.4, 0.265, and 0.35 mm. The annulus with $d_1/d_2 = 0.372$ was used for investigating the effect on heat transfer of a sharp disturbance of previously fully developed flow. The disturbance was produced by placing across the flow an obstacle in the form of a thin ring, which tightly adjoined the inner tube (Fig. 2.12).

When it was impossible to fasten to the inner tube having $d_1 = 3.05$ mm, a sufficient number of thermocouples, we used a special traversing thermocouple probe, introduced through the center of the tube.

The cooling air was supplied to the inlet chamber [(13) in Fig. 2.11] through flow-equalizing meshes [(16) in Fig. 2.11]. It then flowed smoothly through a narrowing down transition piece to the annulus, passed the nonheated hydrodynamic entrance length, and, through the perforated end of the outer tube, was exhausted to the mixing device of the discharge chamber. The air was removed through an outlet, located in the upper part of the discharge chamber. The pressure was measured only in two points—at the annulus inlet and in the vacuum chamber, where it ranged from 3 to 7 Pa.

All the electrical signals from thermocouples and pressure sensors, as well as from instruments measuring the voltage drops along tubes and shunts, were measured by means of an automatic data acquisition system that included a 0.01% precision class high-resistance digital voltmeter, with sensitivity of 1

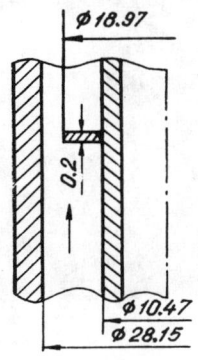

**FIG. 2.12** Geometry of a passage with an obstacle.

**32** HEAT TRANSFER OF GAS COOLANTS

μV. To eliminate the effect of the voltage gradient in the vicinity of thermocouple junctions, all the measurements were performed with the heating current supplied from two opposite sides. The sensor acquisition rate was 2 measurements/second.

Measurement results were recorded automatically, according to a prespecified program, on punched tape, and were then processed on a BESM-4M computer. Components of the experimental setup are shown in Fig. 2.13.

## 2.3 DETERMINATION OF THERMAL AND HYDRODYNAMIC PARAMETERS IN THE BOUNDARY LAYER

Investigations of thermal and hydrodynamic parameters in the boundary layer were performed only in the entrance region of the annulus, where the boundary

**FIG. 2.13** Components of the experimental setup.

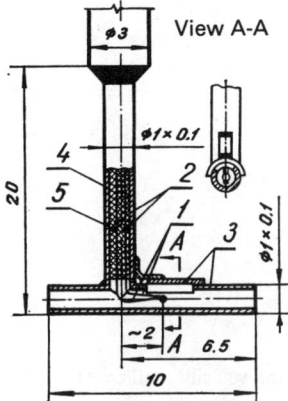

FIG. 2.14 Temperature-field measuring probe. (1) Chromel-alumel thermocouple wires with $d = 0.09$ mm, (2) chromel-alumel wires with $d = 0.3$ mm, (3) thermocouple shield, (4) holder, (5) epoxy compound.

layers forming on the inner and outer walls do not yet merge. In this case we measured the temperature and velocity fields with and without heating of the inner tube. Surface drag and the longitudinal component of the time-dependent velocity fluctuations in the flow core and in the boundary layer of isothermal flows were also measured.

*The temperature profiles* were measured with a shielded thermocouple (Fig. 2.14) made of 0.09-mm-diameter chromel-alumel wire. The thermocouple was insulated from the shield and holder and measured the temperature of the gas flowing along a capillary, which shielded the thermocouple junction from the thermal radiation of the hot wall. Special calibration performed at low temperatures showed that placing the shield has only a slight, negligible effect on the thermocouple readings. However, the presence of the shield significantly limited the portion of the boundary layer in which measurements could be performed. It was impossible to come closer than 0.5 mm from the wall. The thermocouple emf's were measured either by a semiautomatic or by a self-recording potentiometer. The temperature probe was moved across the boundary layer by a special mechanism [66] with a micrometer screw, which allowed measurements of the distance from the wall at minimum increment of 0.05 mm. The initial location of the probes was determined on the basis of electrical contact with the wall.

*The velocity profiles* were measured by Pitot-Prandtl tubes. The small thickness of the boundary layer in which the measurements were performed made it necessary to use a total-pressure tube with a very narrow opening. For this purpose a 1-mm-diameter nickel tube with 0.1-mm wall thickness was flattened and then polished in such a manner that its total span in the narrow part was 0.2 mm, and the span of the open slot was 0.1 mm (Fig. 2.15). The static-pressure tube was placed 4 mm away from the total-pressure tube, closer to the outer wall of the annulus. The total and static-pressure tubes were constructed according to the recommendations of Anisimov et al. [67, 68].

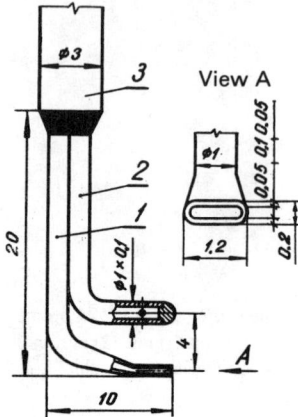

**FIG. 2.15** Probe for measuring velocity profiles. (1) Flattened tube, (2) static-pressure tube, (3) holder.

The velocity in a given location was determined from the measured velocity head:

$$u_f = \xi \sqrt{\frac{2\Delta p}{\rho_f}} \quad (2.1)$$

where $\rho_f$ is the free-stream density, $\Delta p$ is the pressure drop measured by the Pitot-Prandtl tube in newtons per square meter, and $\xi$ is a correction factor.

The freestream density was expressed, as a function of the pressure and temperature, as

$$\rho_f = \frac{348.5 \, P_x}{T_f \times 10^5} \eta \quad (2.2)$$

where $P_x$ is the static pressure at the cross section under measurement, determined from the pressure measured at the start of the test section and from graphs of static pressure drop in the annulus under study; $\eta$ is a correction factor, which is a function of pressure and is determined from tables given by Vargaftik [60]; and $T_f$ is the freestream temperature.

The coefficient $\xi$ in Eq. (2.1) was determined on a calibration stand and was found to be between 0.99 and 1.02. The velocity head was measured by liquid-filled differential manometer, the liquid being either water or bromoform. The velocity was determined to within ±2.5%.

*Displacement thickness and momentum thickness.* The displacement and the momentum thicknesses can be determined from the measured velocity profiles. The expression for these quantities is obtained from the integral momentum equation. For a cylinder this equation is

$$\frac{d}{dx} \int_{r_0}^{r_0+\delta} \rho u (u_\infty - u) \, 2\pi r \, dr$$

$$+ \frac{du_\infty}{dx} \int_{r_0}^{r_0+\delta} (\rho_\infty u_\infty - \rho u) \, 2\pi r \, dr = \tau_w \, 2\pi r_0 \quad (2.3)$$

Since $r = r_0 + y$ and $dr = dy$, then, dividing both sides by $(2\pi r_0)(\rho_\infty u_\infty^2)$ and regrouping, we obtain

$$\frac{d}{dx} \int_0^\delta \left(1 - \frac{u}{u_\infty}\right) \frac{\rho u}{\rho_\infty u_\infty} \left(1 + \frac{y}{r_0}\right) dy$$

$$+ \frac{1}{u_\infty} \frac{du_\infty}{dx} \int_0^\delta \left(1 - \frac{\rho u}{\rho_\infty u_\infty}\right)\left(1 + \frac{y}{r_0}\right) dy = \frac{c_f}{2} \quad (2.4)$$

where by definition the first integral is the momentum thickness

$$\delta^{**} = \int_0^\delta \left(1 - \frac{u}{u_\infty}\right) \frac{\rho u}{\rho_\infty u_\infty} \left(1 + \frac{y}{r_0}\right) dy \quad (2.5)$$

and the second is the displacement thickness

$$\delta^* = \int_0^\delta \left(1 - \frac{\rho}{\rho_\infty} \frac{u}{u_\infty}\right)\left(1 + \frac{y}{r_0}\right) dy \quad (2.6)$$

*Determination of skin friction.* With the exception of long circular tubes, in which the skin friction is easily determined from the pressure drop along the tube, the measurement of skin friction in the general case of flow over bodies involves significant difficulties. Figure 2.16 summarizes the majority of currently available methods for determining surface friction. These methods can be subdivided into two groups. The first is comprised of direct measurements of the surface drag force acting on a surface element (direct methods). In the methods of the second group (indirect methods), the magnitude of the skin friction is determined from values of other quantities, of which skin friction is a function. More details on these methods and instruments are available in papers by Sakiadis [69], Liepman and Skinner [70], Konstantinov [71], Hool [74], Preston [75], the staff of the NPL Aerodynamics Division [76], Dean [77, 78], Konstantinov and Dragnysh [79], Rechenberg [80], and Clauser [81], and in books by Pankhurst with Holder [72], and Gorlin with Slezinger [73].

It appears from surveying the methods for determining skin friction that the most reliable of them is that of direct measurement of the friction force, acting on some surface element of the stream-washed body. This method is suitable for any flow and is not related to its structure or the physical properties of the fluid. Irrespective of whether the flow or the boundary layer is laminar or turbulent, sub- or supersonic, this method allows direct measurements of the force component in the flow direction.

But direct measurement of forces by the balance method is useful only in the absence of longitudinal pressure drop. When the latter is present, it is very difficult and sometimes impossible to use this method due to slot leakage, elimination of which requires special, difficult-to-fabricate devices.

Since the dimensions of the cylinders under study and also the presence of a significant longitudinal pressure drop made it impossible to use direct methods for measurement of the surface drag force, we used only indirect methods to

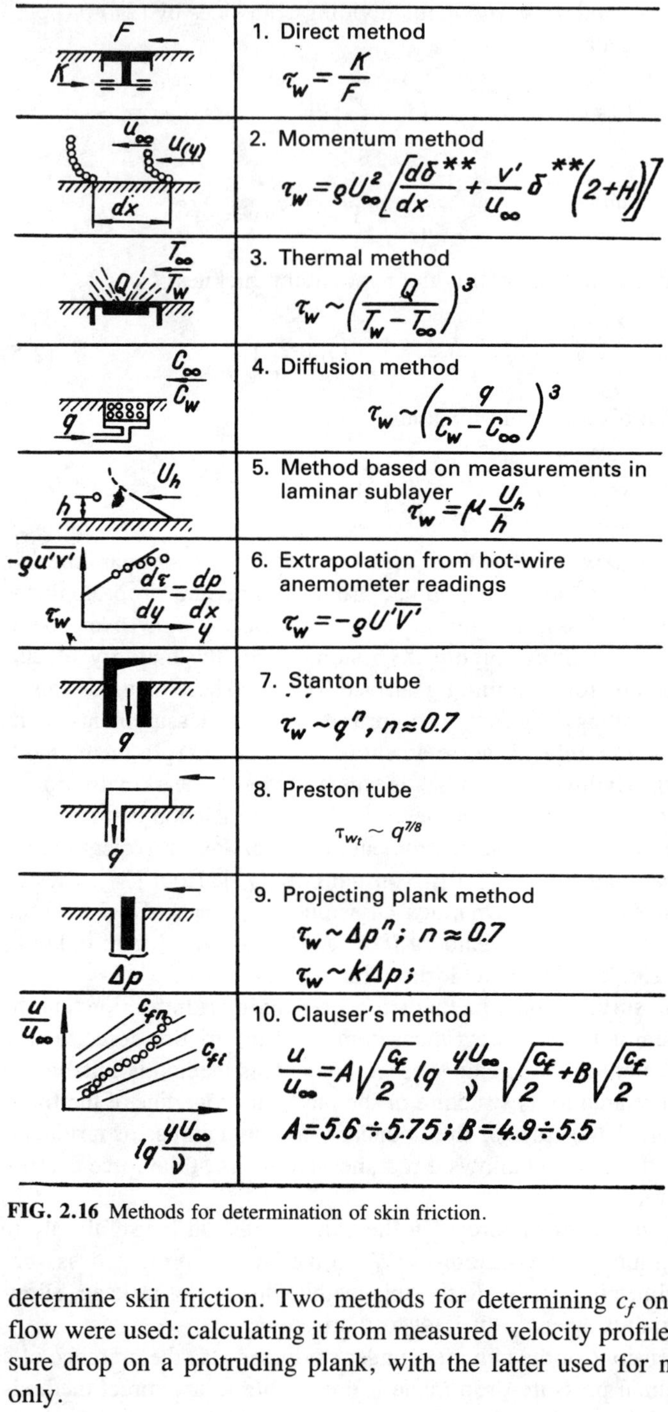

| | Method | Formula |
|---|---|---|
| | 1. Direct method | $\tau_w = \dfrac{K}{F}$ |
| | 2. Momentum method | $\tau_w = \varrho U_\infty^2 \left[ \dfrac{d\delta^{**}}{dx} + \dfrac{v'}{U_\infty} \delta^{**}(2+H) \right]$ |
| | 3. Thermal method | $\tau_w \sim \left( \dfrac{Q}{T_w - T_\infty} \right)^3$ |
| | 4. Diffusion method | $\tau_w \sim \left( \dfrac{q}{C_w - C_\infty} \right)^3$ |
| | 5. Method based on measurements in laminar sublayer | $\tau_w = \mu \dfrac{U_h}{h}$ |
| | 6. Extrapolation from hot-wire anemometer readings | $\tau_w = -\varrho \overline{u'v'}$ |
| | 7. Stanton tube | $\tau_w \sim q^n,\ n \approx 0.7$ |
| | 8. Preston tube | $\tau_{w_t} \sim q^{7/8}$ |
| | 9. Projecting plank method | $\tau_w \sim \Delta p^n;\ n \approx 0.7$<br>$\tau_w \sim k\Delta p;$ |
| | 10. Clauser's method | $\dfrac{u}{U_\infty} = A\sqrt{\dfrac{c_f}{2}}\, lg\, \dfrac{yU_\infty}{\nu}\sqrt{\dfrac{c_f}{2}} + B\sqrt{\dfrac{c_f}{2}}$<br>$A = 5.6 \div 5.75;\ B = 4.9 \div 5.5$ |

**FIG. 2.16** Methods for determination of skin friction.

determine skin friction. Two methods for determining $c_f$ on a cylinder in axial flow were used: calculating it from measured velocity profiles, or from the pressure drop on a protruding plank, with the latter used for nonheated cylinders only.

The local skin friction was determined from the measured velocity profile of a turbulent boundary layer by the Clauser method [81] [method (10) in Fig. 2.16], which has come into extensive use during the past several years. The Clauser method is based on approximating the velocity profile over the logarithmic portion of the boundary layer using the well-known law of the wall.

$$\frac{u}{u_*} = \frac{1}{\varkappa} \ln \frac{yu_*}{\nu} + B \qquad (2.7)$$

Equation (2.7) can be written as

$$\frac{u}{u_\infty} = A \sqrt{\frac{c_f}{2}} \lg\left(\frac{yu_\infty}{\nu}\sqrt{\frac{c_f}{2}}\right) + B\sqrt{\frac{c_f}{2}} \qquad (2.8)$$

If values of the function $u/u_\infty = f(yu_\infty/\nu)$ are recorded for known values of $A$ and $B$ at different $c_f$, they can be used for determining the local skin friction coefficient. Constant $A$, obtained by various investigators for smooth surfaces, ranges from 5.6 to 5.75, whereas $B$ for such surfaces varies from 4.9 to 5.5. The method is simplified if the law of the wall [Eq. (2.7)] is supplemented by another equation, that is, the excess-velocity law,

$$\frac{u_\infty - u}{u_*} = f\left(\frac{y \cdot u_*}{\delta^* \cdot u_\infty}\right) \qquad (2.9)$$

In the latter case one obtains by trial and error a value of the friction velocity $u_*$, which would satisfy the measured velocity profile $u = f(y)$ and Eqs. (2.7) and (2.9) at constant $\varkappa$ (for the cylinder we assumed $\varkappa = 0.43$).

The accuracy of skin friction values determined by the Clauser method was checked using the momentum method [method (2) in Fig. 2.16]. Using $u_\infty(x)$ and $u(y)$ values measured in several locations on the surface along the tube, we can obtain $\delta^*(x)$, $\delta^{**}(x)$, $d\delta^{**}/dx$, $V' = du_\infty/dx$, and $H(x)$. Substituting the values thus found into the momentum equation

$$\frac{d\delta^{**}}{dx} + \frac{V'}{u_\infty}\delta^{**}(2+H) = \frac{c_f}{2} \qquad (2.10)$$

we can determine $c_f(x)$ [63]. However, the accuracy in determining the drag coefficient depends entirely on the accuracy of numerical differentiation, which involves significant error, and hence this method was used only on a supplementary basis. This shortcoming can be eliminated by using Eq. (2.10) for obtaining the total drag coefficient $C_f$ over a given length of the cylinder. Integrating Eq. (2.10) from the initial (1) to the final (2) points of the selected length, Povkh [63] found

$$\frac{2}{L_{1-2}}\left[(\delta_2^{**} - \delta_1^{**}) + \int_1^2 \frac{\delta^{**}(2+H)}{u_\infty} du_\infty\right] = \frac{1}{L_{1-2}} \int_1^2 \frac{\tau_w}{\rho u_\infty^2} dx = C_{f_{1-2}} \qquad (2.11)$$

The above methods for determination of the skin friction from a measured velocity profile can be used for both smooth and rough surfaces with moderate freestream turbulence intensity. At high turbulence intensity of the free stream, the determination of skin friction by the aforementioned methods is made difficult by distortion of velocity profiles. In this case we used the projecting plank method [method (9) in Fig. 2.16]. The principle underlying this method is measuring the pressure difference at the wall upstream and downstream of a plate, protruding from the surface of the body by an amount not exceeding the thickness of the laminar sublayer. This problem was solved by Dean [77, 78] based on the following assumptions: (1) the velocity distribution far from the projection is linear, that is, $u = ky$, where $y$ is the distance from the body under study, and $k$ is a proportionality factor; (2) the liquid moves slowly, that is, it is possible to linearize the equations by dropping inertial terms. Dean obtained the approximate result

$$\Delta p = 2.90 \frac{\mu V}{h} \qquad (2.12)$$

where $\Delta p$ is the pressure difference in the flow up- and downstream of the step, $h$ is the height of the step, and $V$ is the undisturbed flow velocity at $y = h$. The last equation yields

$$\tau_w = k\Delta p \qquad (2.13)$$

where, according to the analytic solution, $k = 0.345$. This solution can be used for experimental determination of the friction forces.

Konstantinov and Dragnysh [79] developed an instrument for determining friction by the above method and checked the limits of applicability of Eq. (2.13). They determined the numerical value of the proportionality constant under different operating conditions of the sensor. They concluded on the basis of direct analysis of results that the proportionality factor $k$ is highly dependent on the step height. An analogous conclusion was drawn also by Rechenberg [80], but since the sensor operating conditions in the protruding-plank system may differ greatly from Dean's assumptions, he suggested that the wall stresses be calculated from the expression

$$\tau_w = \Delta p^n \qquad (2.14)$$

where $n \approx 0.7$.

We constructed sensors for measuring the skin friction by the protruding plank method, placed them on a cylinder, and calibrated them. Calibration was performed at different pressures (in accordance with the planned experiment). It was found that a power law relationship exists between $\tau_w$ and $\Delta p$, that is, a relationship expressed by Eq. (2.14), in which $n$ for individual sensors ranged from 0.61 to 0.67.

The protruding plank method can be used only without heat transfer at the surface. The existence of high-rate heat transfer, accompanied by changes in physical properties in the boundary layer, significantly complicates the problem. The skin friction at the cylinder surface during a variable-property flow of the coolant was determined by us from an integral momentum equation, written for an incompressible fluid flowing isothermally over a cylinder, and, from an analogous equation, written for the case when heat transfer is present, that is,

$$\frac{d\delta^{**}}{dx} + \frac{\delta^*}{u_\infty}\frac{du_\infty}{dx} = \frac{c_f}{2} \qquad (2.15)$$

$$\frac{d\delta_q^{**}}{dx} + \frac{\delta_q^*}{u_\infty}\frac{du_\infty}{dx} = \frac{c_{fq}}{2} \qquad (2.16)$$

In our case the second terms on the left-hand sides of Eqs. (2.15) and (2.16) can be neglected, since they are small compared to the first term (Table 2.2); then, dividing Eq. (2.16) by Eq. (2.15), we obtain

$$\frac{d\delta_q^{**}}{d\delta^{**}} = \frac{c_{fq}}{c_f} = K \qquad (2.17)$$

Equation (2.17) was used for determining the skin friction in the presence of heat transfer. For this we first measured velocity profiles and calculated the values of $c_f$ and $\delta^{**}$ in the absence of heat transfer. Then the cylinder was

**TABLE 2.2** Mean Values of the First and Second Terms in Eq. (2.15)

| No. | $x$, m | $\dfrac{d\delta^{**}}{dx}$ | $\dfrac{\delta^*}{u_\infty}\dfrac{du_\infty}{dx}$ |
|---|---|---|---|
| 1 | 0.325 | 0.00236 | 0.00008 |
|   | 0.525 | 0.00223 | 0.000049 |
| 2 | 0.325 | 0.00227 | 0.000073 |
|   | 0.525 | 0.00212 | 0.00004 |
|   | 0.825 | 0.00206 | 0.000009 |
| 3 | 0.325 | 0.00222 | 0.000058 |
|   | 0.525 | 0.00203 | 0.000032 |
|   | 0.825 | 0.00198 | 0 |
| 4 | 0.325 | 0.00199 | 0.000045 |
|   | 0.525 | 0.00189 | 0.000021 |
|   | 0.825 | 0.00183 | 0 |
| 5 | 0.325 | 0.00196 | 0.000033 |
|   | 0.525 | 0.00182 | 0.00001 |
|   | 0.825 | 0.00174 | 0 |

heated, and $\delta_q^{**}$ was calculated from the measured velocity profile of the hot cylinder boundary layer. Once several velocity profiles were measured along the cylinder with and without heat transfer, it was possible to determine the ratio $d\delta_q^{**}/d\delta^{**}$ graphically. Such a "relative" approach allowed a rather accurate determination of the skin friction coefficient in the presence of heat transfer [82].

The *turbulence intensity* of the flow is originally determined by the design of the wind tunnel and by the operating conditions. A device that reduced vibrations and equalized the velocity field was placed upstream of the test section, providing turbulence intensity of 0.8–2%. The flow turbulence intensity could be raised to 6.3% by placing special turbulence-generating grids upstream of the test section. We used square grids of two types: grid type 1 was made of tubes with $d = 10$ mm, spaced at 30 mm; type 2 was made of tubes with $d = 20$ mm, spaced at 40 mm. To increase the velocity fluctuations and to transport them to the test section, 150-mm-long caprone strips, having a width equal to that of the opening in the grid, were fastened to the latter. As the gas flowed, the tips of the strips reached the narrow part of the inlet section.

The turbulence intensity was determined by the tensometric and hot-wire anemometer techniques. Since no hot-wire anemometer was available at the initial stage of our studies, the bulk of the experiments were performed using the tensometer technique. After an anemometer was obtained, it was used for a careful check of the results obtained by the other method. The turbulence intensity determined by the two instruments differed by not more than 5%.

According to the tensometric method, the turbulence intensity was determined from the longitudinal component of velocity fluctuations that were calculated from the measured longitudinal fluctuations in the dynamic pressure. The longitudinal velocity fluctuation can be represented as the difference between the instantaneous ($u$) and mean ($\bar{u}$) velocity:

$$u' = u - \bar{u} \qquad (2.18)$$

or

$$u' = \sqrt{\frac{2}{\rho}}\left(\sqrt{\Delta H \pm \Delta h} - \sqrt{\Delta H}\right) \qquad (2.19)$$

where $\Delta H$ is the velocity head, corresponding to the mean velocity, whereas $\Delta h$ is the fluctuating component of the velocity head. Transforming Eq. (2.19) on the assumption that $\Delta H \gg \Delta h$, we obtain the following relationship between longitudinal velocity fluctuation $u'$ and the longitudinal fluctuation in the velocity head $\Delta h$:

$$u' = \frac{\Delta h}{\rho \bar{u}} \qquad (2.20)$$

The square of the root mean square (rms) of the fluctuating velocity component is

$$\overline{u'^2} = \frac{1}{n}(u_1'^2 + u_2'^2 + \ldots + u_n'^2) \tag{2.21}$$

Substituting this expression into Eq. (2.20) yields

$$\overline{u'^2} = \frac{1}{\rho^2 \bar{u}^2} \cdot \frac{1}{n}(\Delta h_1^2 + \Delta h_2^2 + \ldots + \Delta h_n^2) \tag{2.22}$$

or

$$\overline{u'^2} = \frac{1}{\rho^2 \bar{u}^2} \cdot \frac{1}{n} \sum_{i=1}^{n} \Delta h_i^2 \tag{2.23}$$

The value of the longitudinal component of velocity head $\Delta h_i$ was measured by tensometric sensors whose signal was amplified by the strain-reading unit and then recorded by a loop oscillograph. The rms longitudinal velocity component $\overline{u'^2}$ was calculated from the measured values of $\Delta h_i$, and this made possible calculating the turbulence intensity using the expression

$$Tu = \frac{\sqrt{\overline{u'^2}}}{\bar{u}} \cdot 100\% \tag{2.24}$$

We used two tensometric sensors, their design and dimensions are shown in Fig. 2.17. One sensor was used under atmospheric, and the other at elevated pressures. The latter differed from the former only by the fact that a hole with $d = 0.2$ mm was drilled in the sensor housing, in order to equalize the static pressure on both sides of the membrane.

The sensors were subjected to static calibration before the experiments, and the fluctuating component of the velocity head was determined from the expression

$$\Delta h_i = K \cdot m_i \tag{2.25}$$

where $m_i$ is the ordinate of $\Delta h_i$ fluctuations as seen on the oscillograph tape, and $K$ is the calibration constant of the sensor.

Results of the turbulence intensity measurements are shown in Table 2.3 and in Fig. 2.18. As seen in the figure, the turbulence intensity of the free stream was a function not only of $Re_x$, but also of $Re_{r_0}$. A rise in $Re_{r_0}$ (at $r_0 =$ constant) reduces the turbulence intensity. At constant $Re_{r_0}$ the natural turbulence intensity of the flow created by the test equipment increases insignificantly along the annulus (from 1.7 to 2% and from 0.85 to 1.4%, respectively, at $Re_{r_0} = 3.98 \times 10^4$ and $1.69 \times 10^5$) approaching Tu of fully developed pipe flow [83].

When turbulence-generating grids and caprone tapes were used, the turbu-

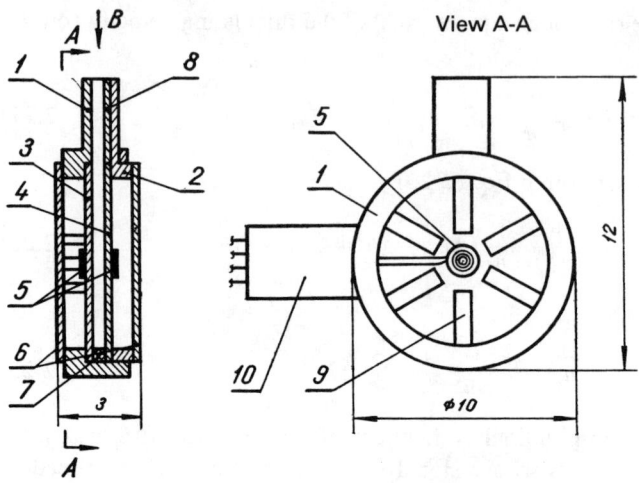

**FIG. 2.17** Tensometer probe for measuring turbulence intensity. (1) Left half of sensing-element housing, (2) right half of sensing-element housing, (3) thermal compensation membrane 0.05–0.1 mm thick, (4) working membrane 0.05–0.1 mm thick, (5) strain gauges, (6) housing covers, (7) spacer, (8) complementing component with thickness equal to that of the membrane, (9) stiffness ribs of thermal compensation membrane, (10) attachment to holder and leadout to the strain-reading unit. $B$ is the direction of the turbulent flow.

lence increased sharply at the test section inlet, that is, immediately downstream of the tapes. Then it rapidly decreased along the flow axis (from 6.28% at the start of the annulus to 4.16% at its end, and from 3.37 to 2.58% at $Re_{r_0} = 4 \times 10^4$ and $1.49 \times 10^5$, respectively). All the above data were used for construct-

**TABLE 2.3** Results of Turbulence Intensity Measurement

| No. | $x$, m | $Re_x \cdot 10^{-5}$ | $\bar{u}$, m/sec | $Re_{r_9} \cdot 10^{-5}$ | $T_f$, K | $u'^2$ | $Tu$, % |
|---|---|---|---|---|---|---|---|
| | | Natural turbulence (without turbulence-generating grid) | | | | | |
| 1 | 0.225 | 6.4 | 48.8 | 0.398 | 305.7 | 0.685 | 1.7 |
| 2 | 0.225 | 8.5 | 46.9 | 0.525 | 308.2 | 0.469 | 1.4 |
| 3 | 0.225 | 27.1 | 47.1 | 1.69 | 310.4 | 0.165 | 0.8 |
| 4 | 0.825 | 25.3 | 49.9 | 0.398 | 306.2 | 1.04 | 2.0 |
| 5 | 0.825 | 103.0 | 48.9 | 1.69 | 309.9 | 0.475 | 1.4 |
| | | Elevated turbulence (with grid 2) | | | | | |
| 6 | 0.225 | 6.44 | 44.7 | 0.4 | 301.2 | 7.91 | 6.28 |
| 7 | 0.225 | 7.49 | 43.0 | 0.462 | 310.0 | 5.1 | 5.27 |
| 8 | 0.225 | 24.0 | 43.3 | 1.49 | 310.1 | 2.11 | 3.37 |
| 9 | 0.825 | 23.3 | 44.8 | 0.4 | 302.7 | 3.46 | 4.16 |
| 10 | 0.825 | 95.9 | 45 | 1.49 | 309.4 | 1.35 | 2.58 |

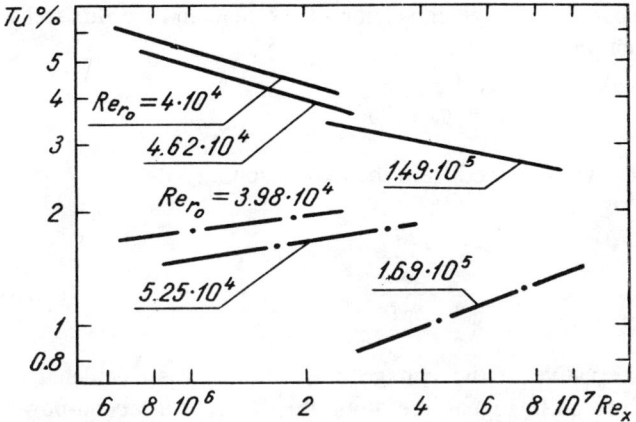

**FIG. 2.18** Turbulence intensity distribution along the annulus test section as a function of $Re_{r_0}$. The dash-dotted lines represent the flow turbulence excited naturally by the test unit. The solid lines were obtained at elevated turbulence.

ing a working nomogram of Tu along the test section as a function of $Re_{r_0}$ for both cases of flow turbulence.

The turbulence intensity along the flow near the wall was determined by a DISA 55D01 constant-temperature hot-wire anemometer, the operating principle of which is described in detail by Žukauskas and Šlančiauskas [52]. The sensing element in our studies was the 55A22, which was calibrated in a special wind tunnel.

## 2.4 TECHNIQUE FOR PROCESSING DATA ON HEAT TRANSFER AND HYDRAULIC DRAG

*Determination of heat transfer in the entrance region of an annulus.* Generation of heat flux directly on the test surface by passing current through it makes it possible to easily and rather accurately determine all the principal quantities in the Newton-Richman law, defining the heat transfer coefficient as

$$dQ = \alpha(T_w - T_f)dF \qquad (2.26)$$

or

$$\alpha = \frac{dQ}{(T_w - T_f)\,dF} = \frac{q}{T_w - T_f} \qquad (2.27)$$

Here $q$ is the heat flux in the section of the cylinder under study, transferred to the cooling gas by convection and defined in our case as the difference between the entire heat flux released in the given section of the tube and the flux transmitted by thermal radiation

$$q = q_{el} - q_r$$

The local heat flux was obtained from the expression

$$q_{el} = \frac{I \dfrac{dU}{dx}}{\pi d} \qquad (2.28)$$

where $dU/dx$ is the derivative of the voltage drop, which was calculated by approximating the values of $U_x$, measured along the tube, by a second-power polynomial using the least squares method.

The radiation heat flux was determined experimentally during special calibration runs. For this purpose the test section was removed from the tunnel cycle and made airtight. The inner space was evacuated to 10–20 Pa, and electric current was passed through the inner cylinder. In such low pressure, heat is transmitted from the cylinder primarily by radiation. The experiments were performed over the same temperature range of the inner and outer annulus walls and in the same sequence of their variation as in the main experiment. The experiments were performed with an inner cylinder whose dimensions, steel brand, and surface finish were the same as those of the inner tube used in the main experiments. The temperature of the outer cylinder was maintained at the specified level by cooling it with water. The relation between the radiant thermal flux and the inner cylinder wall temperature at two different values of the outer-tube temperature $T_e$ is plotted in Fig. 2.19. Both calibration and main experiments were performed with new inner tubes. The measurements started at the lowest temperatures and were terminated upon attaining the highest temperatures. This made it possible to uniquely account for the variation in the radiant flux as a function of the change in surface absorbance, resulting from the formation of oxides on the surface of the experimental tube.

The temperature $T_w$ of the cooled surface of the outer tube was calculated from the measured temperature of its inner surface $T_w'$ and the calculated temperature drop across the wall with allowance for internal heat sources:

$$T_w = T_w' - \frac{q_{el} d_1}{4\lambda}\left(1 - \frac{2 d_{1\,in}^2}{d_1^2 - d_{1\,in}^2} \ln \frac{d_1}{d_{1\,in}}\right) \qquad (2.29)$$

The heat fluxes and wall temperatures were calculated with allowance for variation in linear dimensions (length and diameter) of the cylinder, due to changes in its temperature. The tube was made of 1Kh18N10T stainless steel, whose physical properties are given by Andreyev et al. [84].

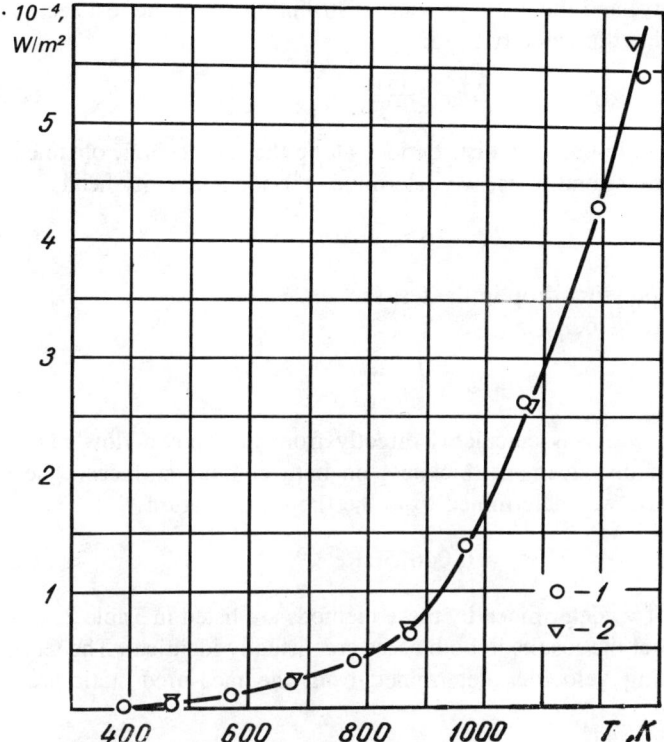

**FIG. 2.19** Graph for determining heat losses due to radiation from tubes. (1) at $\overline{T}_e$ between 320 and 335 K; (2) at $\overline{T}_e$ between 280 and 300 K.

The flow temperature $T_f$ outside the boundary layer was measured by a traversing thermocouple probe in several locations along the annulus. These measurements served as a basis for a graph of $T_f = f(x)$, from which the value of $T_f$ could then be determined for any $x$.

The air flow rate was calculated from the measured pressure drops across the orifice and the static pressure upstream of the orifice:

$$G = A\left(1 - B\frac{\Delta p}{P}\right)\sqrt{\Delta p\, \rho} \qquad (2.30)$$

where $A$ and $B$ are calibration constants.

Assuming constant mass flow rate of air in the tunnel cycle and knowing the inlet pressure and temperature, we determined the velocity at the test annulus inlet. As was shown experimentally this velocity, was constant over the entire cross section.

According to Petukhov and Mukhin [85], the total pressure in the flow core of short ducts does not change axially. Thus, knowing the inlet velocity $u_0$, the freestream velocity, and the static pressure distribution along the test section, it is easy to determine the velocity head:

$$\Delta h_0 = \frac{u_0^2 \, \rho_0}{2} \qquad (2.31)$$

Then using static-pressure distributions along the test section, obtained for each set of operating conditions, we determined the local velocity head,

$$\Delta h_x = \Delta h_0 + \Delta p_x \qquad (2.32)$$

where $\Delta p_x$ is the pressure drop in the test section.

The flow core velocity is

$$u_{\infty x} = \sqrt{\frac{2\Delta h_x}{\rho_x}} \qquad (2.33)$$

The velocity $u_{\infty x}$ was calculated directly from the known flow rate with correction for the displacement thickness on both annulus surfaces. The displacement thickness was determined from the flat plate relation

$$\delta_x^* = 0.04625 \, x \mathrm{Re}_x^{-0.2} \qquad (2.34)$$

The values of $u_\infty$ determined by these methods are listed in Table 2.4, from which it is seen that the results in both cases are virtually identical. The test data were reduced using velocities determined from the measured static-pressure drops $\Delta p_x$.

**TABLE 2.4** Comparison of Flow Core Velocity Values Determined from the Static Pressure Drop ($u_\infty$) with Values Corrected for the Displacement Thickness ($u'_\infty$)

| x, mm | P, MPa | $T_f$, K | $u_\infty$, m/sec | $\delta_x^* \cdot 10^2$, mm | $u'_\infty$, m/sec |
|---|---|---|---|---|---|
| 20.0 | 0.14 | 308.0 | 49.0 | 9.6 | 49.0 |
| 40.5 | 0.14 | 307 | 49.1 | 16.7 | 49.2 |
| 100.5 | 0.14 | 307 | 49.6 | 34.8 | 49.6 |
| 251.0 | 0.14 | 307 | 50.6 | 72.3 | 50.6 |
| 499.5 | 0.139 | 307 | 52.2 | 126.1 | 51.9 |
| 859.0 | 0.139 | 307 | 54.0 | 194.5 | 53.8 |
| 20.0 | 2.33 | 308.1 | 45.4 | 5.70 | 45.5 |
| 40.5 | 2.33 | 308.1 | 45.5 | 10.04 | 45.6 |
| 100.5 | 2.33 | 308.0 | 45.9 | 20.75 | 45.8 |
| 251.0 | 2.33 | 308.0 | 46.7 | 43.14 | 46.3 |
| 499.5 | 2.33 | 307.9 | 47.3 | 74.88 | 47.1 |
| 850.0 | 2.33 | 308.0 | 48.7 | 115.52 | 48.0 |

The reduced data were then expressed in terms of $Nu_x$ and $Re_x$, where the reference dimension is the distance from the start of heating to the cross section under study, that is, the axial coordinate $x$.

*Determination of heat transfer and hydraulic drag coefficients at the hydrodynamic fully developed region of the annulus.* It was established by many investigators that the effect of variable physical properties of gaseous coolants on heat transfer in circular tubes at moderate heating rates is virtually independent of Re. If it is assumed that the same applies also for annuli (which is confirmed experimentally by our results), it is possible to divide the determination of heat transfer at variable physical properties into two parts. One part consists of determining the effect of Re on heat transfer at constant physical properties, and the other part is determining the effect of variability of physical properties. Here the error in determining the expression

$$Nu = f(Re, Pr, x/d) \tag{2.35}$$

has virtually no effect on the accuracy of determining the expression

$$\frac{Nu}{Nu_{\Psi=1}} = f(\Psi, x/d) \tag{2.36}$$

The above considerations dictated the selection of experimental techniques and data-processing methods. All the experiments were performed in sets, with each of the latter performed at 10–40 different experimental conditions. There were two types of sets. In the first type, the surfaces of the tubes forming the annulus were maintained at a constant temperature, and the conditions differed by the values of Re and of the heat flux. In sets of the second type, the flow rate of the coolant, and thus $Re_{in}$, were kept constant. The operating conditions differed by the surface temperatures of the walls and by the heat flux. Two to six sets of experiments were performed with each tube. This technique made it possible to not only perform experiments at optimal (from the point of view of accuracy of the experiment) relations between operating conditions, but also to obtain data in a form most suitable for analysis and correlation.

Since the experiments were performed with two different test sections, the data had to be reduced by two techniques, which differed somewhat from one another. Here we describe the technique for processing the data obtained with test section 2 (with both walls heated), where a single program was used for calculating the heat transfer through the inner and outer tubes at any ratio of heat fluxes. The data-processing procedure used for results from test section 1 will be described separately.

Preliminary reduction of test data was conducted to determine the values of Nu, Re, and Pr and to find elementary relationships such as Eqs. (2.35) and (2.36). The reference temperature was the bulk temperature of the flow $T_f$, the reference dimension was the equivalent diameter $d = 4F/\pi = d_2 - d_1$, whereas the reference velocity was the bulk velocity.

The local heat transfer coefficient was calculated from Eq. (2.27), in which

$$q = \frac{q_l}{\pi d_{1,2}} \tag{2.37}$$

The convective component of heat flux $q_l$ was calculated with correction for the radiant exchange between the inner and outer tubes, losses to the surroundings, and heat dissipating among the annulus walls:

$$q_{l1} = q_{l\,el1} - q_{lr1} - \Delta q_{l1} \tag{2.38}$$

$$q_{l2} = q_{l\,el2} - q_{lr2} - q_{lfp} - \Delta q_{l2} \tag{2.39}$$

where $q_{l\,el} = I(du/dx)$ is the flux of electrical heat released, $q_{lr}$ is the radiant component of the heat flux, $q_{lsh}$ is the heat flux lost through the screens, and $\Delta q_l$ is the change in the convective component of the heat flux as a result of heat dissipation along the wall.

The radiant exchange between the tubes and also between the outer tube and the shield was calculated according to recommendations of Isachenko et al. [86] for both coaxial and infinitely long cylinders:

$$q_{lr1} = \frac{5.67 \cdot 10^{-8} \left(T_{w1}^4 - T_{w2(1)}^4\right) \pi d_1}{\dfrac{1}{\varepsilon_{w1}} + \dfrac{d_1}{d_2}\left(\dfrac{1}{\varepsilon_{w2(1)}} - 1\right)} \tag{2.40}$$

$$q_{lr2} = \frac{5.67 \cdot 10^{-8} \left(T_{w2}^4 - T_{w1(2)}^4\right) \pi d_1}{\dfrac{1}{\varepsilon_{w1(2)}} + \dfrac{d_1}{d_2}\left(\dfrac{1}{\varepsilon_{w2}} - 1\right)} \tag{2.41}$$

$$q_{lfp} = \frac{5.67 \cdot 10^{-8} \left(T_{w2}^4 - T_{sh(2)}^4\right) \pi d_{2out}}{\dfrac{1}{\varepsilon_{w2}} + \dfrac{d_{2out}}{d_{sh}}\left(\dfrac{1}{\varepsilon_{sh(2)}} - 1\right)} \tag{2.42}$$

where $d_{2out}$ is the outer diameter of the outer tube, $d_{sh}$ is the inner diameter of the first reflecting shield, and $\varepsilon$ is the emissivity of the surface. The subscript in parentheses refers to the thermocouple location where the quantity in question is obtained. The values of $T_{w1(2)}$, $T_{w2(1)}$ and $T_{sh(2)}$ were determined by quadratic interpolation of experimental values of $T = f(x)$ obtained in tabulated form.

On the basis of data analysis by Blokh [87], the emissivity of the tubes was determined from the expression

$$\varepsilon_w = 0.58 + 0.17 \cdot 10^{-3}\, T_w \tag{2.43}$$

and for polished shields from the equation

$$\varepsilon_{el} = 0.12 + 0.32 \cdot 10^{-3}\, T_{el} \tag{2.44}$$

The variation of the convective component of the heat flux due to longitudinal heat dissipation within the tube walls was determined from the expression

$$\Delta q_l = -\lambda_w \frac{\Delta\left(\frac{\Delta T_w}{\Delta x}\right)}{\Delta x} \pi \bar{d} \delta \qquad (2.45)$$

where $\lambda_w$ is the thermal conductivity of the tube material, $\bar{d}$ is the mean tube diameter, and $\delta$ is the wall thickness. The value of $\Delta(\Delta T_w/\Delta x)/\Delta x$ was determined by differentiating the tabulated values of $T_w = f(x)$ twice.

In all the calculations a correction was made for changes in linear dimensions due to thermal expansion.

The temperature of the cooled surface of the tubes was calculated from the measured temperature of the noncooled surface $T'_w$ and the temperature difference across the wall.

The maximum airflow velocity did not exceed 70 m/s, which made it possible to neglect compressibility effects. The bulk flow temperature $T_f$ was determined at given locations along the annulus from the local total adiabatic stagnation enthalpy $i$, the value of which, was calculated from the expression

$$i = i_{in} + \frac{Q_1 + Q_2}{G} \qquad (2.46)$$

where

$$Q_1 = I_1 U_1 - Q_{l1} \qquad (2.47)$$

$$Q_2 = I_2 U_2 - Q_{l2} - Q_{sh} \qquad (2.48)$$

Here $Q_l$ is the amount of heat flowing through the section under study due to longitudinal conduction in the walls, and $Q_{sh}$ is the quantity of heat lost through the shields in the section under study.

From the Fourier law,

$$Q_l = -\lambda_w \frac{\Delta T_w}{\Delta x} \pi \bar{d} \delta \qquad (2.49)$$

The value of $T_f$ was obtained by quadratic interpolation of the enthalpy and linear interpolation of the enthalpy pressure, taken from tables of physical properties of air and fed to the computer.

The surface of the nonheated tube was not adiabatic due to radiative heat transfer, axial heat fluxes in the walls, and heat losses. Thus values of Nu calculated for a heated surface do not, strictly speaking, conform to the case of heating from one side. When only one wall was heated, Nu values of the heated surface were corrected for heat losses from the opposite wall by modified Petukhov and Royzen correlations [17]:

$$\text{Nu}_{ad} = \frac{\text{Nu}_1}{1 - \dfrac{q_2}{q_1}\Theta_1 \text{Nu}_1} \qquad (2.50)$$

$$\text{Nu}_{ad} = \frac{\text{Nu}_2}{1 - \dfrac{q_1}{q_2}\Theta_2 \text{Nu}_2} \qquad (2.51)$$

where $\Theta$ is the dimensionless adiabatic wall temperature.

According to Petukhov and Royzen, in the fully developed heat-transfer region,

$$\Theta_1 = 22\left[0.27\left(\frac{d_1}{d_2}\right)^2 - 1\right]\text{Re}^{-0.87}\text{Pr}^{-1.05}$$

$$\Theta_2 = \Theta_1 \frac{d_1}{d_2} \qquad (2.52)$$

Under the most unfavorable conditions ($T_{w\max}$, $x_{\max}$, and $\text{Re}_{\min}$), the maximum difference between Nu and $\text{Nu}_{ad}$ did not exceed 2%. The effect of heat losses from the nonheated wall was negligible in the entrance region.

The results of measurements obtained with test section 1 were processed according to a somewhat simplified technique. Longitudinal heat fluxes in the thermocouple locations were insignificant and were neglected. The annulus surface was not preroasted, and the emissivity was determined from the graph (Fig. 2.20) constructed on the basis of the data of Blokh [87]. The heat flux lost through insulation of the outer tube was calculated from the measured temperature differences across a layer of asbestos insulation of a given thickness.

The local aerodynamic drag coefficients for adiabatic and nonadiabatic flows were calculated from the formula

$$\xi = \frac{[dp/dx + d(\rho u^2)/dx]\, d}{\rho u^2/2} \qquad (2.53)$$

The static pressure $P$ was determined on the basis of the inlet pressure $P_{in}$ and the static pressure difference $\Delta p$. The value of $dp/dx$ was found by graphic differentiation of measured values of $\Delta p$. The local velocity was calculated from the equation

$$u = \frac{G}{F\rho} \qquad (2.54)$$

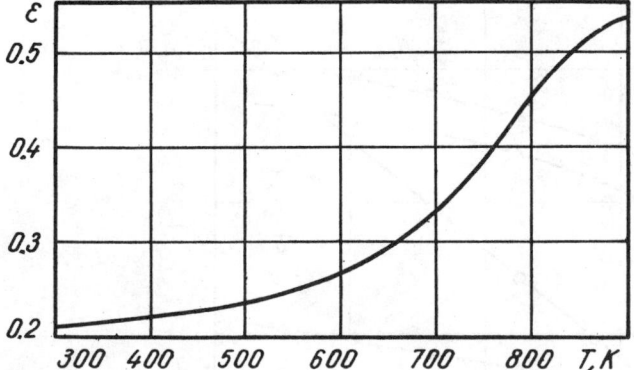
**FIG. 2.20** Graph for determining the emissivity of polished 1Kh18N10T steel.

and the local density was obtained from the Clayperon-Mendeleyev (ideal gas) equation

$$\rho = \frac{P}{RT_f} \qquad (2.55)$$

## 2.5 CALIBRATION AND RUNNING OF EXPERIMENTS

Calibration and initial experimental tests included selection of blanks for the tubes, calibration of thermocouples, and experiments for determining heat losses and for determining heat transfer with a circular tube.

The chromel-alumel wire was roasted prior to experiments. The chromel-alumel thermocouples were calibrated in a specially constructed tubular furnace by comparing the readings of the thermocouples under calibration with readings of a reference platinum—platinum-rhodium thermocouple, or a reference platinum resistance thermometer. Several thermocouples were calibrated from each wire batch. The copper-constantan thermocouples were calibrated in an oil-type constant-temperature bath against a reference mercury thermometer calibrated in 0.1°C divisions.

The heat losses in the annulus (test section 2) were determined by heating the tubes without flow of cooling air through them. For this purpose, the heating power needed for maintaining a constant preset temperature of the outer tube was determined without flow of air through the annulus. Experiments performed at several temperature levels showed that directly measured heat fluxes are in satisfactory agreement with those calculated from the temperature difference in the shields. The absolute values of heat losses through the shields and at the ends of the tubes were insignificant and, in the most unfavorable cases (at high temperatures and low flow rates), did not exceed 5% of the total amount of heat released in the annulus.

**52** HEAT TRANSFER OF GAS COOLANTS

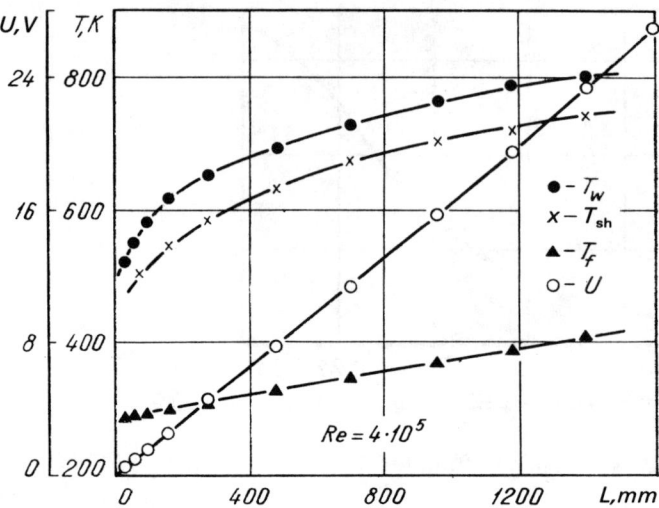

**FIG. 2.21** Distribution of principal measurement parameters along the heated part of the tube.

To comprehensively check the experimental equipment and technique for reduction of experimental data, we performed three sets of experiments for determining the effect of the temperature factor on heat transfer in a circular tube, i.e., with the inner tube removed. The sets differed by the value of Re, whereas the test runs within each set differed by the value of temperature factor $\psi$.

Figure 2.21 shows the manner in which the principal quantities under study ($T_w$, $T_f$, $T_{sh}$, and $U$) change along the channel. The very small scatter of experimental points shows that the random error of measurements is insignificant.

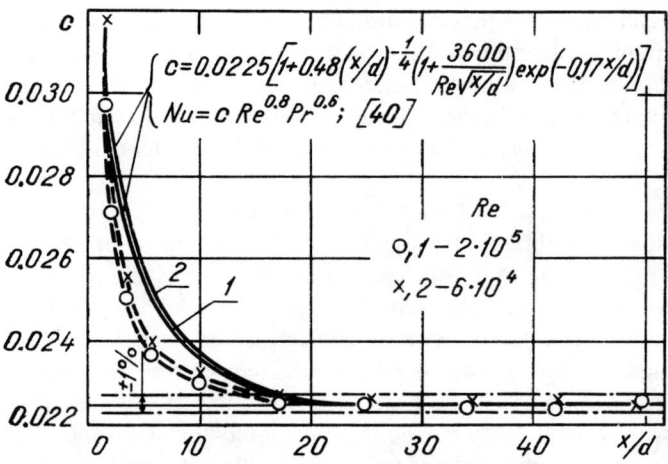

**FIG. 2.22** Heat transfer data in a circular tube, adjusted to conditions of constant physical properties.

Eliminating the effect of the temperature factor, test results at different values of $x/d$ were correlated in the form

$$\mathrm{Nu} = c\mathrm{Re}^{0.8}\,\mathrm{Pr}^{0.6} \tag{2.56}$$

Values of $c$ obtained by us in the fully developed heat-transfer region ($x/d > 20$) at $\mathrm{Re} \leq 2 \times 10^5$ agree within 1% with those calculated by Petukhov et al. [40]. However, in our experiments, fully developed conditions were reached more rapidly in the entrance region (Fig. 2.22).

Analysis and estimate of experimental error performed by standard techniques showed that the systematic rms error in determining Re is 0.8–0.9%, whereas that of Nu ranges from 0.5 to 1.2%.

The information obtained during test runs was used for finally adjusting the experimental technique.

CHAPTER
# THREE

## VELOCITY AND TEMPERATURE DISTRIBUTIONS IN THE BOUNDARY LAYER OF A CYLINDER IN AXIAL FLOW

To investigate the relations governing the flow and heat transfer, we measured the velocity and temperature profiles in the boundary layer on a cylinder in axial flow. The self-similar nature of the boundary layer development was observed by comparing the measured velocity profiles with universal relations of velocity distributions in accordance with the law of the wall and flow retardation. We simultaneously determined the variation of velocity and temperature distributions as a function of density and viscosity in the boundary layer at high heat fluxes. We also measured the variation of these profiles as a function of the freestream turbulence intensity and that due to the effect of a single turbulence promoter. The results of direct measurements are given in the appendixes.

## 3.1 VELOCITY PROFILES WITHOUT HEAT TRANSFER

The boundary layer of a cylinder in axial flow, when the cylinder is short and the boundary layer thickness is significantly smaller than the cylinder radius, corresponds to the boundary layer on a cylinder with a small transverse curvature and can be approximated by equations for flow in the boundary layer on a flat plate. When the cylinder is sufficiently long or when it has a small radius (when $\delta \geq r_0$), a correction for the transverse curvature must be made. This case is the most general and hence shall be considered in detail.

## 56 HEAT TRANSFER OF GAS COOLANTS

The governing equations of a steady-state boundary layer on a long cylinder in axial flow can be derived from the Navier-Stokes and continuity equations:

$$\frac{\partial u}{\partial t} + u\frac{\partial u}{\partial x} + v\frac{\partial u}{\partial r} = -\frac{1}{\rho}\frac{\partial p}{\partial x} + \nu\left(\frac{\partial^2 u}{\partial x^2} + \frac{\partial^2 u}{\partial r^2} + \frac{1}{r}\frac{\partial u}{\partial r}\right)$$

$$\frac{\partial v}{\partial t} + u\frac{\partial v}{\partial x} + v\frac{\partial v}{\partial r} = -\frac{1}{\rho}\frac{\partial p}{\partial r} + \nu\left(\frac{\partial^2 v}{\partial x^2} + \frac{1}{r}\frac{\partial v}{\partial r} - \frac{v}{r^2}\right) \quad (3.1)$$

$$\frac{\partial}{\partial x}(ru) + \frac{\partial}{\partial r}(rv) = 0$$

If we drop the second equation (assuming $\partial p/\partial r = 0$) and the term $\partial^2 u/\partial x^2$ on the right-hand side of the first equation (since it is small compared to other terms), then, replacing $\partial/\partial r$ by $\partial/\partial y$ and introducing the value of velocity at the outer edge of the boundary $u_\infty(x)$, we obtain the following set of equations for the steady-state boundary layer:

$$u\frac{\partial u}{\partial x} + v\frac{\partial u}{\partial y} = u_\infty\frac{du_\infty}{dx} + \nu\left(\frac{\partial^2 u}{\partial y^2} + \frac{1}{r}\frac{\partial u}{\partial y}\right)$$
$$\frac{\partial}{\partial x}(ru) + \frac{\partial}{\partial y}(rv) = 0 \quad (3.2)$$

with the boundary conditions

$$u = 0, \ v = 0 \text{ at } y = 0$$
$$u \to u_\infty(x) \text{ as } y \to \infty \quad (3.3)$$

The integration of Eqs. (3.2) with the boundary conditions of Eqs. (3.3) is very difficult even in the simple case of a uniform flow over a circular cylinder with constant radius $r_0$ and no longitudinal pressure gradient ($du_\infty/dx = 0$). In this case, $r = r_0 + y$ and Eq. (3.2) becomes

$$u\frac{\partial u}{\partial x} + v\frac{\partial u}{\partial y} = \nu\left(\frac{\partial^2 u}{\partial y^2} + \frac{1}{r_0+y}\frac{\partial u}{\partial y}\right) \quad (3.4)$$

$$\frac{\partial u}{\partial x} + \frac{\partial v}{\partial y} + \frac{v}{r_0+y} = 0 \quad (3.5)$$

or

$$\rho\left(u\frac{\partial u}{\partial x} + v\frac{\partial u}{\partial y}\right) = \frac{\partial \tau}{\partial y} + \frac{1}{r_0+y}\tau \quad (3.6)$$

$$\frac{\partial u}{\partial x} + \frac{\partial v}{\partial y} + \frac{v}{r_0+y} = 0 \quad (3.7)$$

with the boundary conditions $u = 0, \ v = 0$ at $y = 0$ and $u = u_\infty$ at $y = \infty$.

The main difficulty in solving this problem is due to the presence of two quantities ($u_\infty$ and $r_0$) that govern the motion. It is not possible to reduce the set of Eq. (3.4) and (3.5) to a single equation.

Near the upstream end of the cylinder, where the boundary layer thickness is still small compared to the radius of the cylinder, the effect of the transverse curvature can be neglected. Then, the development of the boundary layer will not differ from the boundary layer development on a flat plate. In fact, as $r_0 \to \infty$, Eqs. (3.4) and (3.5) become ordinary differential equations of a flat plate boundary layer if there is no pressure gradient in the mainstream.

The first corrections for the transverse curvature in the problem under study, for the case of laminar flow, are given by Seban and Bond [88] and in a somewhat more refined form by Kelly [89]. It was shown that the Blasius expression for wall shear stress $\tau_w = 0.332 \mu u_\infty \sqrt{u_\infty / \nu x}$ should be corrected by the factor $0.7 \; \mu u_\infty / r_0$, which at $\nu x / u_\infty \; r_0^2 = 0.001$ gives a correction of 7%.

The shear stress distribution $\tau$ for turbulent flow is given by the expression

$$\tau = \mu \frac{du}{dy} - \rho \overline{u'v'} \qquad (3.8)$$

with the boundary conditions

$$\begin{aligned} \tau &= \tau_w \text{ at } y = 0 \\ \tau &= 0 \text{ at } y = \infty \end{aligned} \qquad (3.9)$$

Here $u'$ and $v'$ are the fluctuating velocity components.

Equations (3.6)–(3.8) serve as the starting point for analytic studies of turbulent axial flow over a cylindrical surface. Note that these equations can be used to calculate the mean velocity only if the relation between $u'$, $v'$, the time-averaged values of $\overline{u'v'}$, and the average velocity is known. Such relations can be established only empirically. Then, the mean parameters of the boundary layer must be calculated using integral equations, obtained from the momentum-loss balance [3].

Let us analyze the flow in a region close to the surface of the cylinder. In the thin flow region in the immediate vicinity of the wall, the transverse velocity component can be neglected (i.e., $v = 0$). Then it follows from the continuity equation [Eq. (3.7)] that $\partial u / \partial x = 0$ also. In this case the equation of motion, Eq. (3.6), becomes

$$\frac{\partial \tau}{\partial y} + \frac{1}{r_0 + y} \tau = 0 \qquad (3.10)$$

The solution of this equation with the boundary conditions of Eq. (3.9) is

$$\tau = \tau_w \frac{r_0}{r_0 + y} \qquad (3.11)$$

Combining Eqs. (3.8) and (3.11) we have the following expression for the flow in the immediate vicinity of the wall:

$$\tau_w \frac{r_0}{r_0 + y} = \mu \frac{du}{dy} - \overline{\rho u' v'} \qquad (3.12)$$

The turbulent transport can be expressed in terms of the mixing length and its relation with the velocity gradient. The mixing length itself is expressed by Prandtl theory as $l = \kappa y$. Assuming that molecular and turbulent transport interact within the sublayer, Van Driest [90] introduced a damping function into the mixing length equation. Then

$$l = \kappa y (1 - e^{-y^+/A}) \qquad (3.13)$$

where $A$ is an empirical constant.

Converting to dimensionless quantities, we use the following scaling factors for velocity and length:

$$u_* = \sqrt{\tau_w/\rho_f}, \text{ m/sec} \qquad (3.14)$$

$$\mu_f/\rho_f u_*, \text{ m} \qquad (3.15)$$

Then Eq. (3.12) becomes

$$\frac{r_0^+}{r_0^+ + y^+} = \frac{du^+}{dy^+} + \kappa^2 (1 - e^{-y^+/A})^2 y^{+2} \left(\frac{du^+}{dy^+}\right)^2 \qquad (3.16)$$

Solving this equation for the nondimensional velocity $u^+$, we find

$$u^+ = \int_0^{y^+} \frac{2 dy^+}{\frac{r_0^+ + y^+}{r_0^+} + \sqrt{\left(\frac{r_0^+ + y^+}{r_0^+}\right)^2 + 4\left(\frac{r_0^+ + y^+}{r_0^+}\right) \kappa^2 y^{+2} (1 - e^{-y^+/A})^2}} \qquad (3.17)$$

This equation describes the velocity in the region governed by the law of the wall, that is, in the region of one-dimensional flow in the immediate vicinity of the cylinder surface. The solution of this equation is plotted in Fig. 3.1.

Analysis of Fig. 3.1 shows that, in the region governed by the law of the wall, $u^+$ not only is a function of the universal variable $y^+$ but also depends on $r_0^+$. The upper most (dashed) curve corresponds to a flat plate, and the successive curves correspond to decreasing radius of cylinder $r_0^+$. If the velocity profiles were represented in terms of actual physical variables $u$ and $y$, then at a given length $x$ the velocity profile near the wall would become increasingly steep with reduction in $r_0$.

Figure 3.2 shows several characteristic velocity profiles, expressed in terms of variables $u^+$ and $y^+$, using the technique of Sparrow et al. [55]. They obtained the following differential equations for calculating the velocity profile near the wall:

$$dy^+ = \left(1 + \frac{y^+}{r_0^+}\right)(1 + n^2 u^+ y^+) du^+ \qquad (3.18)$$

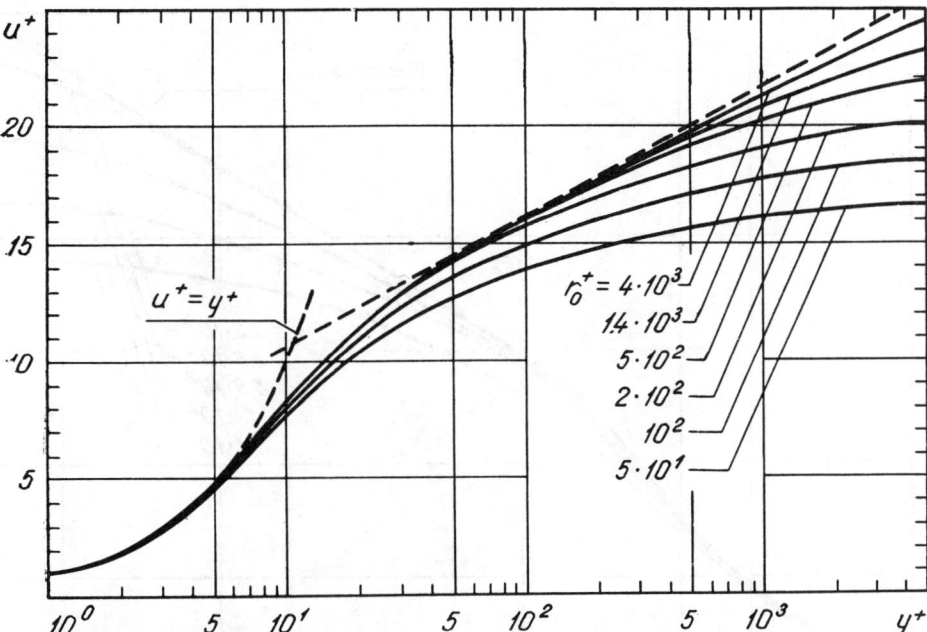

**FIG. 3.1** Law of the wall for a cylinder in axial flow. The dashed line represents a flat-plate velocity profile (according to Clauser).

and far from the wall:

$$dy^+ = \frac{2}{3} \varkappa r_0^+ \left[\left(1 + \frac{y^+}{r_0^+}\right)^{3/2} - 1\right] du^+ \qquad (3.19)$$

Here $n$ and $\kappa$ are empirical constants, assumed to be 0.109 and 0.36, respectively.

Sparrow et al. [55] integrated Eqs. (3.18) and (3.19) taking $y^+ = 26$ and $u^+ = 12.9$ as boundaries between regions "near" and "far" from the wall (on the basis of Deissler's data on flow in tubes). At very low $r_0^+$ the value of $y^+$ corresponding to $u^+ = 12.9$ was found to be somewhat higher than 26. To check how this could affect the results at low values of $r_0^+$, Sparrow et al. [55] performed special calculations in which the dividing point was varied such that the ratio of turbulent to laminar viscosity at this point was the same as for a flat plate. The shear stresses calculated from auxiliary profiles did not differ from stresses calculated from profiles obtained when the dividing point was specified by the condition $u^+ = 12.9$. This confirmed the validity of the above assumption.

Comparison of Figs. 3.1 and 3.2 confirms the similarity of their shapes. The only exception is that the corresponding curves in Fig. 3.1 lie somewhat lower than those of Fig. 3.2. This rather insignificant discrepancy is due to different methods of determining the turbulent transport coefficient.

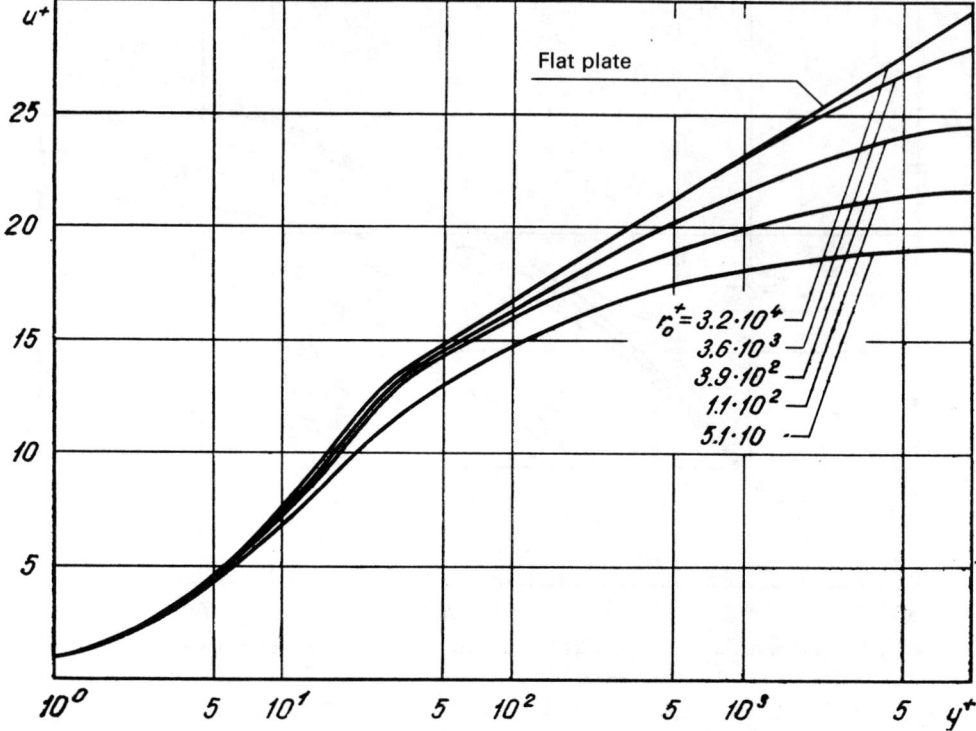

**FIG. 3.2** Characteristic velocity profiles according to Eqs. (3.18) and (3.19).

The above law of the wall [Eq. (3.17)] was used extensively in our studies for comparison with experimental results.

The experimental data were expressed in universal coordinates $u^+ = f(y^+)$. Figures 3.3 and 3.4 show measurement results of the velocity profile for isothermal flow over a cylinder with $d = 15.5$ mm and for static pressures of 0.13 and 0.435 MPa. It is seen from these figures that all measured velocity profiles give similar values in the region where the law of the wall is obeyed. Together with the experimental profiles, Figs. 3.3 and 3.4 show velocity profiles calculated analytically from Eq. (3.17). The experimental points obtained in the law of the wall region are in satisfactory agreement with the velocity distribution obtained from analytic Eq. (3.17). They also confirm the conclusions drawn from analysis of Fig. 3.1.

The quantity $u_\infty^+$ can be interpreted as the value of $u^+$ at the edge of the boundary layer. Since the shear stress at the wall $\tau_w$ changes with location along the cylinder, the value of $u_\infty^+$ changes also. At the start of the boundary layer (at $x = 0$), we have $\tau_w \to \infty$ and $u_\infty^+ = 0$. Further downstream along the cylinder (with increasing $x$), the boundary layer thickness and $u_\infty^+$ increase, whereas $\tau_w$ decreases.

Study of velocity profiles in experiments without heat transfer, in the re-

## VELOCITY AND TEMPERATURE DISTRIBUTIONS OF A CYLINDER 61

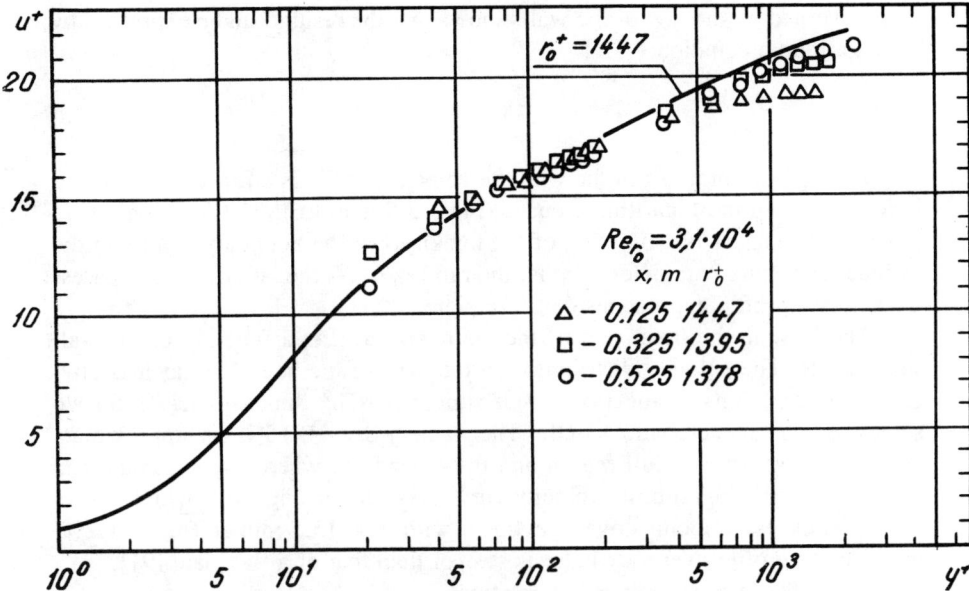

**FIG. 3.3** Velocity distribution in universal coordinates for a cylinder with $d = 15.5$ mm at $P = 0.134$ MPa and $\overline{Tu} \approx 1.2\%$. The solid line represents Eq. (3.17).

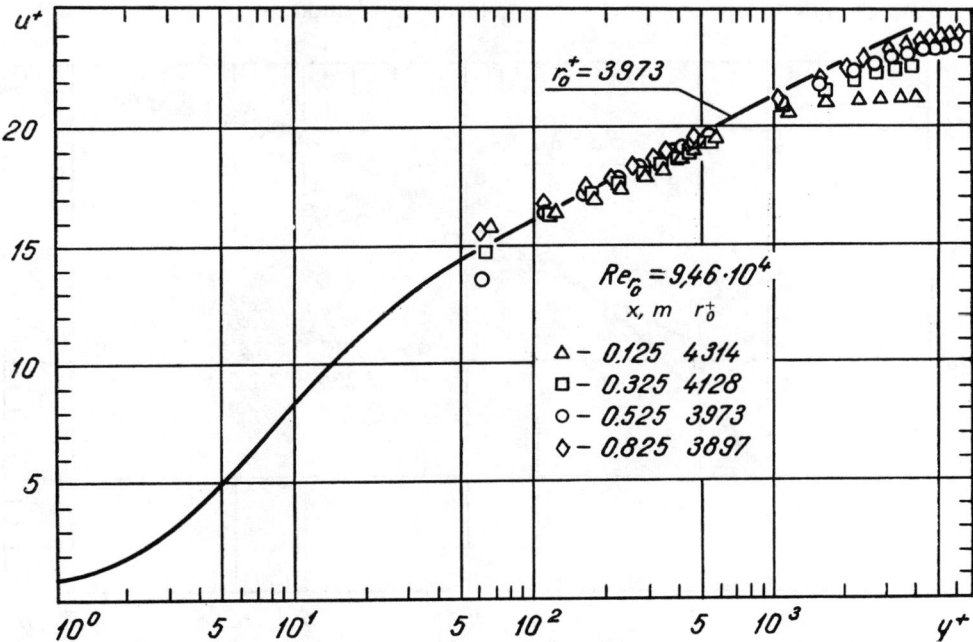

**FIG. 3.4** Velocity distribution in universal coordinates for a cylinder with $d = 15.5$ mm at $P = 0.435$ MPa and $\overline{Tu}_\infty \approx 1.2\%$. The solid curve represents Eq. (3.17).

gion governed by the law of the wall, shows that the results can be expressed by the logarithmic equation:

$$u^+ = 5.35 \lg y^+ + 5.2 \qquad (3.20)$$

At large distances from the wall the velocity profile is a function of Re and it deviates from the logarithmic curve (Figs. 3.5 and 3.6). The location where this deviation starts is a function of the thickness of the boundary layer and the cylinder curvature parameter, that is, the ratio $r_0/\delta$. As the latter ratio increases, the velocity profile approaches that of a plate.

The flow in the outer part of the boundary layer is a function of the wall shear stress. The velocity distribution in this part of the boundary layer is governed by conditions of the flow's self-similarity. The dimensionless ratio $y/\delta$ serves as the characteristic length. The value $y/\delta = 0.15$ is an approximate boundary between the wall region and the outer part, where experimental data deviate from the logarithmic velocity curve. As seen in Fig. 3.7, which shows measured excess velocities over a cylinder with $d = 15.5$ mm at $\overline{Tu}_\infty = 1.2\%$, the velocity profile over a cylinder is steeper than that of a flat plate [91].

Using the familiar power-law relation

$$\frac{u}{u_\infty} = \left(\frac{y}{\delta}\right)^n \qquad (3.21)$$

we can find for the universal excess velocity distribution

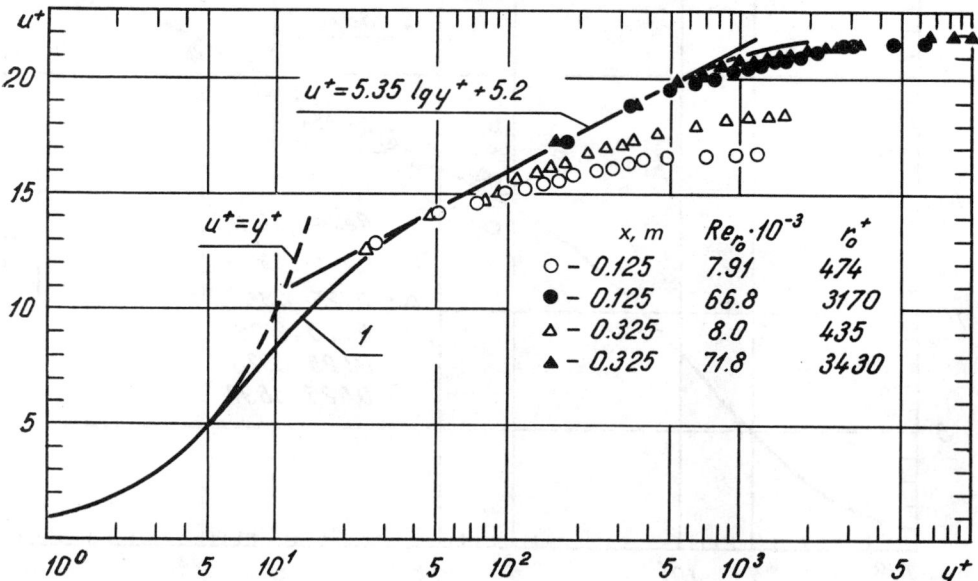

**FIG. 3.5** Velocity distribution in universal coordinates for a cylinder with $d = 4$ mm and $\overline{Tu}_\infty \approx 1.2\%$. Solid curve 1 is from Eq. (3.17) for $r_0^+ = 3430$.

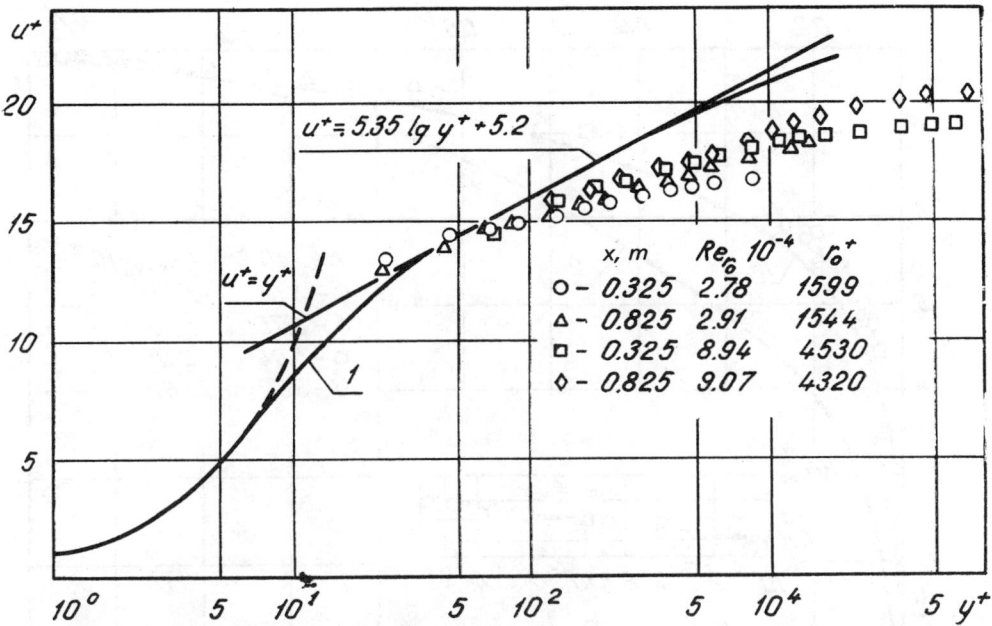

**FIG. 3.6** Experimental velocity profile over a cylinder with $d = 15.5$ mm at $\overline{Tu}_\infty \approx 5\%$. Solid curve 1 is plotted from Eq. (3.17) for $r_0^+ = 4320$.

$$\frac{u_\infty - u}{u_*} = f\left(\frac{y}{\delta}\right) \qquad (3.22)$$

the following expression:

$$\frac{u_\infty - u}{u_*} = \sqrt{\frac{2}{c_f}}\left\{1 - \left(\frac{y}{\delta}\right)^n\right\} \qquad (3.23)$$

Equation (3.23) is shown by a solid curve in Fig. 3.7; for a specific case ($c_f = 0.00473$ and $n = 1/8$ obtained from our measurements) it becomes

$$\frac{u_\infty - u}{u_*} = 21.18\left\{1 - \left(\frac{y}{\delta}\right)^{0.125}\right\} \qquad (3.24)$$

It is seen by comparing Figs. 3.3 and 3.6 that the distribution of excess velocity in the outer part of the boundary layer is a function of not only the cylinder radius $r_0$ but also the freestream turbulence.

The velocity field in the outer part of the boundary layer is governed by Eqs. (3.4) and (3.5), which were used to obtain the law of the wall. Farther from the wall the equations are solved only with the convective terms. The solution is not given here (for a flow over a flat surface it is presented by Šlančiauskas and Žukauskas [52]), and the outer part of the boundary layer is analyzed on the basis of calculated profiles of the velocity defect, which are of a form close to self-similar

**64** HEAT TRANSFER OF GAS COOLANTS

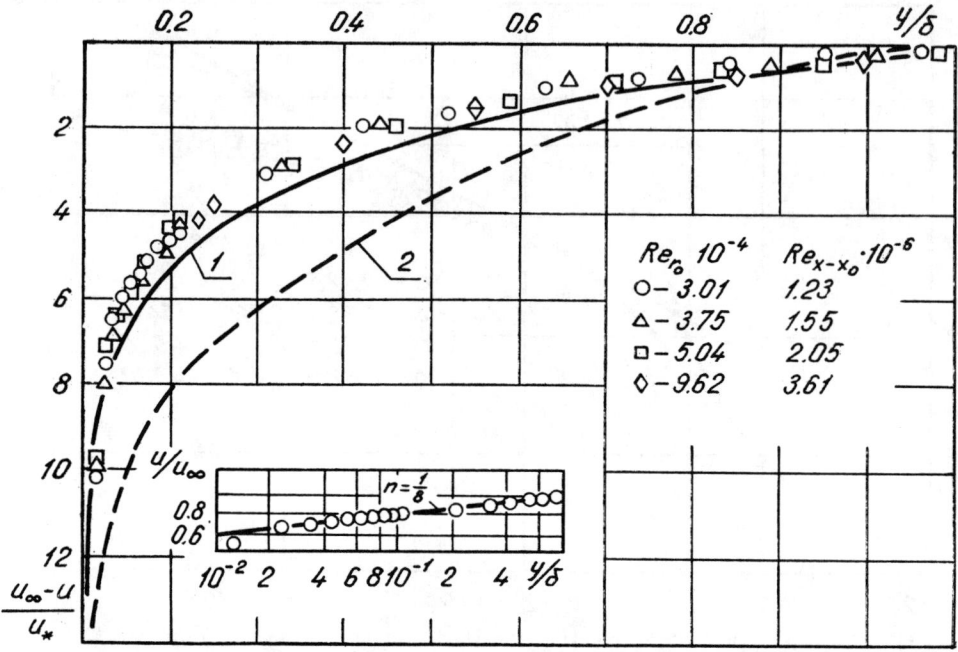

**FIG. 3.7** Excess velocity distribution. The points represent our experiments with a 15.5-mm-diameter cylinder at $\overline{Tu}_\infty \approx 1.2\%$ and $x = 0.525$ m: (1) plot of Eq. (3.24) under the above conditions; (2) flat-plate curve according to Clauser [91].

$$\frac{u_\infty - u}{u_*} = f\left(\frac{yu_*}{\delta^* u_\infty}\right) \tag{3.25}$$

This data processing technique, proposed by Rotta [92] and verified by Šlančiauskas et al. [93], provides a better description of experimental data (Fig. 3.8). The excess velocity distribution in this case can be expressed in the following form:

$$\frac{u_\infty - u}{u_*} = -\frac{1}{\varkappa} \ln \frac{yu_*}{\delta^* u_\infty} + B \tag{3.26}$$

In the given case of a cylinder with $d = 15.5$ mm at $Tu_\infty \approx 1.2\%$ and $x = 525$ mm, it becomes

$$\frac{u_\infty - u}{u_*} = -5.35 \lg \frac{yu_*}{\delta^* u_\infty} - 3.25 \tag{3.27}$$

Increasing the surface curvature (i.e., reducing the diameter of the cylinder) or increasing the turbulence intensity produces a steeper profile in the outer part of the boundary layer (Figs. 3.9 and 3.10). In these cases the constant $B$ changes and the velocity defect profiles form a single-parameter family. The constant $B$, in Eq. (3.26), is a function of only a single parameter [92].

$$I = \int_0^\infty (u_\infty^+ - u^+) \, d\frac{y}{\delta^* u_\infty^+} \tag{3.28}$$

## 3.2 DISTORTION OF VELOCITY AND TEMPERATURE FIELDS IN THE CASE OF VARIABLE PHYSICAL PROPERTIES OF THE COOLANT

Our study and determination of the distortions of velocity and temperature profiles due to changing density and viscosity over the span of the boundary layer made it possible to follow the change in the flow pattern and to obtain a deeper insight into the mechanism of momentum and heat transfer across the boundary layer under conditions of large heat fluxes. The resultant velocity profiles were expressed in terms of the universal variables $u_q^+ = u/u_{*q}$ and $y_{fq}^+ = yu_{*q}/\nu_f$, whereas the temperature profiles were expressed in terms of dimensionless variables $\vartheta^+ = (T_w - T_f)/\vartheta_*$ and $y_{fq}^+$. The friction velocity $u_{*q}$ was determined from the skin friction coefficient $c_{fq}$. The kinematic viscosity was obtained from conditions outside the boundary layer. Certain characteristic velocity profiles for variable physical properties are shown in Fig. 3.11.

Figure 3.12 is a plot of distorted temperature profile measured with the same boundary conditions as in measuring the velocity distortion.

Our study of turbulent boundary layers made it possible to establish that the velocity profiles are universal. This aided in more completely analyzing the

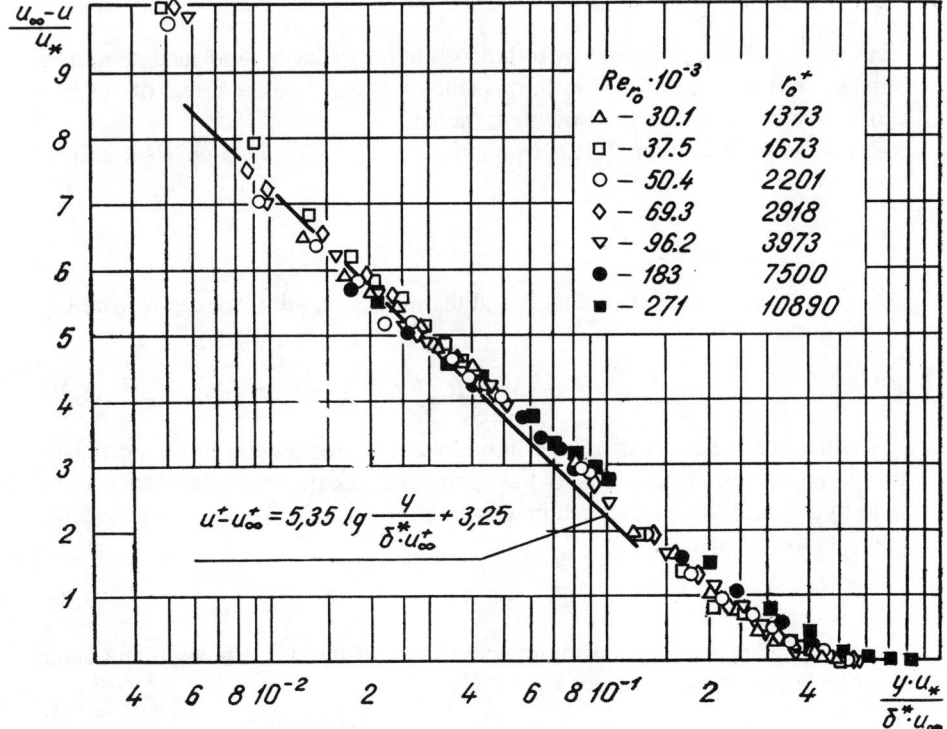

**FIG. 3.8** Velocity defect for a cylinder with $d = 15.5$ mm at $\overline{Tu}_\infty \approx 1.2\%$, $x = 0.525$ m.

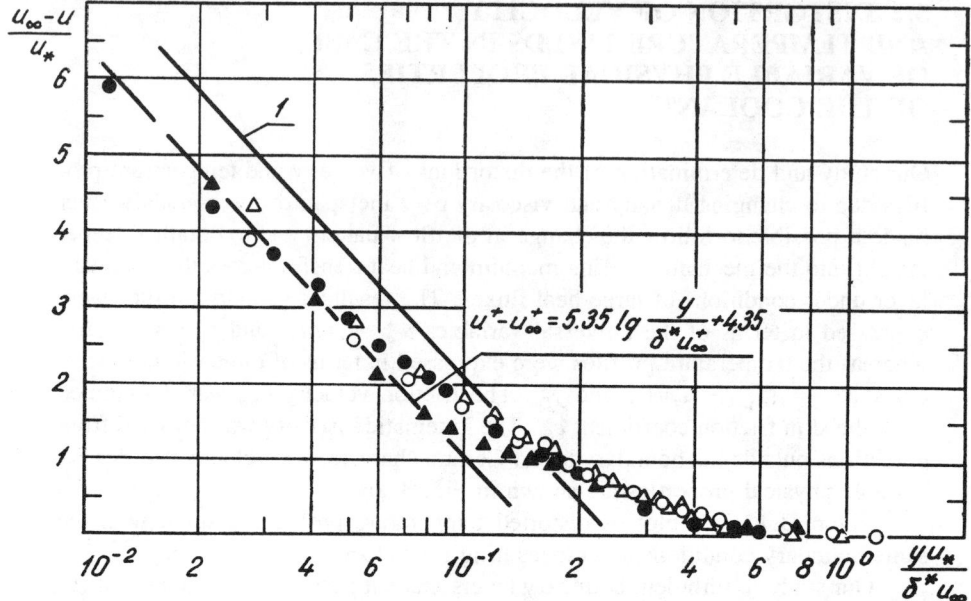

**FIG. 3.9** Velocity defect for a cylinder with $d = 4$ mm and $\overline{Tu}_\infty \approx 1.2\%$. Symbols are defined in Fig. 3.5. Line (1) is the logarithmic velocity defect distribution for a cylinder with $d = 15.5$ mm.

relations governing the flow in turbulent boundary layers and in determining turbulent transport coefficients. In this study we analyzed the effect of variable density and viscosity on the flow near the wall.

As shown in Section 3.1, the law of the wall for variable physical properties flow can be written as

$$\tau = \mu_y \frac{du}{dy} - \rho_y \overline{u'v'} \tag{3.29}$$

Using correlation coefficient $R$ and the mixing length $l$, we can rewrite Eq. (3.29) in the form

$$\tau = \mu_y \frac{du}{dy} - \rho_y R_{uv} l_u l_v \left(\frac{du}{dv}\right)^2 \tag{3.30}$$

Since the existence of a logarithmic velocity distribution in the portion of the boundary layer where $l = \kappa y$ has been verified experimentally, the velocity field in the transition region can be described by introducing the damping function $n$ to the constant $\kappa$:

$$n = 1 - e^{-y^+/23} \tag{3.31}$$

Eq. (3.30), simplified by assuming some value of the mixing length and using Eq. (3.31), becomes

$$\tau = \mu_y \frac{du}{dy} + \rho_y \kappa^2 y^2 n^2 \left(\frac{du}{dy}\right)^2 \tag{3.32}$$

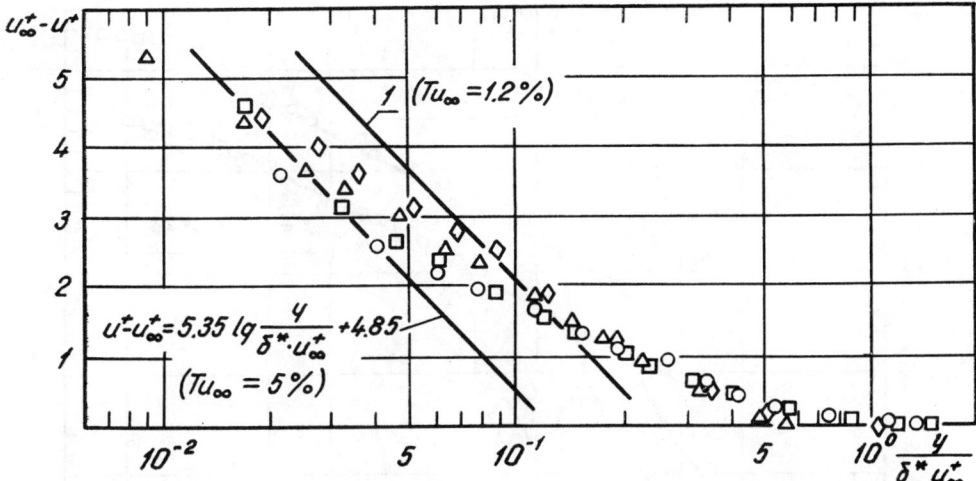

**FIG. 3.10** Effect of freestream turbulence on the excess velocity distribution for a cylinder with $d$ = 15.5 mm (symbols defined in Fig. 3.6). The points represent our experimental data at $\overline{Tu}_\infty \approx$ 5%; line (1), logarithmic velocity defect distribution at $\overline{Tu}_\infty \approx 1.2\%$.

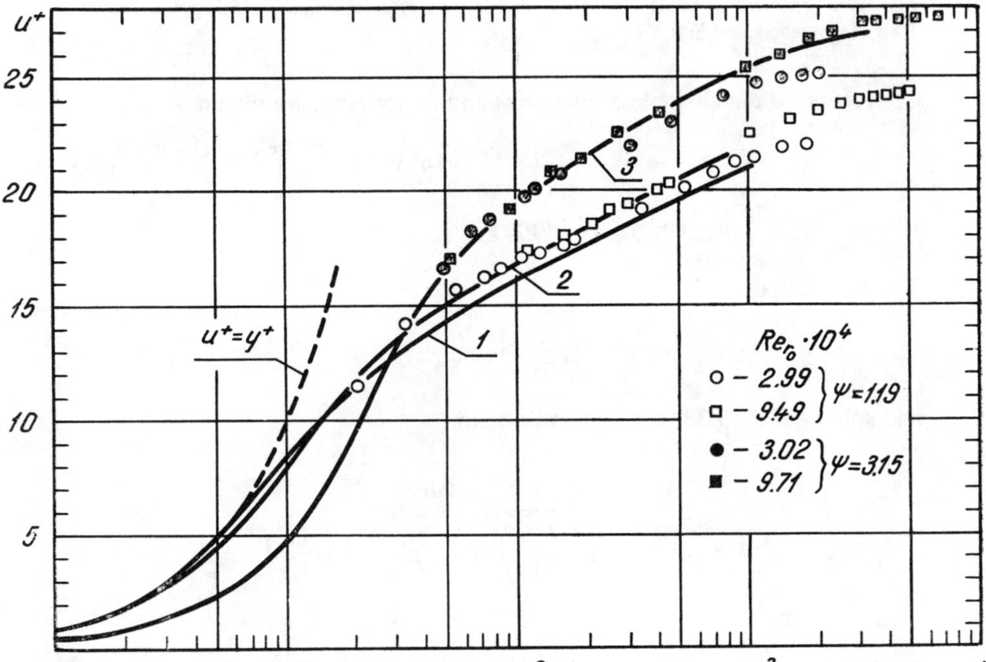

**FIG. 3.11** Distorted universal velocity profile. The solid curves were plotted from Eq. (3.34): (1) without heat transfer; (2) $\psi = 1.19$; (3) $\psi = 3.15$. The damping factor $n$ was determined from the local value of $y_{fq}^+ = yu_{*q}\nu_y$. The data points are from our experiments with a cylinder with $d$ = 15.5 mm at $\overline{Tu}_\infty \approx 1.2\%$.

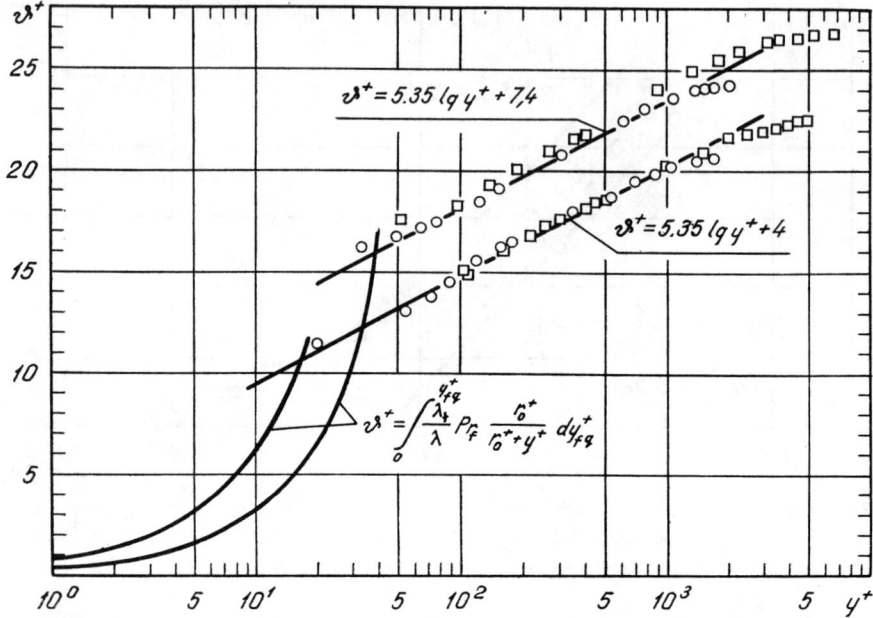

**FIG. 3.12** Distorted universal temperature profile for a cylinder with $d = 15.5$ mm at $\overline{\mathrm{Tu}}_\infty \approx 1.2\%$ (for symbols, see Fig. 3.11).

Converting to dimensionless quantities and regrouping, we obtain

$$\frac{\tau}{\tau_w} = \frac{\mu_y}{\mu_f} \frac{du_q^+}{dy_{fq}^+} + \frac{\rho_y}{\rho_f} \varkappa^2 n^2 y_{fq}^{+2} \left(\frac{du_q^+}{dy_{fq}^+}\right)^2 \tag{3.33}$$

where $u_q^+ = u/u_{*q}$ and $y_{fq}^+ = (y/\nu)u_{*q}$.

Given that

$$\frac{\tau}{\tau_w} = \frac{r_0^+}{r_0^+ + y_{fq}^+}$$

and solving Eq. (3.33) for $u_q^+$, we obtain

$$u_q^+ = \int_0^{y_{fq}^+} \frac{2\,dy_{fq}^+}{\dfrac{\mu_y}{\mu_f}\left(\dfrac{r_0^+ + y_{fq}^+}{r_0^+}\right) + \sqrt{\left(\dfrac{\mu_y}{\mu_f}\right)^2 \left(\dfrac{r_0^+ + y_{fq}^+}{r_0^+}\right)^2 + 4\,\dfrac{\rho_y}{\rho_f}\left(\dfrac{r_0^+ + y_{fq}^+}{r_0^+}\right) \varkappa^2 n^2 y_{fq}^{+2}}} \tag{3.34}$$

Under conditions of variable physical properties, $n$ is a function of the shear stress and of the temperature profile. Thus, it is very difficult to determine it on the basis of the parameter $y^+ = yu_{*q}/\nu_f$. It is possible to determine $n$ from the value of $u/u_{*q}$ since the universal relationship $u/u_* = f(\overline{y u_*}/\nu_f)$ is unique. It is assumed then that turbulent transport, in the case of variable physical conditions, is determined by the local value of $n$, which is based on the dynamic parameter $u/u_{*q}$ obtained from the universal pattern of $n(u/u_{*q})$ for constant physical properties [52]. The additional effect of variability of physical properties on the universal transport pattern is neglected in such calculations. The universal velocity profiles, calculated in our case using this technique lie somewhat lower than experimental profiles. If $n$ is determined from the local value of $y_{fq}^+ = yu_{*q}/v_f$, the velocity profiles calculated analytically from Eq. (3.34) (Fig. 3.13) in the zone governed by the law of the wall are in satisfactory agreement with the experimental profiles (Fig. 3.11). The physical properties of the flow were determined from tables on the basis of the experimental temperature profile (Fig. 3.12).

Due to lack of measured temperature values near the wall, the temperature profile in this region was calculated analytically. The temperature field in the region where the law of the wall is obeyed can be described based on the fact that the heat flux is constant this region:

$$q_w = \lambda \frac{d\vartheta}{dy} - c_p \rho \overline{v' \vartheta'} \qquad (3.35)$$

where $v'$ and $\vartheta'$ are the pulsatile velocity and temperature components.

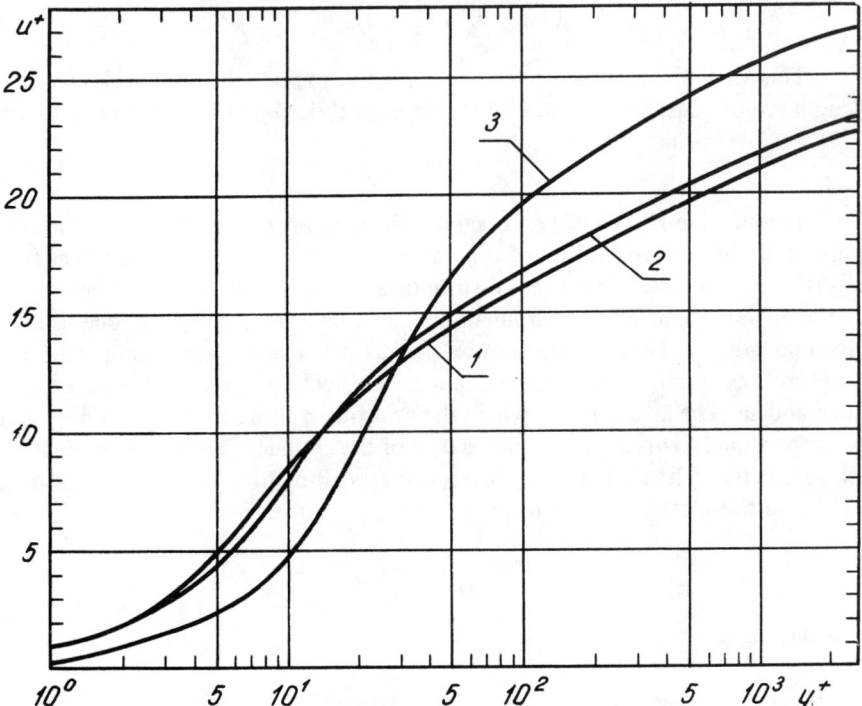

**FIG. 3.13** Velocity profiles calculated from Eq. (3.34). (1) $\psi = 1$, (2) $\psi = 1.24$, (3) $\psi = 3.24$. (The damping function $n$ was determined from the dynamic parameter $u/u_{*q}$ obtained from the universal pattern of $n(u/u_{*q})$ at constant physical properties.)

Turbulent flow parameters will be described based on the mixing length

$$-\overline{u'v'} = l^2 \left(\frac{du}{dy}\right)^2 \tag{3.36}$$

with correction for the difference between the transfer of momentum and heat. For this, following Žukauskas and Šlančiauskas [52], we introduce the turbulent Prandtl number,

$$\mathrm{Pr}_t = \frac{\overline{u'v'}}{\overline{v'\vartheta'}} \frac{\partial \vartheta/\partial y}{\partial u/\partial y} \tag{3.37}$$

The turbulent heat flux in this case will correspond to the expression

$$-\overline{v'\vartheta'} = \frac{l^2}{\mathrm{Pr}_t} \left(\frac{du}{dy}\right) \frac{d\vartheta}{dy} \tag{3.38}$$

and the heat flux equation [Eq. (3.35)] becomes

$$q_w = \lambda \frac{d\vartheta}{dy} + \rho c_p \frac{\varkappa^2 n^2}{\mathrm{Pr}_t} y^2 \frac{du}{dy} \frac{d\vartheta}{dy} \tag{3.39}$$

Converting to dimensionless quantities, we obtain

$$\frac{q}{q_w} = \frac{\rho_y c_{py}}{\rho_f c_{pf}} \left(\frac{a_y/a_f}{\mathrm{Pr}_f} + \varkappa^2 \frac{1}{\mathrm{Pr}_t} y_{fq}^{+2} n^2 \frac{du_q^+}{dy_{fq}^+}\right) \frac{d\vartheta^+}{dy_{fq}^+} \tag{3.40}$$

This conversion was performed with the previously cited velocity and length scales [Eqs. (3.14) and (3.15), respectively], and the scaling temperature

$$\vartheta_{*q} = q_w/\rho_f c_{pf} u_{*q}. \tag{3.41}$$

To integrate Eq. (3.40) and determine the temperature profile, it is necessary to know the distribution of $q/q_w$ across the boundary layer. The heat flux distribution, like the shear stress distribution, is usually specified. To determine the temperature profile of a turbulent boundary layer on a flat plate, one usually assumes $q/q_w = 1$. Heat transfer coefficients calculated by means of temperature profiles found in this manner are in satisfactory agreement with experimental data. The area through which the heat flux $q$ flows in the radial direction is proportional to distance $r$ from the axis of the cylinder. Thus, the distribution of $q/q_w$ in the cylinder boundary layer, corresponding to the distribution of $q/q_w = 1$ in a flat plate boundary layer, is

$$\frac{q}{q_w} = \frac{r_0}{r} = \frac{r_0}{r_0+y} = \frac{r_0^+}{r_0^+ + y_{fq}^+} \tag{3.42}$$

We thus have

$$\vartheta^+ = \int_0^{y_{fq}^+} \frac{\dfrac{r_0^+}{r_0^+ + y_{fq}^+} \, dy_{fq}^+}{\dfrac{\rho_y c_{py}}{\rho_f c_{pf}} \left(\dfrac{a_y/a_f}{\mathrm{Pr}_f} + \varkappa^2 y_{fq}^{+2} \dfrac{1}{\mathrm{Pr}_t} n^2 \dfrac{du_{fq}^+}{dy_{fq}^+}\right)} \tag{3.43}$$

Due to the significant effect of enhanced flow turbulence, the temperature profile calculated with Eq. (3.43) does not agree with experimental results. This may be due to the effect of turbulence on the shear stress (see Chapter 4), which is a component of the parameter $\vartheta_{*q}$ and can distort the universal pattern. Accordingly, calculation of the temperature profile from Eq. (3.34) is possible under turbulence conditions only in the laminar sublayer region (Fig. 3.12).

The variability of density and viscosity across the boundary layer span distorts the boundary layer profiles by sharply reducing the velocity near the wall (Fig. 3.11). Profile distortion occurs over the entire region of logarithmic velocity distribution. In Fig 3.11 the distorted profiles are compared with curve 1, calculated from Eq. (3.34) for a constant-property flow of the coolant.

This means that the extent of velocity profiles distortion under conditions of variable physical properties is determined by the thickness of the viscous sublayer and by distortion of the transition region, which result in changes in the $u^+$ range of the logarithmic part.

The pattern change in the temperature and velocity profiles with changes in the physical properties of the coolant is similar.

The logarithmic velocity and temperature distributions

$$u^+ = 5.35 \lg y^+ + c \qquad (3.44)$$

$$\vartheta^+ = 5.35 \, Pr_t \cdot \lg y_{fq}^+ + c_1 \, (Pr) \qquad (3.45)$$

where $c$ and $c_1$ are determined experimentally at distance $y_{fq}^+ > 30$. They depend on the surface curvature, freestream turbulence, and temperature factor.

The velocity distribution in the outer part of the boundary layer (Fig. 3.14) can be described by the equation

$$\frac{u_\infty^+ - u^+}{u_*} = -5.35 \lg \frac{yu_*}{\delta^* u_\infty} + B \left( \frac{yu_*}{\delta^* u_\infty} \right) \qquad (3.46)$$

and the temperature profile by

$$d\vartheta^+ = Pr_t du^+ \qquad (3.47)$$

The correction factor $B$ is found experimentally and is a function of the aforementioned quantities. It is seen from Fig. 3.14 that the excess velocity profile becomes steeper with increasing heat flux through the boundary layer.

The relation between the velocity or temperature profile and the coefficients of turbulent transfer is given by Eqs. (3.29) and (3.35) or Eqs. (3.34) and (3.43). Analysis of these relations on the basis of the previously described universal velocity and temperature distributions gives an idea about changes in the eddy diffusivity $\epsilon_\tau$. In the viscous sublayer $\epsilon_\tau$ tends to zero in the logarithmic part of the boundary layer $\epsilon_\tau = u_* \kappa y$, whereas in the outer part of the boundary layer it takes a constant value that is a function of the freestream velocity [81].

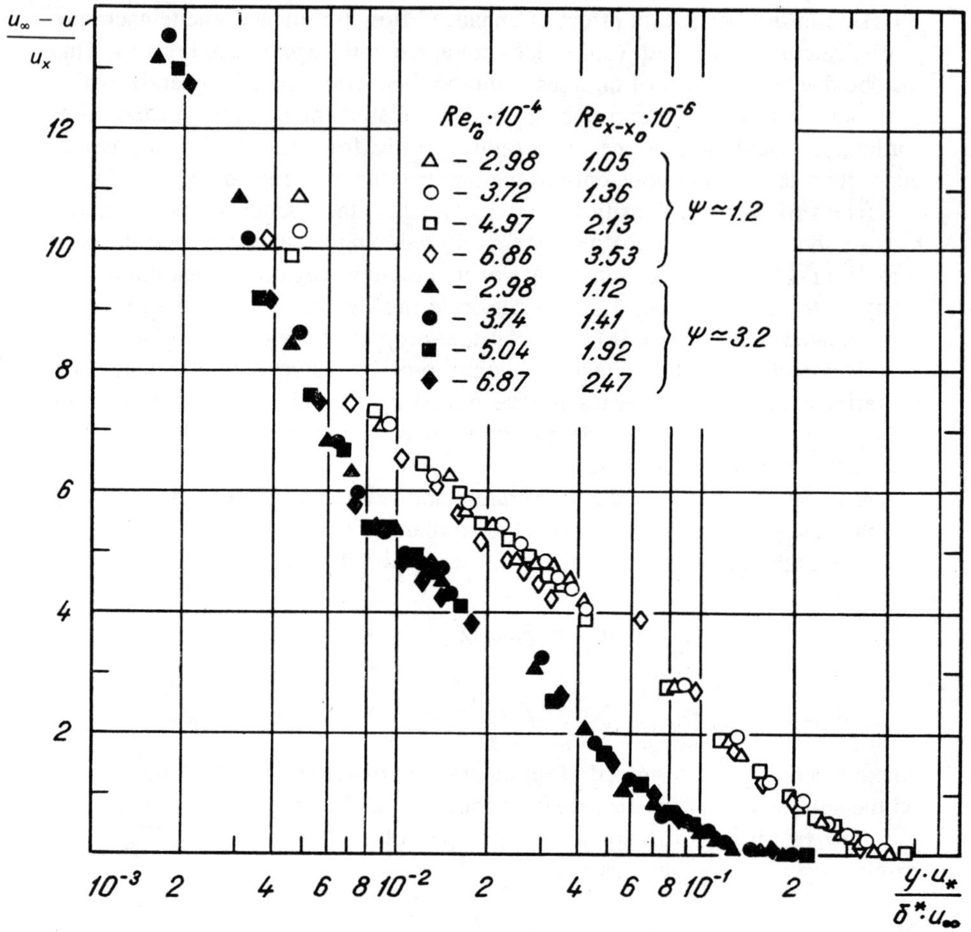

# 72 HEAT TRANSFER OF GAS COOLANTS

**FIG. 3.14** Excess velocity distribution as a function of the temperature factor for a cylinder with $d = 15.5$ mm at $\overline{Tu}_\infty \approx 1.2\%$.

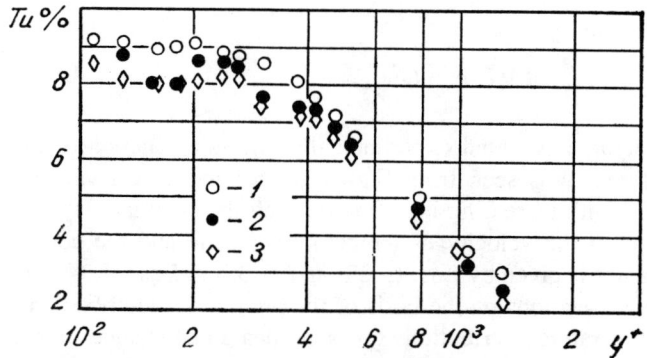

**FIG. 3.15** Turbulence intensity distribution in the boundary layer 125 mm downstream of the turbulence promoter. (1) Data for a cylinder with turbulence promoter 1; (2) data for a cylinder with turbulence promoter 2; (3) data for a smooth cylinder.

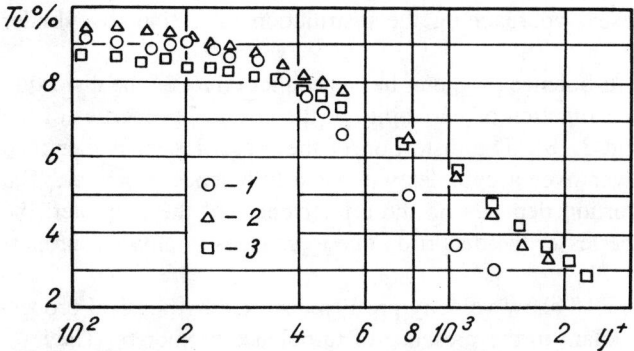

**FIG. 3.16** Turbulence intensity distribution in the boundary layer downstream of turbulence promoter 2 as a function of distance $x$ from the promoter. (1) $x = 125$ mm, (2) $x = 325$ mm, (3) $x = 525$ mm.

As seen from Figs. 3.11 and 3.12, all these conclusions are confirmed by measured velocity and temperature profiles.

## 3.3 DISTORTION OF THE VELOCITY PROFILE DUE TO VARIABILITY OF PHYSICAL PROPERTIES OF A FLOW WITH SURFACE-TYPE TURBULENCE-GENERATING MECHANISM

A more profound analysis of experimental results showed that distortion of velocity fields depends not only on physical properties variation due to changes in the temperature factor but also on the intensity of boundary layer turbulence. To clarify this problem we performed experiments using surface-type turbulence promoters with rectangular and triangular threads (turbulence promoters 1 and 2), placed near the upstream end of the cylinder (Fig. 2.5).

Data on the distribution of turbulence intensity across the boundary layer, measured in the flow over a cylinder with $d = 15.5$ mm are plotted in Figs. 3.15 and 3.16. Figure 3.15 shows the distribution of Tu in the boundary layer as a function of the type of turbulence promoter. The measurements were performed downstream of the promoter at a distance of 125 mm from it. The results are compared with data on the distribution of Tu in the boundary layer of a cylinder without a turbulence promoter. As seen in Fig. 3.15, turbulence promoter 1 is more effective than turbulence promoter 2: it shifts the turbulence intensity peak to the left, that is, closer to the wall. The absolute values of Tu at the same distance $y^+$ from the wall also depend on the type of turbulence promoter. Due to the latter, the turbulence intensity also increases somewhat outside of the boundary layer, in the flow core.

At large downstream distances from the turbulence promoter, the turbulence intensity decreases, approaching the distribution of a smooth cylinder (Fig. 3.16).

The surface-type turbulence promoter has a peculiar effect on the distortion of velocity profiles as a function of variability of the physical properties of the coolant (Figs. 3.17 and 3.18). The distortion of these profiles is smaller than that of a boundary layer over a cylinder without a turbulence promoter. The magnitude of the distortion depends on the effectiveness of the promoter: the more effective it is, the lesser the distortion due to changes in physical properties.

Figures 3.17 and 3.18 show velocity profiles distorted by physical properties variation of the coolant in the presence of turbulence promoters. They are compared with a profile obtained at identical operating conditions over a cylinder without a turbulence promoter. It is seen from the figures that the distortion increases with reduction in the turbulence promoter effectiveness, and it gradually approaches the pattern of velocity profile distortion seen in the boundary layer of a cylinder without a turbulence promoter. At high $x/d$ (in our case at $x/d > 35$ for turbulence promoter 1 and $x/d > 14$ for promoter 2), the

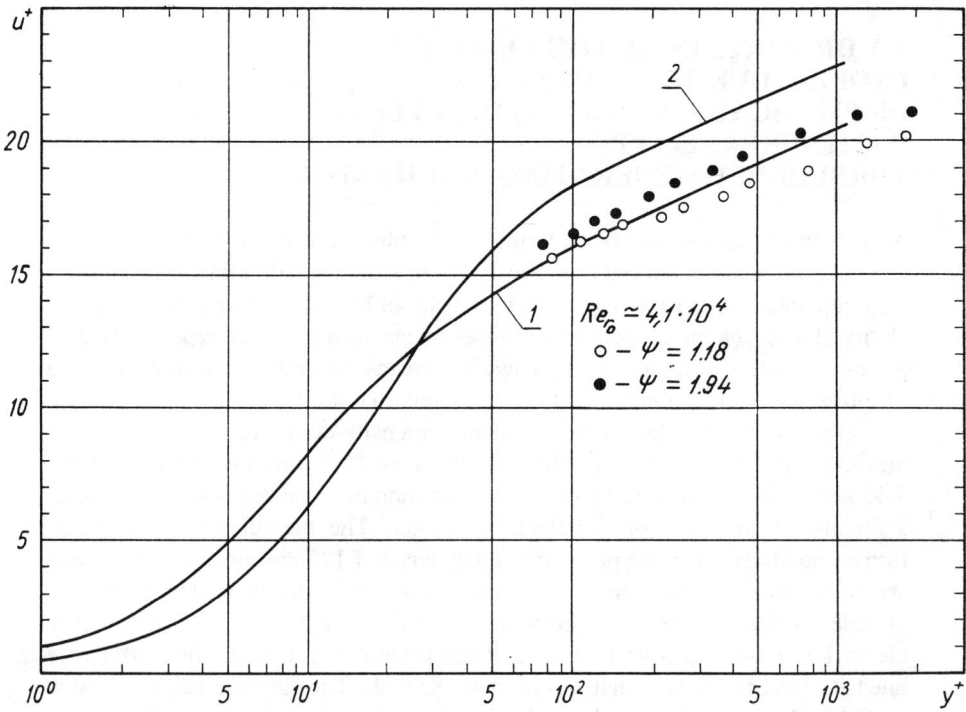

**FIG. 3.17** Distortion of the velocity profile 125 mm downstream of turbulence promoter 1 as a function of variability of physical properties. The solid curves were plotted from Eq. (3.34) for a smooth cylinder: (1) without heat transfer, (2) at $\psi = 2$. The points represent our experimental data.

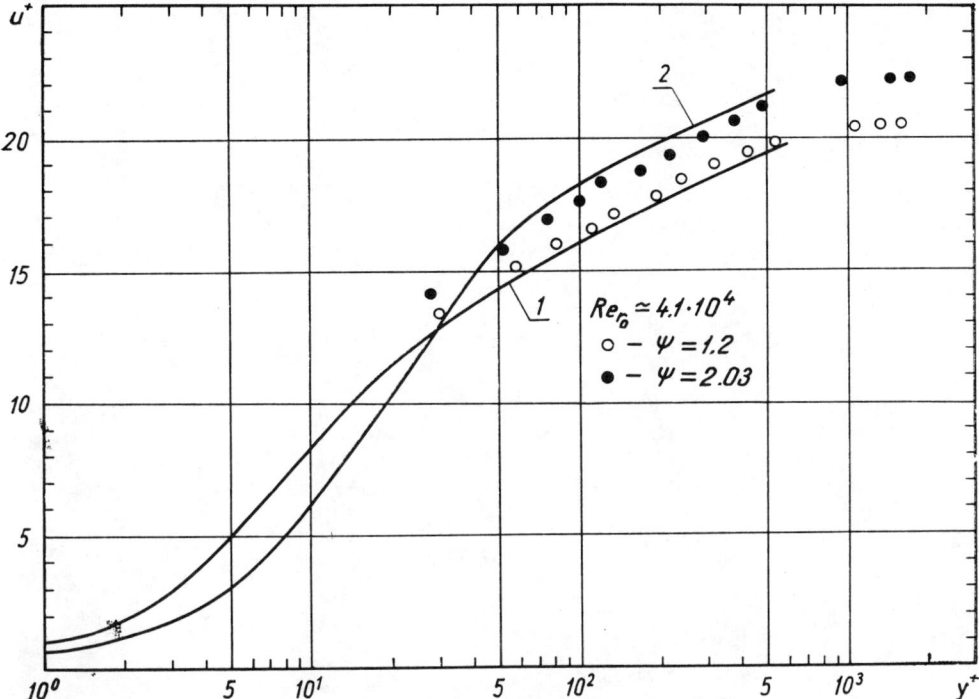

**FIG. 3.18** Distortion of the velocity profile 125 mm downstream of turbulence promoter 2 as a function of variability of physical properties. The solid curves were plotted from Eq. (3.34) for a smooth cylinder: (1) without heat transfer, (2) at $\psi = 2$. The points represent our experimental data.

effect of the temperature factor, in the presence of a promoter, is to increase the distortion, which becomes virtually identical to that of a smooth cylinder boundary layer. This is natural, since the turbulence generated by a turbulence promoter decays with increasing distance from the latter [95].

CHAPTER
# FOUR
## SKIN FRICTION OF A CYLINDER IN AXIAL FLOW

A large number of analytic and experimental studies have been performed to determine the effect of physical properties variation on skin friction in turbulent pipe flows. Despite slight discrepancies between the results of various investigators, this question can be regarded as sufficiently explored. However, due to differences in the velocity and temperature profiles, the results obtained in pipe flows cannot provide a clear picture of the processes occurring in the entrance regions of annuli or in external flow over bodies.

Analytic calculation of turbulent momentum transfer involves insurmountable difficulties, since the set of equations of motion and heat flux in the boundary layer is incomplete. Using semiempirical solution methods of turbulent transfer, new results were obtained by Petukhov with Popov [26] and Deissler [96]; however, the condition of variable physical properties required additional assumptions on the relation between the viscosity and the turbulence transport coefficient, which greatly limited the applicability of computational methods.

The currently available experimental data are insufficient for obtaining a complete picture of the pattern of skin friction in external axial flows of incompressible fluids at high temperature differences. Even if we use compressible flow results, which are correlated by Chi and Spalding [97], the temperature factor range is limited to $0.2-1.7$.

The effect of transverse curvature of the cylinder surface on friction was investigated only analytically, at constant physical properties of the fluid, by Eckert et al. [54, 55] and by Solodkin and Ginevskiy [98]. The data on the effect of surface curvature on friction obtained by these investigators are not in agreement.

The calculated skin friction values in [55] are represented graphically, but although they are very important for estimating the effect of cylindrical curvature, they are inconvenient for practical calculations. This study also requires an experimental check concerning the effect of transverse curvature on skin friction.

As previously noted, we performed an experimental study of the local skin friction on a cylinder in axial turbulent flow of air over a wide range of Re and over the largest possible range of the temperature factor. The analysis, described below, is based on results of direct measurements, which are tabulated in the appendixes.

In correlating results on skin friction on the basis of $Re_x$, the leading edge of the cylinder cannot be used as the origin of the turbulent boundary layer since the boundary layer remains laminar over some region downstream of it. The turbulent boundary layer develops gradually, approaching a self-similar behavior, which, as is known, is attained only at $Re_x$ on the order of $10^{10}$. In calculating different boundary conditions in the case of natural transition from laminar to turbulent flow, it is convenient to introduce the concept of virtual origin of the turbulent boundary layer [52]. If this origin is situated at distance $x_0$ from the leading edge, the characteristics of the boundary layer will be described with the same accuracy by the displacement thickness $\delta^*$, momentum thickness $\delta^{**}$, and the length $(x - x_0)$.

The relationship between the friction factor and the distance $(x - x_0)$ was obtained by Cesna et al. [99] by experimentally determining universal velocity profiles and assuming that they develop self-similarity.

Let us consider a control volume, containing the boundary layer between planes $x$ and $(x + dx)$. According to the momentum equation, the overall momentum flux is equal to the work of the forces acting inside the control volume or on its boundaries. Mathematically this is expressed as

$$\tau_w (2\pi r_0 \, dx) = d \left\{ \int_{r_0}^{\delta} \rho u (u_\infty - u) \, 2\pi r \, dr \right\} \tag{4.1}$$

Writing the above expression in terms of the dimensionless parameters $u^+$, $y^+$, and $r_0^+$ and the virtual origin Reynolds number

$$Re_{x-x_0} = \frac{u_\infty (x - x_0)}{\nu} \tag{4.2}$$

we obtain

$$1 = u_\infty^{+2} \frac{d}{d \, Re_{x-x_0}} \left\{ \int_0^{\delta^+} \frac{\rho}{\rho_\infty} u^+ \left(1 - \frac{u^+}{u_\infty^+}\right) \left(1 + \frac{y^+}{r_0^+}\right) dy^+ \right\} \tag{4.3}$$

or

$$Re_{x-x_0} = \int_0^{\{\}} u_\infty^{+2} d \left\{ \int_0^{\delta^+} \frac{\rho}{\rho_\infty} u^+ \left(1 - \frac{u}{u_\infty^+}\right) \left(1 + \frac{y^+}{r_0^+}\right) dy^+ \right\} \tag{4.4}$$

where the expression in braces in the upper limit of the outer integral corresponds to the expression in braces under the differentiation sign.

Equation (4.4) is too complicated for numerical integration. Integrating it by parts and replacing $dy^+$ by $(dy^+/du^+)du^+$, we obtain

$$\text{Re}_{x-x_0} = u_\infty^{+2}\,\text{Re}_{\delta^{**}}(u_\infty^+) - 2\int_0^{u_\infty^+} \text{Re}_{\delta^{**}}(u_\infty^{+\prime})\,u_\infty^{+\prime}\,du_\infty^{+\prime} \qquad (4.5)$$

where

$$\text{Re}_{\delta^{**}} = \frac{u_\infty\,\delta^{**}}{\nu} = \int_0^{y_\infty^+} \frac{\rho}{\rho_\infty} u^+ \left(1 - \frac{u^+}{u_\infty^+}\right)\left(1 + \frac{y^+}{r_0^+}\right) dy^+ \qquad (4.6)$$

The primes in the integral of Eq. (4.5) indicate that this quantity is regarded as an integration variable.

## 4.1 SKIN FRICTION WITHOUT HEAT TRANSFER

The data shown in Fig. 4.1 indicate that the local skin friction coefficients, determined from the measured velocity profiles and by the projecting plank method, are in satisfactory agreement. The mean skin friction coefficients calculated from Eq. (2.11) lie 12% above the local coefficients. It was found that over the $\text{Re}_{x-x_0}$ range of $3 \times 10^5$ to $7.6 \times 10^6$ these values of local skin friction coefficients can be correlated by the equation

$$c_f = 0.455\,(\lg \text{Re}_{x-x_0})^{-2.58} \qquad (4.7)$$

Equation (4.7) describes data obtained at freestream turbulence intensity $\text{Tu}_\infty = 0.8-1.4\%$.

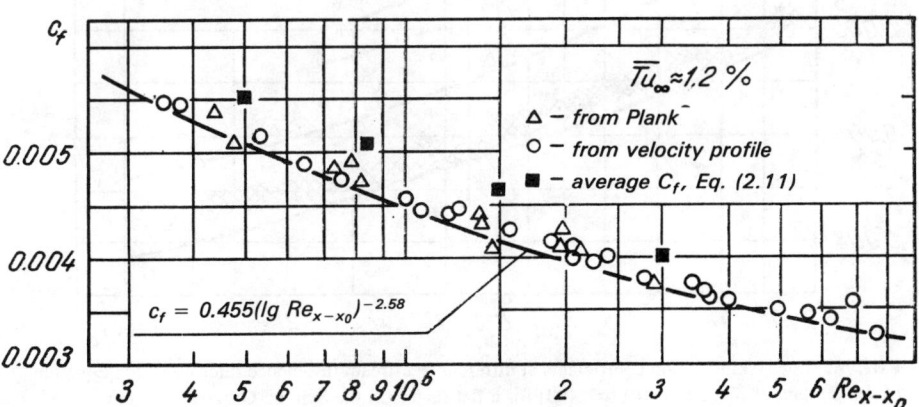

**FIG. 4.1** Local skin friction coefficients for a cylinder with $d = 15.5$ mm.

To determine the power exponent of log $\mathrm{Re}_{x-x_0}$, we constructed curves of $c_f = f(\log \mathrm{Re}_{x-x_0})$ in logarithmic coordinates at all the $\mathrm{Re}_{r_0}$ studied. It was found that for the entire range under study $m = -2.58$.

It was clarified in further analysis of experimental data that the skin friction coefficient is a function of $\mathrm{Tu}_\infty$. To obtain a more complete picture, experiments were performed at elevated $\mathrm{Tu}_\infty$.

It is seen by comparing skin friction coefficients measured at different values of $\mathrm{Tu}_\infty$ by the projecting plank method (Fig. 4.2) that surface friction, increases significantly with $\mathrm{Tu}_\infty$.

To determine the effect of Tu we constructed curves of $c = f(\mathrm{Tu})$, where

$$c = c_f (\lg \mathrm{Re}_{x-x_0})^{2.58} \tag{4.8}$$

For the Tu range from 0.85 to 6.3%, we found that the skin friction coefficient $c_f$ is proportional to $(\mathrm{Tu})^{0.08}$ (Fig. 4.3). The shaded region in Fig. 4.2 shows how the skin friction coefficients would have varied at Tu = 0.5% (assuming this behavior is the same upstream of the region under study and in it). The local skin friction coefficients obtained with this assumption are in satisfactory agreement with the skin friction coefficients of a flat plate at low $\mathrm{Re}_{x-x_0}$ ($\sim 5 \times 10^5$) and lie 10% above data for a plate at $\mathrm{Re}_{x-x_0}$ of about $3 \times 10^6$.

Figures 4.1 and 4.2 also show that the cylindrical shape affects the skin friction (when $\mathrm{Re}_{x-x_0}$ increases, the values of $c_f$ on a cylinder in axial flow increasingly deviate from values of $c_f$ for a flat plate). It is seen from Fig. 4.2

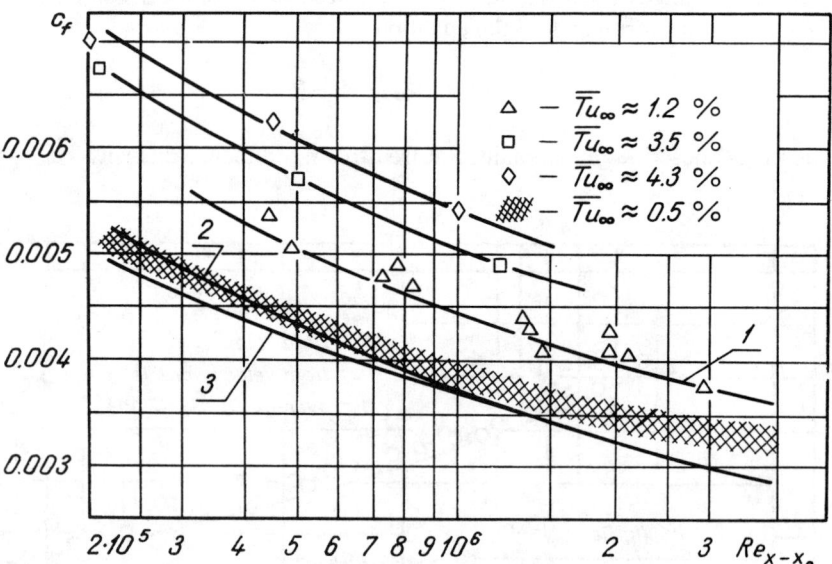

**FIG. 4.2** Local skin friction coefficients at different freestream turbulence intensities. (1) Equation (4.7), (2) data of Šlančiauskas et al. [93] for a flat plate, (3) the Schultz-Grunow drag law.

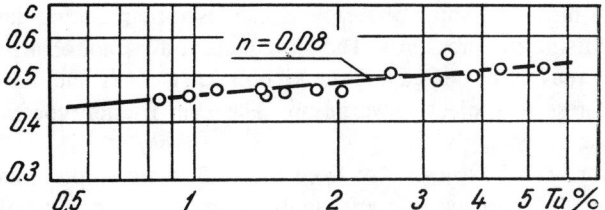

**FIG. 4.3** Effect of turbulence intensity on skin friction.

that the local skin friction coefficient on a cylinder is greater than the corresponding coefficient on a flat plate. The effect of the cylindrical shape increases at large $Re_{x-x_0}$ and at low $Re_{r_0}$ (Fig. 4.4).

The above tendencies become understandable if it is noted that the difference between flows above a cylindrical surface and a flat plate is due to the increase in the flow passage cross section with increasing distance from the cylindrical surface. When the boundary layer thickness is small compared to the radius of the cylinder, the changes in flow passage cross section across the boundary layer are insignificant, and the flow over a cylinder becomes identical to that over a flat plate.

When the boundary layer thickness is of the same order of magnitude as the cylinder radius, the flow passage cross section differs significantly from that for a flat plate, and this results in a significant difference in the friction characteristics of a cylinder and a plate. The boundary layer thickness increases with distance along the cylinder axis. This growth is very steep near the upstream end of the cylinder and it slows down as $x$ increases. Hence, the greater $Re_{x-x_0}$, which is based on the downstream distance from the virtual origin, the thicker the boundary layer and the larger the effect of cylindrical shape.

All the above pertains to the case of a constant $Re_{r_0}$. The points in Fig. 4.4 represent data on the distribution of local skin friction coefficients, determined

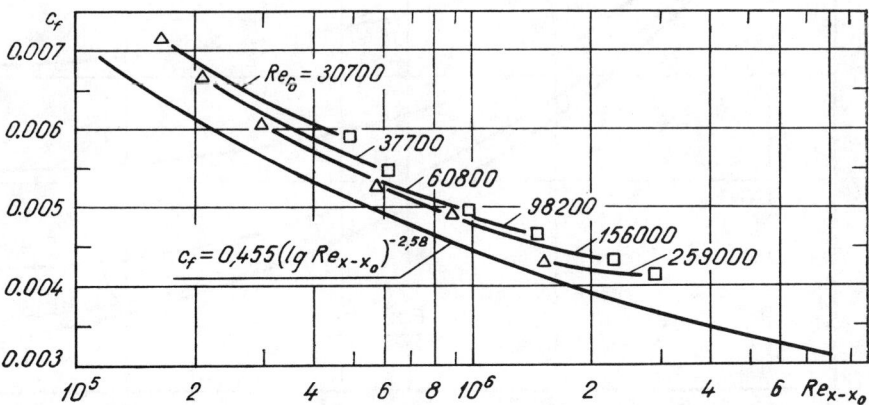

**FIG. 4.4** Distribution of local skin friction coefficients measured on a cylinder with $d = 4$ mm at $Tu_\infty \approx 1.2\%$. The bottom curve is for a cylinder with $d = 15.5$ mm.

for a cylinder with $d = 4$ mm. The value of $\text{Re}_{r_0}$ was increased by reducing the kinematic viscosity (by raising the pressure). The flow pattern does not change when $\text{Re}_{r_0}$ is varied by increasing or decreasing the kinematic viscosity and only the boundary layer thickness is subject to variation. The skin friction coefficients increase or decrease accordingly.

Analogous conclusions can be drawn also from Fig. 4.5, which represents analytically calculated local skin friction coefficients, using the technique of Sparrow et al. [55]. The parameter of these curves is $\text{Re}_{r_0}$, which is based on the cylinder radius, and varies from $10^3$ to $10^6$. To more conveniently use the results for determining the effect of cylindrical geometry, Fig. 4.5 also shows a curve of the local skin friction coefficient for a flat plate.

Figure 4.6 presents experimental data on skin friction over a cylinder with $d = 15.5$ mm as a function of $\text{Re}_{\delta **}$. Such a representation of results made it possible to compare them with results calculated from the Ludwieg–Tillman empirical formula [92]

$$c_f = 0.246 \cdot 10^{-0.678 H} \text{Re}_{\delta **}^{-0.268} \tag{4.9}$$

using the shape factor $H$ and $\text{Re}_{\delta **}$ calculated from experimental velocity profiles.

Equation (4.9) was obtained by Ludwieg and Tillman based on their own

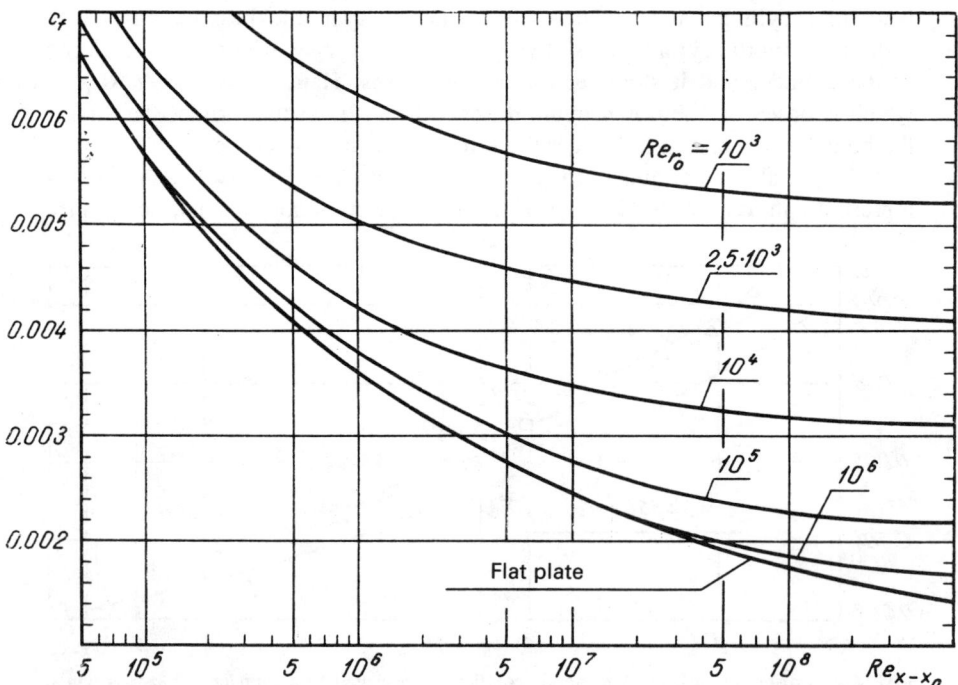

**FIG. 4.5** Calculated local skin friction coefficients [55].

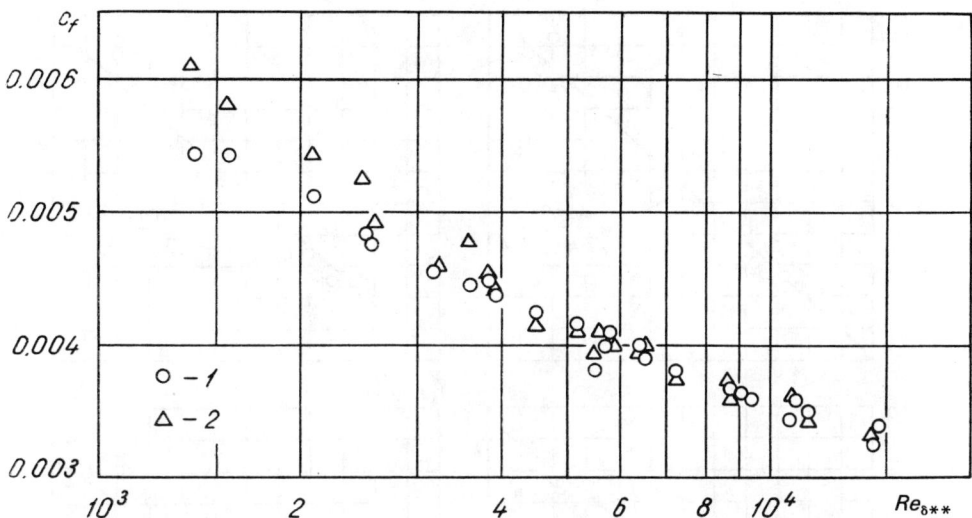

**FIG. 4.6** Comparison of experimental data on skin friction with those calculated from Eq. (4.9) for a cylinder with $d = 15.5$ mm. (1) Results of present study for a cylinder with $d = 15.5$ mm, (2) coefficients of local skin friction, calculated using the Ludwieg-Tillman equation.

experiments assuming the existence of both a universal law of the wall and a single-parameter velocity profile.

It is seen from Fig. 4.6 that correction for the shape factor $H$ results in satisfactory agreement between experimental and analytic results. However, Eq. (4.9) can be used only when the effect of transverse curvature is insignificant, i.e., only for large-diameter cylinders. In the opposite case it is not recommended that skin friction be determined from Eq. (4.9), since it yields results that are much too low and that do not agree with those determined from measured velocity profiles.

## 4.2 EFFECT OF TEMPERATURE FACTOR ON FRICTION

The skin friction coefficient in the presence of heat transfer was determined from the previously obtained equation

$$\frac{d\delta_q^{**}}{d\delta^{**}} = \frac{c_{fq}}{c_f} = K \qquad (2.17)$$

To do this we measured velocity profiles and calculated the values of $c_f$ and $\delta^{**}$ in the absence of heat transfer and obtained $\delta_q^{**}$ in the presence of heat transfer. Using several velocity profiles along the cylinder, we determined $d\delta_q^{**}/d\delta^{**}$ graphically (Fig. 4.7).

The local skin friction coefficients on a cylinder, determined by the above

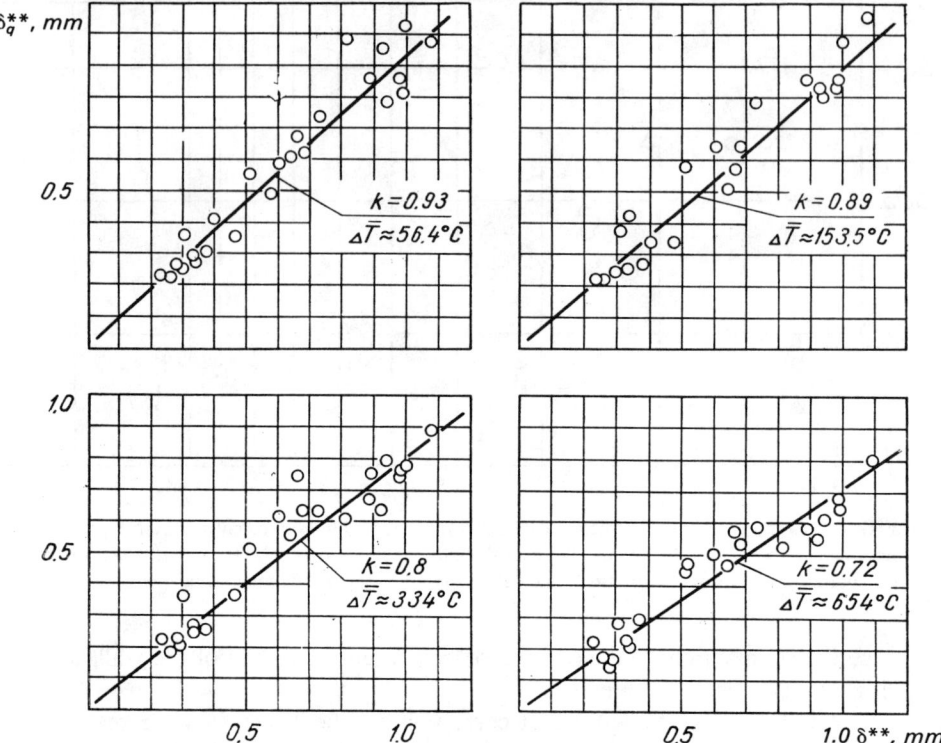

**FIG. 4.7** Determination of the variation in skin friction due to the temperature factor.

method at different temperatures of the cylinder surface, are plotted in Fig. 4.8. It is seen by examining this figure that $c_f$ decreases with increasing temperature factor. Over the range of Re and temperature factor under study, the reduction in $c_f$ is as high as 43%. This happens because the physical properties of the flow (particularly density) are not constant across the boundary layer but change with the temperature. When the temperature of the cylinder surface rises, which also increases the temperature factor, the physical properties in the boundary layer change and its shape factor increases. The latter reduces the skin friction [92].

All cases with approximately the same mean temperature of the cylinder wall can be described by the same dimensionless power-law equation:

$$c_{fq} = c(\lg \text{Re}_{x-x_0})^m \tag{4.10}$$

To obtain a general equation for the skin friction coefficient at all the temperature differences, we must clarify how it is affected by variation of physical properties in the boundary layer.

It is known that changes in physical properties of gases can be expressed by the ratio of their absolute temperature raised to some power, that is, $\rho/\rho_0 =$

$T_{f_0}/T_f$, $\mu/\mu_0 = (T_f/T_{f_0})^n$ etc. Hence it is better to correlate experimental data at different temperature differences by introducing the temperature factor raised to some power into Eq. (4.10). Then the nondimensional equation for the local skin friction coefficient becomes

$$c_{fq} = c(\lg \text{Re}_{x-x_0})^m \psi^n \quad (4.11)$$

where $\psi = T_w/T_f$.

To determine the exponent $m$ of $\log \text{Re}_{x-x_0}$, we assumed that $\psi = 1$ in Eq. (4.11) and, for each pressure, constructed a curve of $c_{fq} = f(\log \text{Re}_{x-x_0})$ in logarithmic coordinates using values at different $x/d$ (at constant $x/d$ the temperature of cylinder surface is approximately constant), and we graphically determined the value of $m$ for each level of $\psi$. It was found that for $\text{Re}_{x-x_0}$ from $2.5 \times 10^6$ to $7 \times 10^6$ and $\psi$ from 1 to 3.1, $m = -2.56$.

It is seen that $m$ was found to be approximately the same as that for isothermal flow. To simplify subsequent correlations and analysis of results, we assumed that $m = -2.58$. After constructing the function $c = f(\psi)$, where

$$c = c_{fq}(\lg \text{Re}_{x-x_0})^{2.58} \quad (4.12)$$

we determined the power exponent $n$ of $\psi$ (Fig. 4.9). It was found that at all the values of $\text{Re}_{x-x_0}$ and irrespective of the value of $m$, the value of $n$ fluctuates about $-0.3$. This value was used for the entire range of $\text{Re}_{x-x_0}$ under study.

Despite the simplifications made in determining $c_{fq}$, $m$, and $n$, the experimental data are satisfactorily described by an equation such as Eq. (4.11). The majority of the test points deviates from the approximating curve by not more than $\pm 6\%$ (Fig. 4.10).

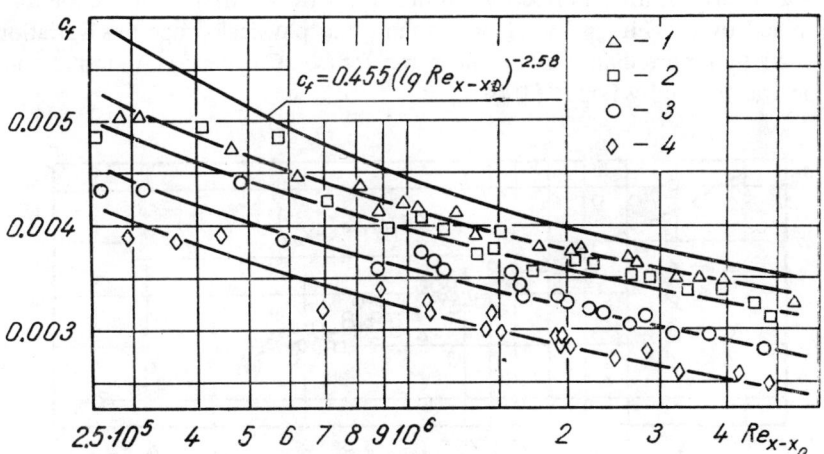

**FIG. 4.8** Local skin friction coefficients at different values of the temperature factor $\psi$ (cylinder with $d = 15.5$ mm, $\text{Tu}_\infty \approx 1.2\%$). (1) $\psi = 1.24$, (2) $\psi = 1.57$, (3) $\psi = 2.24$, (4) $\psi = 3.24$. The top curve is for conditions without heat transfer.

**86** HEAT TRANSFER OF GAS COOLANTS

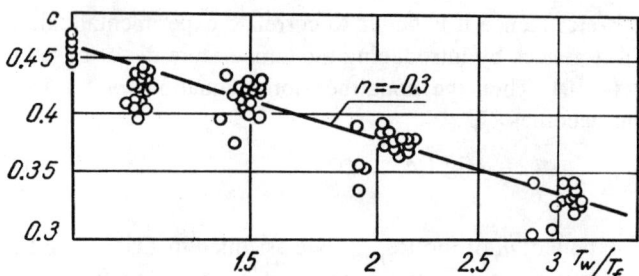

**FIG. 4.9** Effect of temperature factor on friction.

This means that the local skin friction coefficient on a cylinder with $d = 15.5$ mm, at $\mathrm{Re}_{x-x_0}$ from $2.5 \times 10^5$ to $7 \times 10^6$ and $\psi$ from 1 to 3.2 can be expressed as

$$c_{fq} = 0.455 \, (\lg \mathrm{Re}_{x-x_0})^{-2.58} \psi^{-0.3} \quad (4.13)$$

Since no data on the distribution of $c_f$ on a cylinder as a function of $\psi$ are available, the effect of $\psi$ on skin friction can be compared to previous studies only when $r_0 \to \infty$, i.e., for a flat plate. The effect of $\psi$ on friction, calculated on the basis of analytic results of Popov [50, 100] from the equation

$$\frac{c_{fq}}{c_f} = 1 + A \left[ \left( \frac{\rho_w}{\rho_\infty} \right)^{0.2} - 1 \right] \quad (4.14)$$

where

$$A = 1.54 \left[ \mathrm{Re}_x \frac{\mu_\infty}{\mu_w} \left( \frac{\rho_w}{\rho_\infty} \right)^{0.2} \right]^{0.009} \quad (4.15)$$

points to a less significant effect of physical properties variation on friction than determined by us (Fig. 4.11). It was found that physical properties variation affects friction more than it affects heat transfer (see Chapter 5). A similar conclusion was drawn by Popov [100].

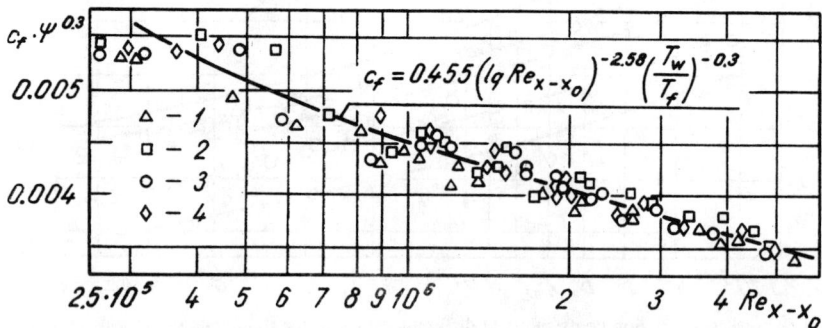

**FIG. 4.10** Local skin friction coefficients with correction for the temperature factor (cylinder with $d = 15.5$ mm, $\overline{\mathrm{Tu}}_\infty \approx 1.2\%$). (1) $\overline{\psi} = 1.24$, (2) $\overline{\psi} = 1.57$, (3) $\overline{\psi} = 2.24$, (4) $\overline{\psi} = 3.24$.

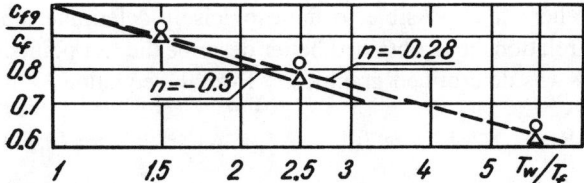

**FIG. 4.11** Comparison of data on the effect of physical properties variation on skin friction. The solid curve represents our data, and the dashed curve (drawn along the points) represents the data by Popov [100].

## 4.3 EFFECT OF THE CURVATURE OF THE CYLINDER ON FRICTION

As previously noted, skin friction depends not only on $\mathrm{Re}_{x-x_0}$, Tu, and $T_w/T_f$ but also on the transverse curvature of the surface, whose influence increases with the ratio of boundary layer thickness to the cylinder radius ($\delta/r_0$).

It can be assumed that the effect of surface curvature on skin friction will change with $\delta/r_0$, irrespective of the manner in which the latter is varied, that is, by changing either one or both of the terms in the ratio $\delta/r_0$. In our case, the effect of surface curvature can be determined using experimental data on a single cylinder, that is, using the change in boundary layer thickness $\delta$ as a function of $\mathrm{Re}_x$ at constant Tu.

It can be stated upon analyzing the variation in $\delta/r_0$ (at $r_0$ = constant and Tu = constant), that it decreases with increasing $\mathrm{Re}_{x-x_0}$ due to reduction in the kinematic viscosity, and increases if $\mathrm{Re}_{x-x_0}$ increases due to increase in the reference dimension. The higher the value of $\delta/r_0$, the stronger the effect of curvature on skin friction. The effect of curvature on friction was approximately the same at all the temperature differences under study.

Experimental data can be correlated with sufficient accuracy by introducing a separate factor into the power-law expression for skin friction, which in the case of variable physical properties in the boundary layer, becomes

$$c_f = c(\lg \mathrm{Re}_{x-x_0})^m \psi^n \Omega^k \tag{4.16}$$

Here $\Omega$ is a correction factor for surface curvature.

Since the effect of surface curvature on friction has been explored very little, it is very important to select the value of $\Omega$. Eckert [54] attempted to correct for this effect analytically. Assuming that the velocity distribution obeys the 1/7 law, and that the shear stresses are inversely proportional to $(\delta_{fp}/\delta)^{0.25}$ (Blasius law), he arrived at the conclusion that the skin friction coefficient on a cylinder with a turbulent boundary layer is higher than that on a plate by the amount $\Omega^k = (1 + \delta/3r_0)^{0.2}$.

Jacob and Dow [53] suggested that the effect of curvature on heat transfer be corrected for by the factor $(1 + 0.3\delta_{fp}/r_0)$. In this case it was found best to assume $\Omega^k = (1 + \delta/r_0)^k$ here it is possible to more precisely determine the power exponent $k$, and the relations thus obtained better describe the test points.

The power exponent $k$ was determined graphically from the equation

$$c_f(\lg \text{Re}_{x-x_0})^{-m}\psi^{-n} = f(1 + \delta/r_0) \qquad (4.17)$$

The boundary layer thickness $\delta$ was determined from the value of $\delta^*$, calculated from measured velocity profiles.

To determine $k$ precisely, it is necessary to know the value of $m$ for a flat plate. If the latter is unknown, then $k$ should be determined at $\text{Re}_{x-x_0}$ = constant where it will be independent of the assumed value of $m$. However, if $m$ is close to the power exponent of a plate at the given range of Re, then a small deviation from the condition that $\text{Re}_{x-x_0}$ = constant will not significantly affect the value of $k$. We determined $k$ in the above manner (Fig. 4.12) and found it to vary only little with $\text{Re}_{x-x_0}$, thus we assumed it to be equal to 0.18 over the entire range under study.

Experimental data on skin friction of a cylinder at different temperature differences and flow pressures, with correction for $\psi$ and for the surface curvature, are correlated by the expression

$$c_f = 0.316(\lg \text{Re}_{x-x_0})^{-2.45}\psi^{-0.3}\Omega^{0.18} \qquad (4.18)$$

Equation (4.18) describes experimental data to within $\pm 6\%$ and is convenient because it extends over a rather wide range of $\text{Re}_{x-x_0}$.

The experimental points can be better correlated by expressing $c_f$ as

$$c_f = c\text{Re}_{x-x_0}^m \psi^n \Omega^k \qquad (4.19)$$

Then the experimental points are correlated by two equations (Fig. 4.13):

for $\text{Re}_{x-x_0}$ from $2.5 \times 10^5$ to $2 \times 10^6$,

$$c_f = 0.0593 \, \text{Re}_{x-x_0}^{-0.2} \psi^{-0.3} \Omega^{0.18} \qquad (4.20)$$

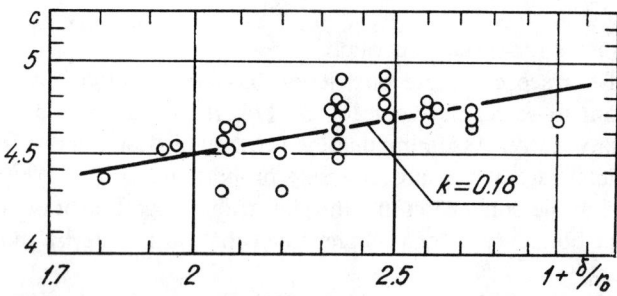

**FIG. 4.12** Determination of the power exponent of the factor correcting for the effect of surface curvature on skin friction.

SKIN FRICTION OF A CYLINDER IN AXIAL FLOW    89

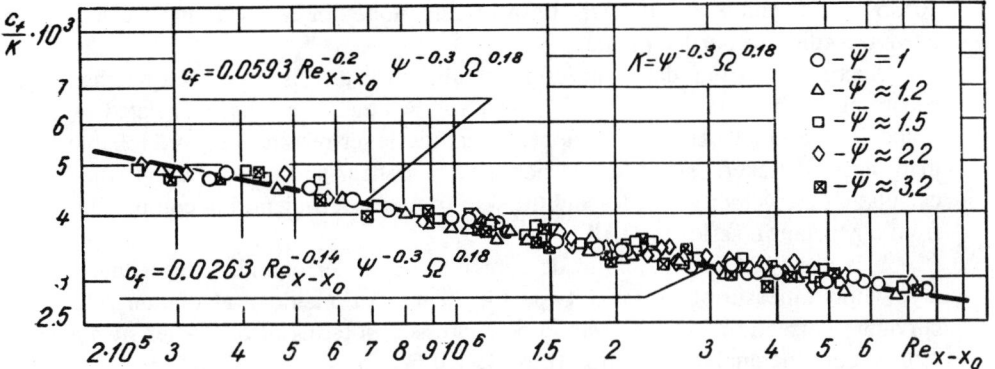

**FIG. 4.13** Correlated experimental data on local skin friction with correction for the temperature factor and surface curvature (cylinder with $d = 15.5$ mm, $\overline{Tu}_\infty \approx 1.2\%$).

and for $Re_{x-x_0}$ from $2 \times 10^6$ to $7 \times 10^6$,

$$c_f = 0.0263 \, Re_{x-x_0}^{-0.14} \psi^{-0.3} \Omega^{0.18} \quad (4.21)$$

Since no other experimental data on the effect of surface curvature on the skin friction with a turbulent boundary layer are available, our data were compared with available analytic results. The effect of the curvature on friction as determined by us is in satisfactory agreement with the analytic solution of Sparrow et al. [55] (Fig. 4.14). The comparison was performed at $Re_{r_0}$ of $2 \times$

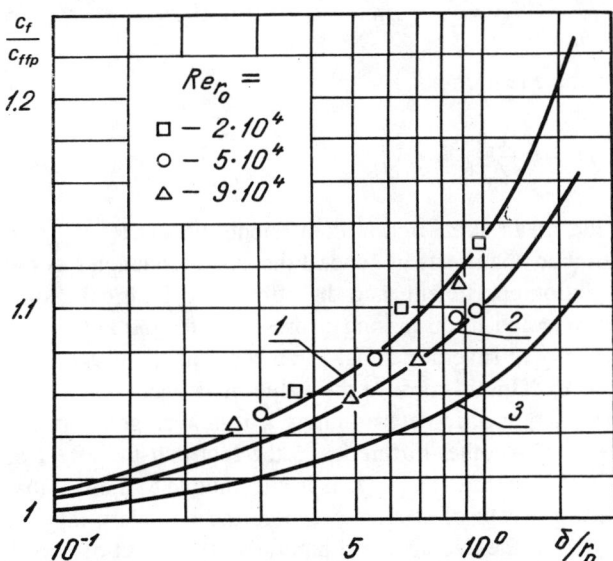

**FIG. 4.14** Comparison of the effect of surface curvature on friction. (1) Correlation curve of our data, (2) of ref. [98], (3) of ref. [54]. The points are calculated according to ref. [55].

# 90 HEAT TRANSFER OF GAS COOLANTS

$10^4$, $5 \times 10^4$, and $9 \times 10^4$ on the basis of data plotted in Fig. 4.5, and recalculated according to our technique.

According to our data, the effect of the curvature is 2–6% larger than according to the data of Eckert [54], and 1–3% larger than that calculated by Solodkin and Ginevskiy [98]. It appears that the latter results are more reliable than those of Eckert, since he did not specify a shear stress distribution when calculating the velocity profiles and the skin friction coefficient, but computed it from a Maclaurin series instead.

It is presently impossible to determine the cylinder boundary layer thickness without measuring the velocity profile; hence, for practical calculations, the curvature correction factor $\Omega$ should be expressed in terms of $\delta_{fp}$. After such a replacement the analytic equation [Eq. (4.18)] becomes

$$c_f = 0.329(\lg \mathrm{Re}_{x-x_0})^{-2.45} \psi^{-0.3} \Omega^{0.14} \qquad (4.22)$$

where $\Omega = 1 + \delta_{fp}/r_0$.

## 4.4 MEAN SKIN FRICTION COEFFICIENT

The results analyzed in this chapter pertain to the local skin friction coefficients. Knowing them, it is possible to determine the mean skin friction coefficient $C_f$. If the drag force acting on surface area $F$ that extends between points 0 and $x$ is expressed as

$$D = \int_F \tau_w \, dF \qquad (4.23)$$

$C_f$ can be calculated from the equation

$$C_f = \frac{2D}{\rho u_\infty^2 F} = \frac{1}{x} \int_0^x c_f \, dx \qquad (4.24)$$

The distribution of the mean skin friction coefficients versus $\mathrm{Re}_{x-x_0}$ (the length-based Reynolds number) manifests in general the same tendencies as the distribution of the local friction coefficient, and the effect of cylindrical geometry is similar to those enumerated in discussing expressions for the latter. Important differences appear only in several details. First, the numerical values of $C_f$ are higher than those of $c_f$. This is because the value of $c_f$ decreases downstream (i.e., with increasing $x$), and, in determining $C_f$, it is averaged between the limits $x$ and $x = x + \Delta x$. Another difference is the fact that the effect of cylindrical geometry on $C_f$ is lesser than on $c_f$. This is attributable to the fact that $c_f$ values at points situated upstream of section $x$, where the boundary layer is thiner than in $x$, are included in the average, and naturally, the effect of curvature is smaller.

More detailed data on the mean skin friction coefficient can be found in the analytic study by Sparrow et al. [55].

CHAPTER
# FIVE

# HEAT TRANSFER IN THE ENTRANCE REGION OF AN ANNULUS

Turbulent flow in the entrance region of a channel differs significantly from fully developed turbulent flow. The development and formation of the flow depends on a large number of factors, such as inlet velocity distribution, inlet design, and the initial turbulence intensity.

The flow conditions in the entrance region of a duct is similar in most cases to the boundary layer pattern in flow over a plate. However, in annuli, in particular at high diameter ratios, the nature of formation of boundary layers on the outer and inner surfaces is significantly different. The formation of the inner cylinder boundary layer is significantly affected by the surface curvature, whereas the boundary layer of the outer cylinder is approximated by familiar relations, obtained for a flat plate in longitudinal flow. For this reason the study of heat transfer was performed by us only for the inner tube of the annulus. Since the moderate axial pressure gradients existing in our test section do not affect the formation of the boundary layer or the heat transfer pattern, the surface of the inner tube can be regarded as a cylinder in axial potential flow.

This chapter presents the results of a study on local heat transfer from a cylinder in axial flow. The effects of physical properties variation, transverse curvature of the surface, initial freestream turbulence, and surface disturbances upstream of the heated length are investigated.

## 5.1 DISTRIBUTION OF HEAT TRANSFER COEFFICIENTS ALONG THE CYLINDER

It is known that heat transfer coefficients in symmetrical axial flow over a cylinder do not vary around the perimeter at a fixed location along the cylinder. We are interested only in the distribution of the heat transfer coefficients along the cylinder. If heat transfer occurs at a constant rate through the wall, the surface temperature of bodies in external flow is a mirror image of the distribution of the heat transfer coefficient. The conditions in our experiments differ somewhat from $q_c(x)$ = constant; however, the axial temperature distribution pattern is analogous to that of the distribution of local heat-transfer coefficients. It is seen from Fig. 5.1 that the surface temperature of cylinder 1 ($d$ = 15.5 mm) changes steeply only near its upstream end. Temperature changes become more and more moderate as $L$ increases, approaching isothermal conditions at low temperature differences.

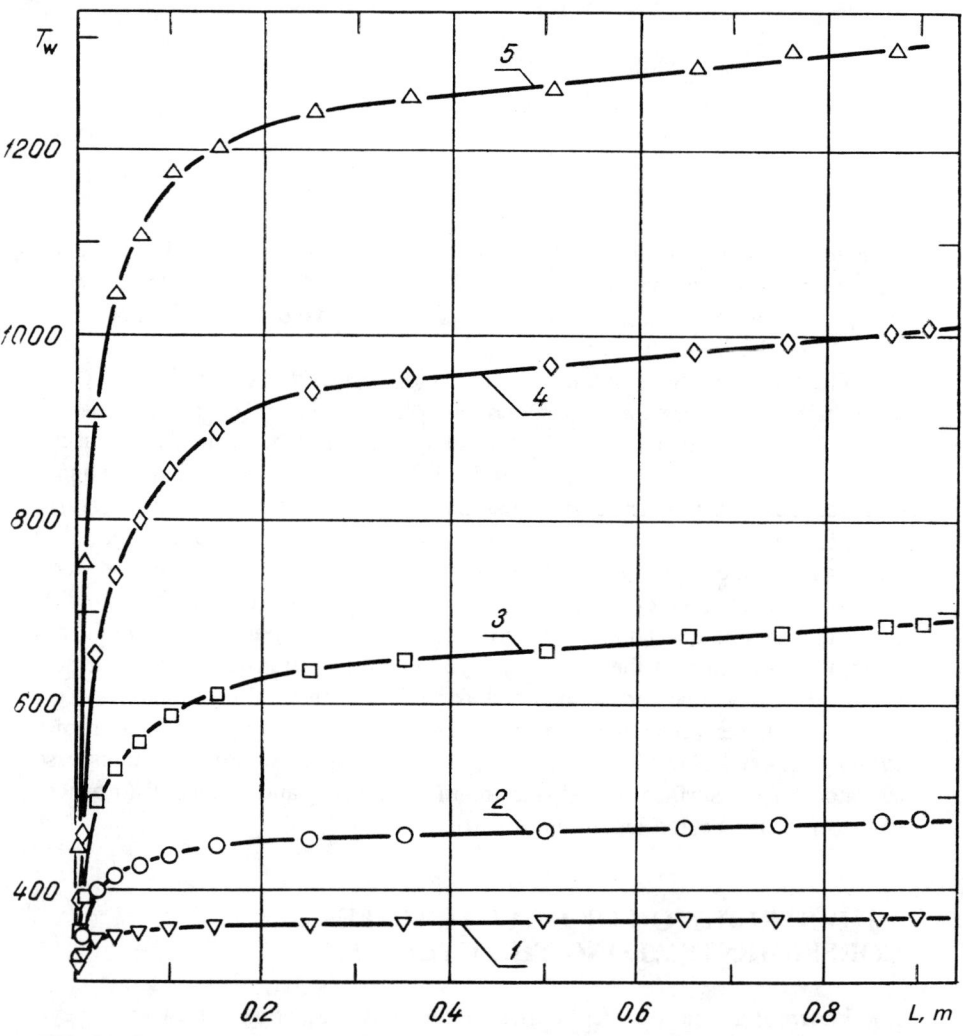

**FIG. 5.1** Temperature distribution along the surface of cylinder 1 at $Re_{r_0} \approx 9 \times 10^4$ and at different $\psi$. (1) $\psi$ = 1.2, (2) $\psi$ = 1.5, (3) $\psi$ = 2.1, (4) $\psi$ = 3.0, (5) $\psi$ = 4.0.

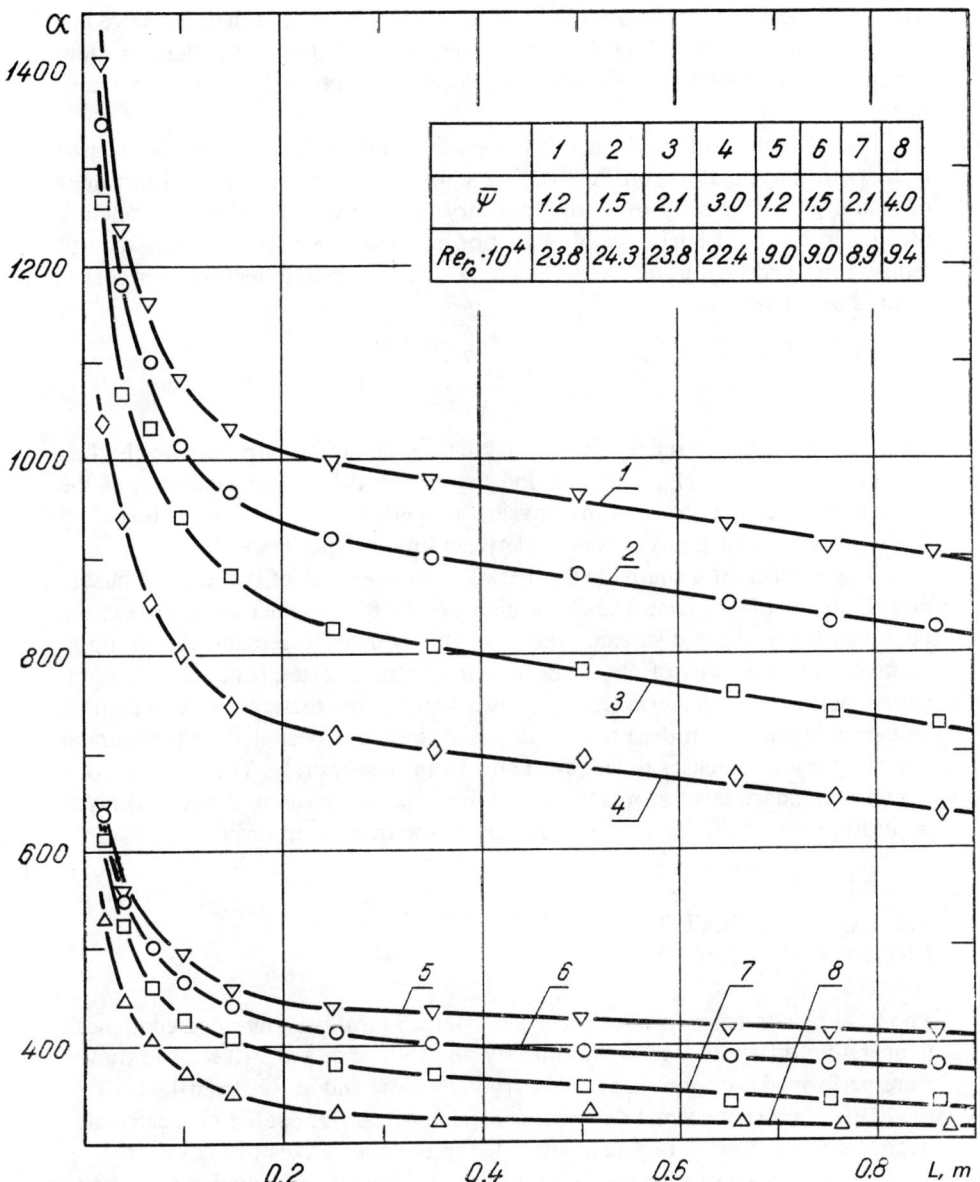

**FIG. 5.2** Variation of local heat transfer coefficients along cylinder 1.

The shape of the surface temperature curves of cylinder 1 is almost independent of the flow variables and of the heat transfer through the wall, provided that the mean temperature difference is constant; however, when $\overline{\Delta T}$ is not constant, the temperature distribution along the cylinder varies significantly (Fig. 5.1).

The convective heat transfer coefficient changes steeply in the upstream portion of the cylinder, as does the surface temperature, and then its curve becomes asymptotic (Fig. 5.2). It is seen from Fig. 5.2 that under identical flow conditions, the heat transfer coefficient decreases with increasing temperature factor $\bar{\psi}$.

The reduction of the heat transfer coefficients with increasing $\bar{\psi}$ is associated with physical properties variation across the boundary layer. The values of $\rho$ and $\mu$ vary greatly over the boundary layer span, distorting the velocity profile. In the final analysis, this variation is responsible for gradients of wall temperature and heat transfer coefficient. If the convective heat transfer coefficient is expressed as

$$\alpha = \frac{-\lambda \left(\frac{\partial T}{\partial y}\right)_{y=0}}{\Delta T} \tag{5.1}$$

then the heat transfer reduction with increasing $\bar{\psi}$ can be attributed to the fact that the rise in $(\partial T/\partial y)_{y=0}$ lags behind the rise in $\Delta T$ due to distortion of the velocity profile brought about by physical properties variations in the boundary layer, whereas $\lambda$ of gases depends relatively little on the temperature.

The absence of a sharp change between the diameter of the forward busbar of cylinder 5 ($d = 4$ mm) and the diameter of the cylinder itself causes the boundary layer in the upstream portion of the cylinder to remain laminar up to relatively large values of $Re_x$. The rise in the surface temperature at the upstream portion of cylinder 5 (Fig. 5.3) is a result of the existence of this laminar boundary layer. When the pressure is raised, the zone occupied by the laminar boundary layer decreases in length and then vanishes entirely. The existence of a laminar boundary layer at relatively high $Re_x$ (the transition region was delayed up to $Re_x$ of $5 \times 10^5$ to $8 \times 10^5$) indicates low freestream turbulence.

## 5.2 LOCAL HEAT TRANSFER FROM A CYLINDER

The local heat transfer from cylinder 1 ($d = 15.5$ mm) was investigated at mean temperature differences of approximately 56, 150, 350, and 940 K. The studies were performed with turbulent boundary layer flow and at $Re_x$ from $1 \times 10^5$ to $5 \times 10^7$. The experimental data on the local heat transfer coefficients calculated over a length $L$ from 2 to 86 cm are listed in the appendixes (Tables 8–12).

Figure 5.4 presents local heat transfer coefficients for cylinder 1. It shows that at any temperature difference, the heat transfer is an identical function of $Re_x$. It can be easily seen by considering an individual case, at a given temperature factor, that the local heat transfer from a cylinder is not rigorously represented by a power-law relation of the form

$$\text{Nu} = c\text{Re}^m \tag{5.2}$$

which is a straight line in logarithmic coordinates, but rather it deviates some-

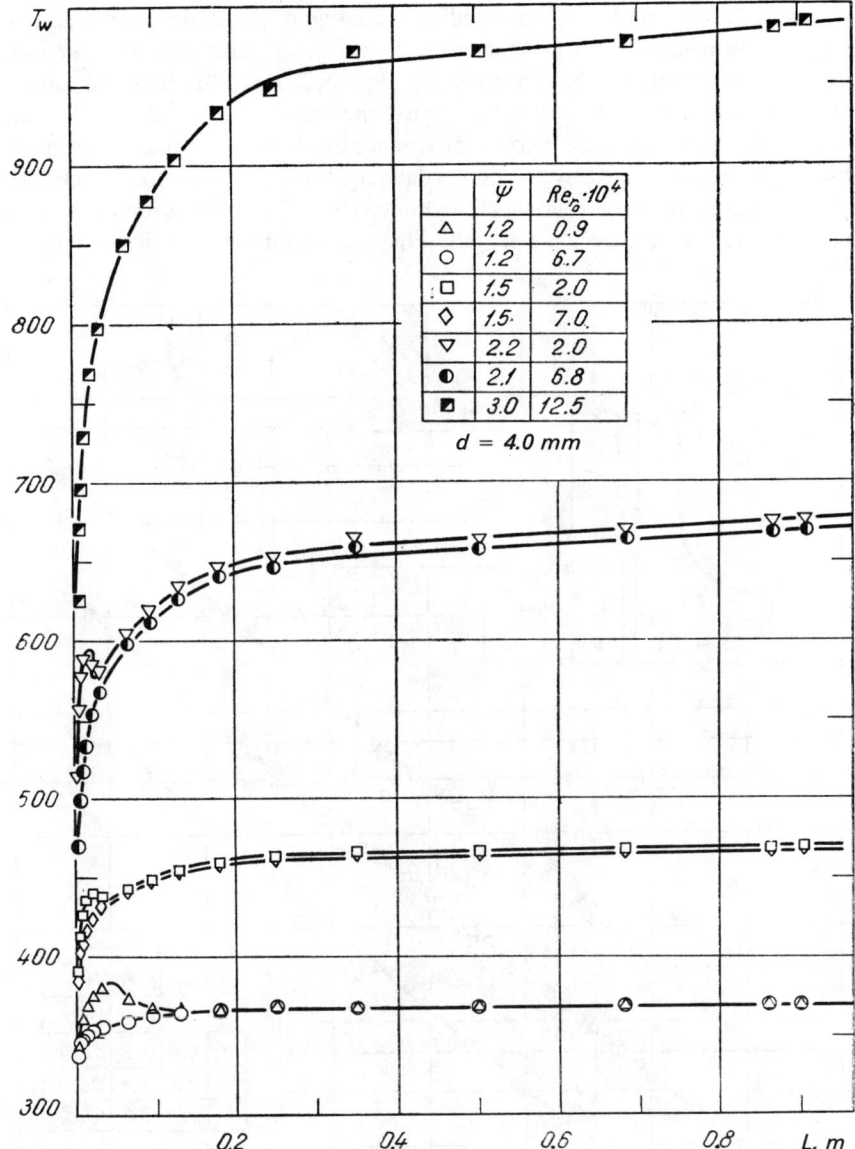

**FIG. 5.3** Temperature distribution along cylinder 5.

what upward from this line, reaching relatively high values of the heat transfer coefficient with increasing Re. In the case at hand, this deviation is due to two reasons: first, enhancement of turbulent transport in the boundary layer with increasing $Re_x$, and second, the effect of the cylinder's curvature. In spite of this, we reduced our experimental results on the basis of Eq. (5.2), since it simplifies the analysis of data in which the effect of the temperature factor on heat transfer is combined with the effect of surface curvature.

**96** HEAT TRANSFER OF GAS COOLANTS

Analysis showed that by subdividing the range of $Re_x$ under study into two intervals, the results on local heat transfer at a given temperature factor can, with sufficient accuracy, be described by Eq. (5.2) with different constants $c$ and $m$ for each range of $Re_x$. It is best to use the value of $Re_x = 2 \times 10^6$ as the dividing line between these two ranges, since then the experimental data are most accurately described by the power-law equations. In our experiments, the surface temperature varied along the cylinder (Fig. 5.1), indicating changes in the temperature factor over the length. Thus, to avoid the error introduced by

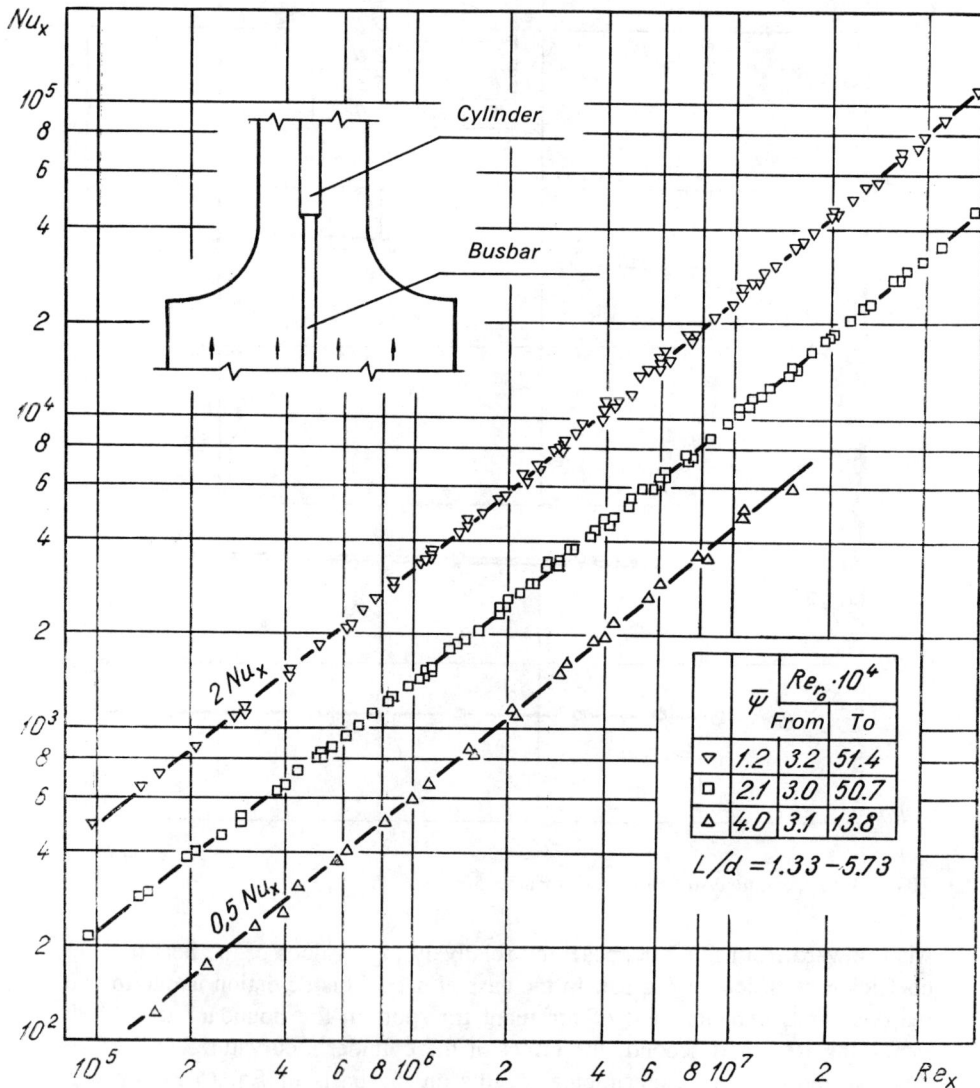

**FIG. 5.4** Local heat transfer from cylinder 1.

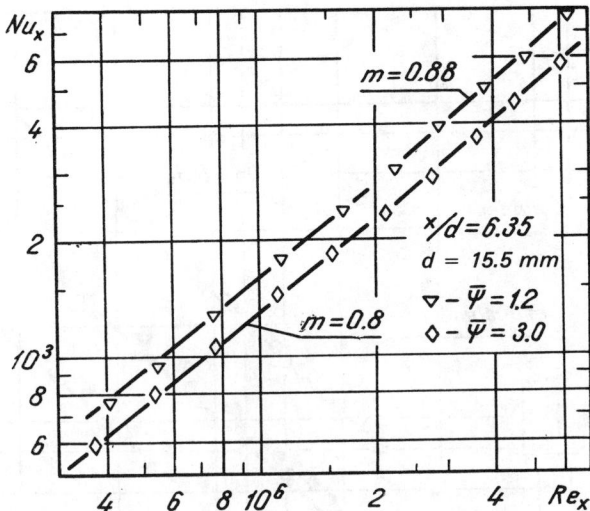

**FIG. 5.5** Heat transfer rate versus $Re_x$ at different temperature differences.

physical properties variation in determining the value of $m$, we selected data at $\psi$ = constant. This corresponds to data at $x$ = constant, with a fixed mean temperature difference. It was found that the local turbulent heat transfer of cylinder 1 can be described by power-law relations such as Eq. (5.2) with $m = 0.8$ for $Re_x < 2 \times 10^6$ and $m = 0.88$ for $Re_x \geqslant 2 \times 10^6$, irrespective of the value of $\psi$ (Fig. 5.5).

The results on local heat transfer in Figs. 5.4 and 5.6 cluster about lengthwise averaged temperature differences. The width of the band occupied by experimental points (at $\overline{\psi}$ = constant) is a function not only of the experimental accuracy, but of the temperature variation along the surface of the cylinder, and also of the different effects of the curvature, which, as shall be shown below depend on the cylinder diameter and on the flow variables. The effect of curvature on heat transfer is shown more clearly in Fig. 5.6, which represents local heat transfer along a cylinder with $d = 4$ mm. It is seen that at certain values of $Re_x$ and $\overline{\psi}$, the heat transfer is higher at lower flow pressures (in these experiments). This effect is magnified with increasing $Re_x$. Unfortunately, our experimental equipment did not allow attaining high $Re_x$ at both low and high pressures.

The effect of physical properties variations on local heat transfer from cylinders 5 ($d = 4$ mm) and 1 ($d = 15.5$ mm) is analogous.

As a result of its higher curvature, the increase in heat transfer from cylinder 5 with increasing $Re_x$ is significantly larger than that for a plate or a cylinder with a large diameter (cylinder 1). If the results on local heat transfer from cylinder 5 are subdivided into two ranges with respect to $Re_x$, as was done in the case of cylinder 1, then the heat transfer rate at high values of $Re_{r_0}$ can be described by Eq. (5.2) with $m = 0.86$ for the lower $Re_x$ range, and 0.94 for the

**98** HEAT TRANSFER OF GAS COOLANTS

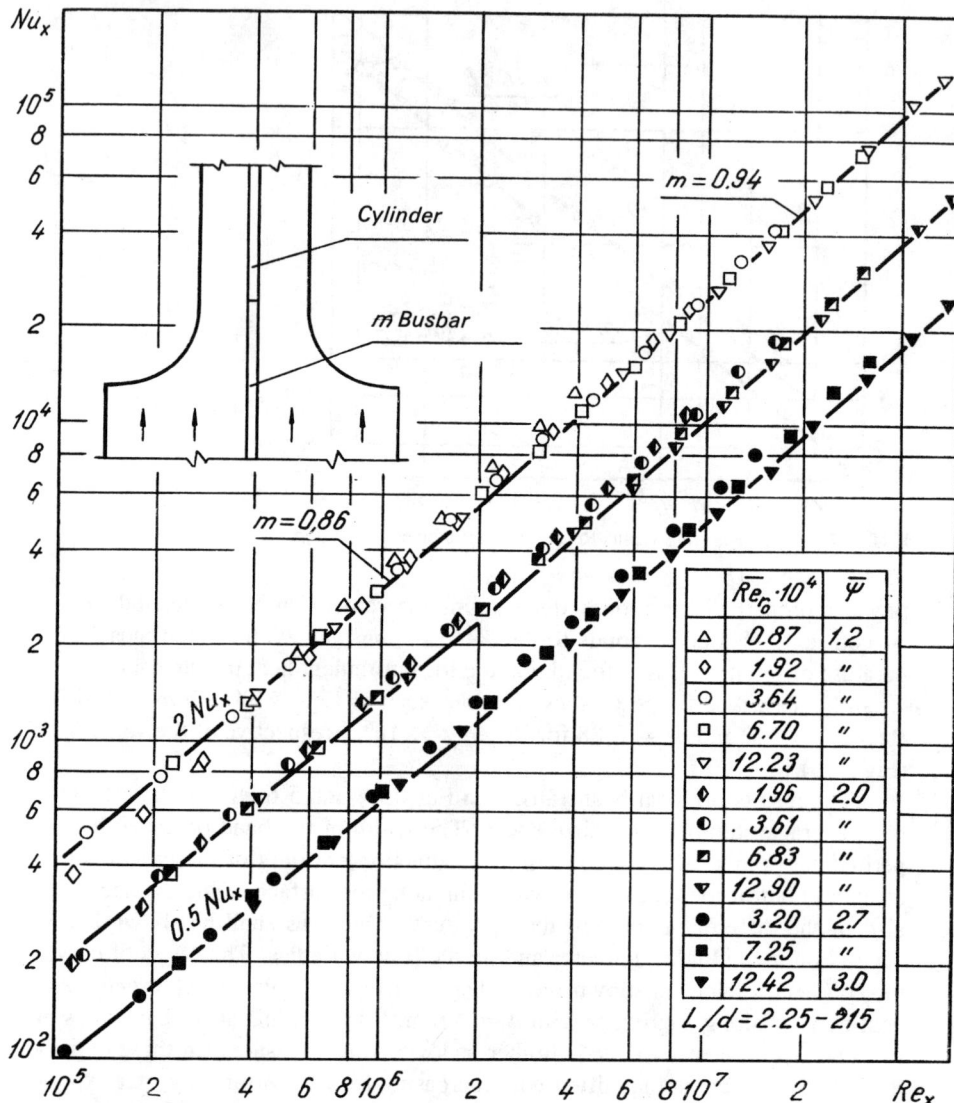

FIG. 5.6 Local heat transfer from cylinder 5.

upper range. At low $Re_{r_0}$ the values of the power exponent $m$ were ~0.95 and ~0.99, for the lower and upper $Re_x$, respectively. Analysis of our results showed that over the lower range of $Re_x$ the $m$ value of the function $Nu_x = f(Re_x)$ is rather on the high side; this occurred because the lower busbar of the cylinder under study had a diameter equal to that of the cylinder itself. However, this is not important, since the results obtained with cylinder 5 are only used for checking the effect of physical properties variations on heat transfer (in such case it is important only that the experimental conditions would not change with the temperature factor), and for determining the effect of curvature. For the latter purpose we used results at high $\text{Re}_x$, exceeding $2 \times 10^6$, where $m$ in Eq. (5.2) is close to unity. Consequently, the error in measuring the location of the boundary layer origin could not introduce a significant error in determining constant $c$ in the heat transfer equation.

It is clear from the above that heat transfer from a cylinder in axial flow depends significantly both on physical properties variation in the boundary layer, and on the surface curvature. We shall determine the magnitude of these effects and their mutual dependence. In the first place we shall attempt to determine the effect of physical properties variation assuming that they are independent of the curvature effect.

## 5.3 EFFECT OF TEMPERATURE FACTOR ON HEAT TRANSFER FROM A CYLINDER

In the general case, heat transfer is a function of a number of factors, such as physical properties of the coolant, flow conditions, direction of heat flux, magnitude of heat flux, unheated length, etc. The above can be written as

$$\text{Nu} = f\left(\text{Re},\ \text{Pr},\ \frac{\mu_f}{\mu_w},\ \frac{\lambda_f}{\lambda_w},\ \frac{c_{p,f}}{c_{p,w}},\ \frac{\rho_f}{\rho_w},\ \frac{x}{L},\ \ldots\right) \qquad (5.3)$$

The physical properties of the gas over the temperature range under study are a power function of the temperature, for example,

$$\frac{\rho_f}{\rho_w} = \frac{T_w}{T_f}\ ;\quad \frac{\mu_f}{\mu_w} = \left(\frac{T_f}{T_w}\right)^n\quad \text{etc.,} \qquad (5.4)$$

Thus, Eq. (5.3) for the case under study (Pr of air is virtually independent of the temperature) can be significantly simplified:

$$\text{Nu} = f\left(\text{Re},\ \frac{T_w}{T_f},\ \frac{x}{L},\ \Omega\right) \qquad (5.5)$$

where $\Omega$ is a parameter correcting for the effect of curvature in axial flow over a cylinder.

In investigating heat transfer from cylinders 1 and 5, the nonheated length, measured from the inlet of the test section (pipe with $d = 100$ mm), was only 2 mm for a total length of 20–860 mm, and hence the reference dimension in calculating Nu and Re was the length $L$ from the start of heating, whereas $x/L$ in Eq. (5.5) was taken to be equal to unity.

In determining the effect of physical properties variation (which is a function of the temperature factor) on heat transfer, we first assume that it is independent of the curvature. Then heat transfer from a cylinder can be described by the power-law equation:

$$\text{Nu}_x = c\ \text{Re}_x^m \psi^n \qquad (5.6)$$

**100** HEAT TRANSFER OF GAS COOLANTS

and the problem reduces to determination of constants $c$, $m$, and $n$. According to Survila and Stasiulevičius [101], the local heat transfer along cylinder 1 ($d = 15.5$ mm) is rather satisfactorily described by Eq. (5.6) with $n = -0.25$ along the entire cylinder and over the entire range of $Re_x$; the value of $m$ for the lower and upper ranges of $Re_x$ is taken to be 0.8 and 0.88, respectively. True, $n$ changes somewhat as a function of the length of the cylinder and the curvature parameter, but the above value of $n$ ensures sufficient accuracy for practical use.

Results on local heat transfer from cylinder 1 with correction for $\psi$ (Fig. 5.7) show that the effect of curvature, for a cylinder with radius $r_0 = 7.75$ mm at $Re_x$ from $10^5$ to $5 \times 10^7$, is not too high. When no correction is made for it, the greatest deviation of experimental data from the approximating curves does not exceed $\pm 10\%$. This means that over the above range of $Re_x$ the local heat transfer from cylinder 1 (corrected for the effect of physical properties variation in the boundary layer at $\psi$ from 1.1 to 4) can be described by two power-law equations, each highly accurate below or above $Re_x = 2 \times 10^6$ (Fig. 5.7). Note that the above equations should be regarded as valid for cylinders with diameters

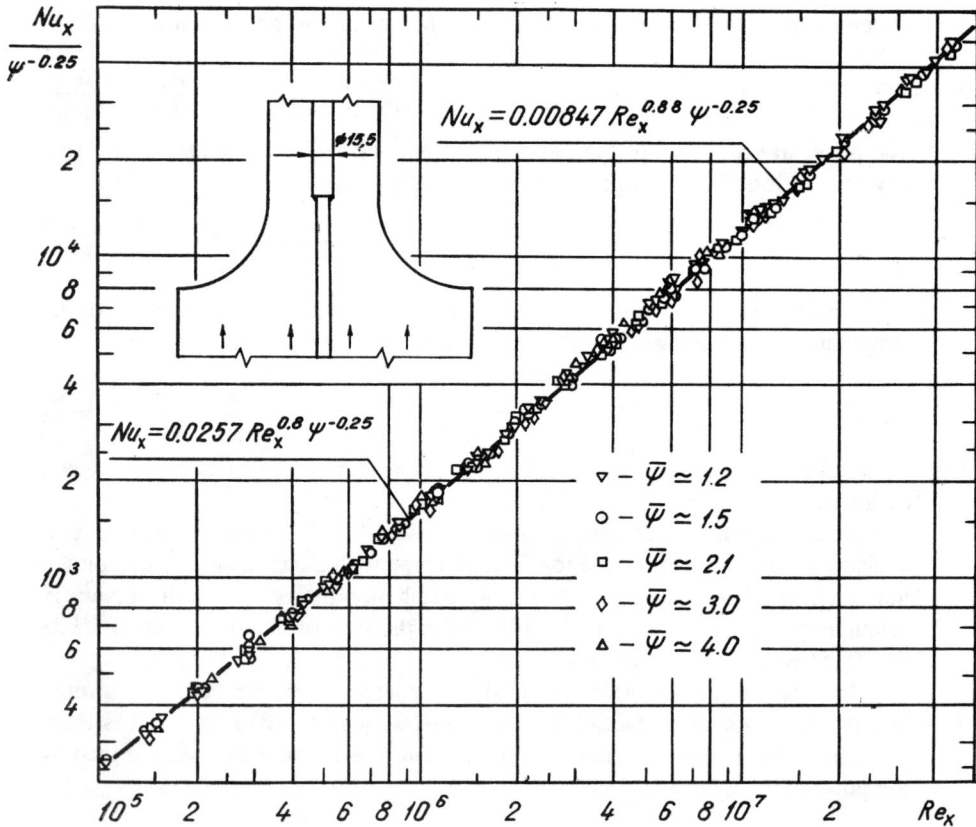

**FIG. 5.7** Local heat transfer from cylinder 1 with allowance for variability of physical properties.

not less than 15.5 mm and length not greater than that investigated (860 mm). Reducing the diameter or increasing the length amplifies the effect of curvature on heat transfer.

Studies of heat transfer from a cylinder with $d = 4$ mm performed at $\psi$ of 1.2, 1.5, 2.1, and 2.7 (3.0) and flow pressures from 0.14 to 2.32 MPa showed that due to variation of flow variables the experimental points stratify as a function of $\overline{\psi}$, and, at high $Re_x$, also as a function of $Re_{r_0}$. Hence the effect of $\psi$ on heat transfer from cylinder 5 was determined from data at maximum pressure over a wide range of $Re_x$. The value of $m$ was taken from results obtained at lowest cylinder-surface temperature ($\overline{T}_w = 365$ K), when temperature variation along the segment under study was only 5%. The variation of cylinder surface temperature increases with the temperature difference. In this case the value of $m$ determined from results at various $x$ locations differs somewhat from the value of $m$ determined at $T_w(x) = $ constant. But, in determining the effect of $\psi$ from data at different $Re_x$, one should assume $m$ to be constant at all the temperature differences.

When the entire range of $Re_x$ was subdivided into two parts as in the case of cylinder 1, the effect of $\psi$ for cylinder 5 was found to be of the same order of magnitude as for cylinder 1 over the entire range of $Re_x$ [102]. This means that the power exponent of $\psi$ can be assumed without reference to the cylinder diameter.

Figure 5.8 presents data on heat transfer from cylinder 5 ($d = 4$ mm) with correction for physical properties variation. It is seen that the experimental points have a greater scatter than in Fig. 5.7. This is due to the fact that no correction was made for the effect of curvature, which is significant in the case of a small-diameter tube. Figure 5.8 also shows a line approximating heat transfer from cylinder 1. It is seen that over the lower range of $Re_x$, up to about $Re_x = 10^6$, heat transfer from cylinder 1 is greater than from cylinder 5. This is due to differences in inlet conditions. The diameter of the forward busbar of cylinder 1, placed in a convergent section upstream of the test section, is smaller than the diameter of the cylinder itself (Fig. 5.4). Thus, the boundary layer forming on the busbar within the convergent section is washed down by the accelerating flow at the step between the busbar and the test section inner cylinder. In the case of cylinder 5, this situation is somewhat modified; hence, one obtains here what appears to be a significant nonheated length. This apparent nonheated length is a function of flow variables, and it is very difficult to take into account. Therefore, we excluded data on cylinder 5 up to $Re_x = 2 \times 10^6$ from subsequent analysis, and used them only for checking the effect of the temperature factor over a wide range of $Re_{r_0}$. Data on local heat transfer from cylinder 5 at $Re_x > 2 \times 10^6$ do not contain a large error due to uncertainty in locating the origin of the reference dimension. This is due to the fact that the heat-transfer rate over this range of $Re_x$ is proportional to the reference dimension raised to a power close to unity. These data, together with results obtained for cylinder 1, were used in subsequent analysis of the effect of surface curvature on heat transfer.

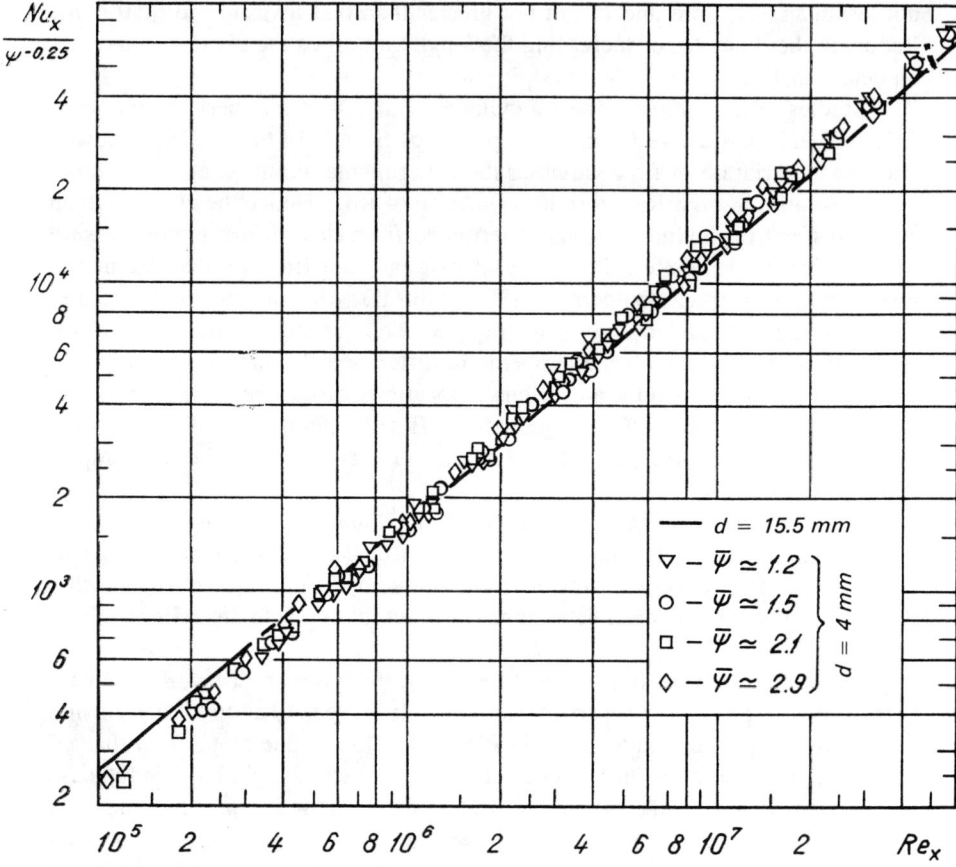

**FIG. 5.8** Local heat transfer along cylinder 5 with correction for the temperature factor.

## 5.4 EFFECT OF THE CURVATURE OF THE CYLINDER SURFACE ON ITS HEAT TRANSFER CHARACTERISTICS

The smaller the diameter of a cylinder, the greater its surface curvature. However, heat transfer from a cylinder in axial flow is not a simple function of this curvature, but depends on $Re_{r_0}$ and $Re_x$. The smaller the Re based on cylinder radius, the greater the difference in heat transfer between the cylinder and a plate. The smallest $Re_{r_0}$ values in our experiments were obtained with a cylinder having $d = 4$ mm at low flow pressure. The heat transfer rate is highest under these conditions (Fig. 5.6). This is observed especially at high $Re_x$. It follows from this that the heat transfer is governed by the ratio of some flow parameter to the cylinder radius. Since the boundary layer thickness is a controlling factor in heat transfer, it is sensible to express this ratio as $\delta/r_0$. It was shown experimentally that the heat transfer is indifferent to the manner in which this ratio is changed (i.e., whether $\delta$ or $r_0$ is varied): an increase in $\delta/r_0$ increases the difference between heat transfer from a flat plate and a cylinder.

It is desirable to have a correlation that would describe heat transfer for different curvatures and would make allowance for physical properties variation. This could be Eq. (5.6) in which $c$ and $m$ are functions of $\delta/r_0$ and $\mathrm{Re}_x$. However, another approach to this correlation is also possible. Jacob and Dow [53] assume that turbulent heat transfer from a cylinder is greater than that from a plate by the amount $(1 + 0.3\,\delta_{fp}/r_0)$. Eckert [54] found that skin friction of a cylinder is greater than that of a plate by the factor $(1 + \delta/3r_0)^{0.2}$. It was found from our experimental data that satisfactory correlation is possible on the assumption that heat transfer from a cylinder differs from that of a flat plate by the amount

$$\Omega^k = (1 + \delta_{fp}/r_0)^k \tag{5.7}$$

where $k = 0.14$ over the entire range of $\mathrm{Re}_x$ under study [102]. The expression for the heat transfer from a cylinder can then be written as

$$\mathrm{Nu}_x = c\,\mathrm{Re}_x^m \left(\frac{T_w}{T_f}\right)^n (1 + \delta_{fp}/r_0)^k \tag{5.8}$$

A more accurate determination of the value of $k$ from experimental data involves certain difficulties. One must know the value of the $\mathrm{Re}_x$ power $m$ for a plate, but for $\mathrm{Re}_x > 2 \times 10^6$ its exact value is unknown, whereas at $\mathrm{Re}_x < 2 \times 10^6$ the curvature effect is relatively small due to the small value of $\delta/r_0$. If the value of $m$ for a flat plate is unknown, then $k$ should be determined at $\mathrm{Re}_x = $ constant, since in this case its value will be independent of the assumed $m$. Given the above, we determined $k$ on the basis of data for cylinders 1 and 5 over the $\mathrm{Re}_x$ range of $2 \times 10^6$ to $5 \times 10^7$; $m$ for the plate was taken, according to the analytic results of Sparrow et al. [55], to be 0.85.

It was found that experimental data obtained over a wide range of flow pressures make it possible to determine the effect of the curvature from data on heat transfer from a cylinder of a single diameter (Fig. 5.9). Data obtained with two different cylinder diameters made it possible to increase the range of values of $\delta_{fp}/r_0$ and to improve the accuracy in determining $k$; it also showed that $k$ is independent of the cylinder diameter. It is also seen that, within the limits of experimental accuracy, $k$ can be taken to be independent of $\mathrm{Re}_x$. Rounding off, we obtained $k = 0.14$ for the entire range of $\mathrm{Re}_x$.

With corrections for the effects of curvature and physical properties variation, the local heat transfer from cylinders 1 and 5 (Fig. 5.10) can be described by two power-law equations such as Eq. (5.8). The correlation of local heat

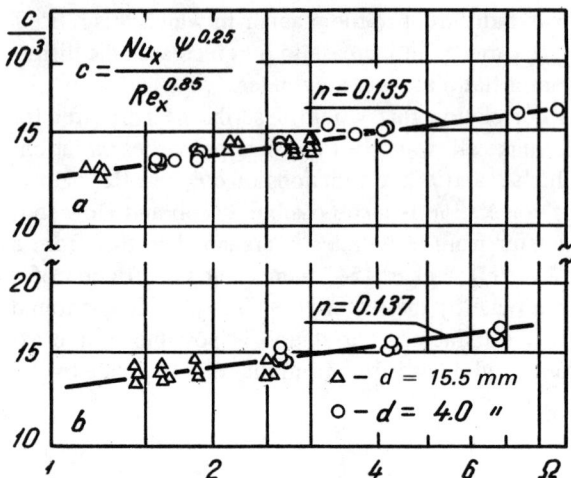

**FIG. 5.9** The power exponent of the curvature parameter as a function of $Re_x$. (a) Data at $Re_x = 3 \times 10^6 - 4 \times 10^6$, (b) data at $Re_x$ from $1.5 \times 10^7$ to $1.8 \times 10^7$.

transfer from a cylinder for the lower range of $Re_x$ ($10^5$ to $2 \times 10^6$) is

$$Nu_x = 0.0253 \, Re_x^{0.8} \psi^{-0.25} \Omega^{0.14} \tag{5.9}$$

For a constant property flow over a flat plate (assuming that when $r_0 \to \infty$, $\Omega = 1$ and $T_w/T_f \to 1$) it yields a result that is 1% lower than that obtained by Isachenko et al. [86] from the familiar relation (for a flat plate):

$$Nu_x = 0.0296 \, Re^{0.8} Pr^{0.43} (Pr_f/Pr_w)^{0.25} \tag{5.10}$$

The flat plate values calculated from Eq (5.9) are 8% lower than the results listed by Žukauskas and Šlančiauskas [52], obtained for a plate with a turbulent boundary layer of heated air. The reason for this difference is apparently the fact that in our experiments $q_w \neq$ constant.

At $Re_x$ of $2 \times 10^6$ to $5 \times 10^7$, heat transfer from a cylinder to air at large temperature differences can be described by the expression

$$Nu_x = 0.0106 \, Re_x^{0.86} \psi^{-0.25} \Omega^{0.14}, \tag{5.11}$$

which, like Eq. (5.9), is suitable at $\psi$ of 1.1 to 4. The power exponent of $Re_x$ used in this equation is close to that assumed in determining $k$ (0.85). Hence, we assume that the error due to this difference is insignificant in determining $k$ (in addition, it was determined from data at close values of $Re_x$). It can be assumed that the heat transfer from a plate in the case of a highly developed turbulent flow will also be proportional to $Re_x$ with a power exponent close to 0.86.

Equations (5.9) and (5.11) describe experimental data to within ±5%.

With somewhat less accuracy, about ±6%, local heat transfer from a cylinder for $Re_x$ from $10^5$ to $5 \times 10^7$ can be described by a single relation,

$$St = 0.183(\lg Re_x)^{-2.45}\psi^{-0.25}\Omega^{0.14} \tag{5.12}$$

which was obtained assuming that for air $Pr = 0.71$.

For comparison, Fig. 5.10 shows an analytic curve of heat transfer from a flat plate [55]. (Experimental data from previous studies on local heat transfer from a plate at high $Re_x$ in incompressible flows are not available.) The divergence of the present results from the analytic solution increases with $Re_x$ and becomes 16% at $Re_x = 5 \times 10^7$.

It should be emphasized that the correction for the curvature is rather significant. In these experiments, for example, it is as high as 16.5% for cylinder 1

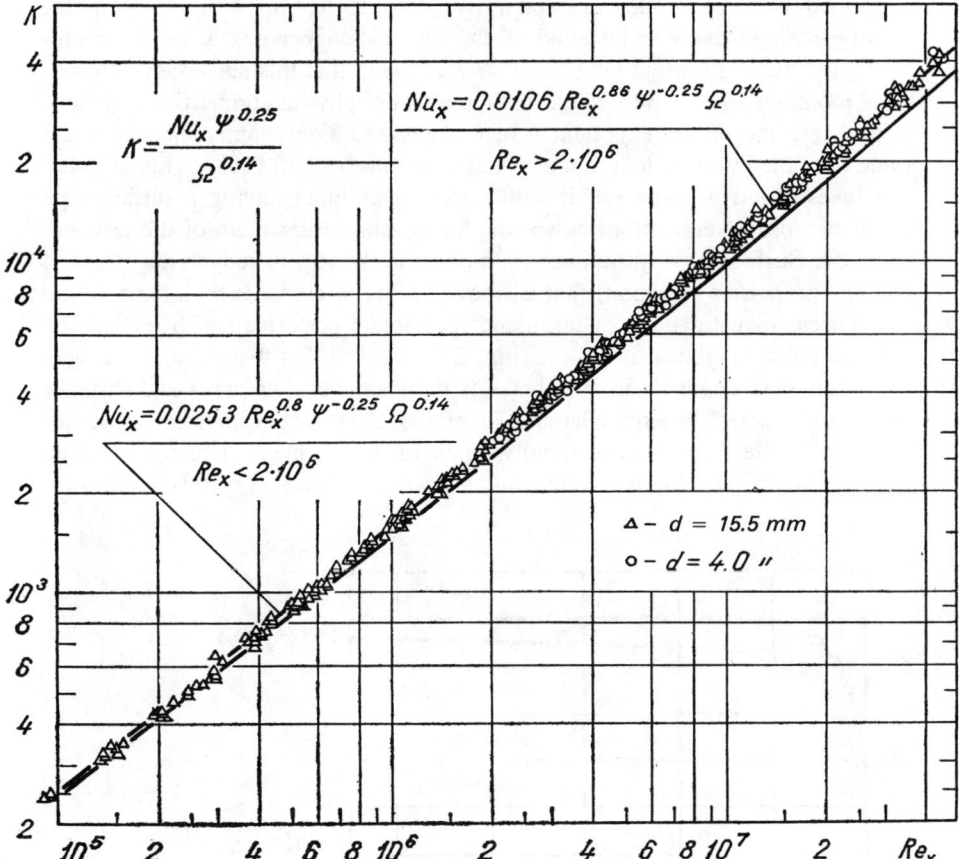

**FIG. 5.10** Correlation of experimental data on local heat transfer from cylinders 1 and 5 with correction for variability of physical properties and for surface curvature. The lower curve represents analytic data for a plate [55].

(obtained at the highest $Re_x$, which was $4 \times 10^6$, at minimum pressure, i.e., at minimum $Re_{r_0}$); for cylinder 5 it is 35% (for the same $Re_x$). These values increase with increasing $Re_x$ and decreasing $Re_{r_0}$; hence, under these conditions, heat transfer from cylinders in axial flow cannot be calculated from equations for a plate.

## 5.5 HEAT TRANSFER WITH ENHANCED TURBULENCE INTENSITY IN THE BOUNDARY LAYER AND FREE STREAM

*Effect of surface-type turbulence promoters.* Experiments in local heat transfer with artificially disturbed boundary layer flowing over a heated surface were performed with a cylinder in axial flow ($d$ = 15.5 mm). For this purpose, sand-type roughness was installed on the forward copper busbar used for supplying the heating current (Fig. 2.5). It was found that this surface-type turbulence promoter significantly decreases the effect of physical properties variation. In this case the power exponent $n$ of $\psi$ decreased significantly, and its mean value over the cylinder length under study became $-0.15$ [103]. This question was investigated in more detail with a similar cylinder having a surface-type turbulence promoter with a known roughness, placed upstream of the test section [95]. Surface-type turbulence promoters are known to reduce the effect of physical properties variation, that is, they become effective at high temperature differences. Investigations of turbulence promoters performed with rectangular and triangular roughness elements (Fig. 2.5) showed that the effect of surface-type promoters depends on the shape of the roughness elements and distance from the promoter. When the latter is increased, the effect of physical properties variation is "restored" more rapidly when finer-roughness elements are used and later when the roughness elements are rough (Fig. 5.11). Results on local

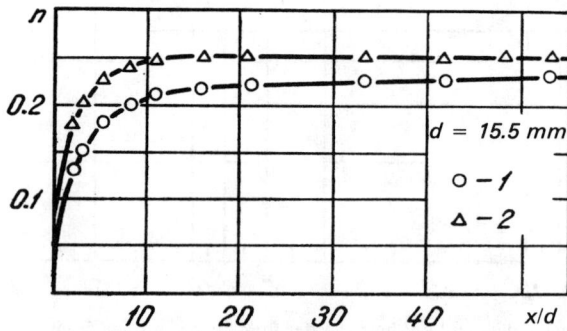

**FIG. 5.11** The power exponent of $\psi$ as a function of the type of surface turbulence promoter and distance from it. (1) Triangular roughness elements, (2) rectangular roughness elements.

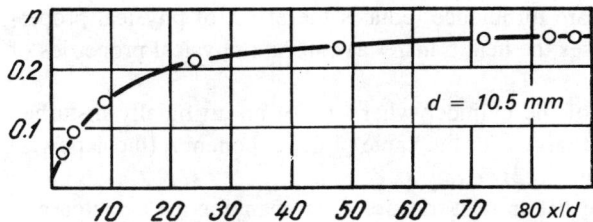

**FIG. 5.12** Plot of $n = f(x/d)$ at elevated freestream turbulence.

heat transfer at $Re_x$ of $10^5$ to $3 \times 10^7$ and $\overline{\psi}$ of 1.2–2.1, obtained with turbulence promoters having triangular and rectangular roughness elements, coincide when a correction is made for physical properties variation (Fig. 5.11). No significant rise in heat transfer is obtained at $\psi \approx 1$ [95].

*Effect of freestream turbulence.* A special experiment for determining the effect of turbulence intensity on local heat transfer from a cylinder in axial flow was performed with the same test section. In the first measurements set, a cylinder with $d = 10.5$ mm and elevated freestream turbulence of 3.2 to 7.5% over the design length of a cylinder was used. In the second set, a cylinder having $d = 15.5$ mm was used at lower freestream turbulence of 0.7 to 1.9%. It is seen that in the first case Tu was damped out along the test section, whereas in the second it increased.

The results obtained at mean temperature factors of 1.2, 2.1, and 2.8 show

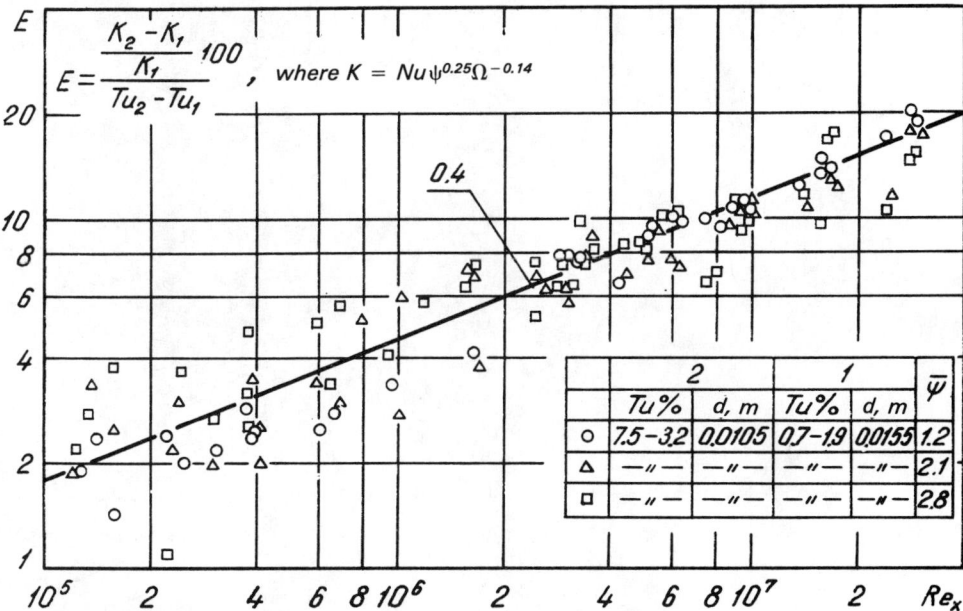

**FIG. 5.13** Effect on heat transfer of increasing freestream turbulence intensity.

that increasing the freestream turbulence reduces the effect of physical properties variation, but it increases the heat transfer at constant physical properties of the flow.

In the upstream part of the cylinder where Tu of an artificially disturbed flow is much higher than at large $L/d$, the value of the exponent $n$ (the $\psi$ power) is significantly smaller (Fig. 5.12).

Figure 5.13 shows the rise in heat transfer accompanying a 1% increase in Tu. According to our experiments, the rise in heat transfer with freestream Tu agrees with analytic data on the effect of Tu on surface friction [104].

Since Tu significantly affects heat transfer only at high $Re_x$, many investigators concerned with heat transfer at $Re_x$ not exceeding $10^6$ could neglect the effect of freestream Tu. Apparently Tu is in part responsible for the increase of the power exponent $m$ in expressions for heat transfer at high $Re_x$.

CHAPTER
# SIX
## HEAT TRANSFER IN ANNULI UNDER FULLY DEVELOPED FLOW CONDITIONS

The effect of physical properties variation of coolants on heat transfer has not been considered in analytic, as well as in the majority of experimental, studies of heat transfer in annuli. Almost all the studies were performed at low temperature differences, and those few investigations that were performed at large temperature differences were performed only with heating of the inner tube over a rather limited range of tube diameters. It was assumed in engineering practice that correction for physical properties variation can be obtained from equations derived for circular tubes. Given the above, and also the fact that in order to design modern heat exchangers with annular ducts or their equivalents, one must have rather precise and reliable relationships, it follows that a detailed study of the governing equations of heat transfer in such ducts is timely and necessary.

The present chapter describes the results of an extensive experimental study of local heat transfer coefficients in turbulent gas flows in annuli. Large temperature differences are employed and hydrodynamic fully developed conditions are reached upstream of the heated length. Under these conditions we investigated the effect on heat transfer of physical properties variation, the ratio of tube diameters, and the relative length of the annulus heated both from one and from both sides with different ratios of heat fluxes.

The effect of a sharp disturbance of the flow on the local heat transfer over both surfaces was investigated by placing a ring onto the inner tube. This made it possible to determine ways for eliminating the harmful effect of high heat fluxes in gas-cooled assemblies of fuel elements releasing heat at a high rate.

The experimental results are correlated and presented in the form of engi-

**110** HEAT TRANSFER OF GAS COOLANTS

neering equations and graphs. Due to the limits on the scope of the book, the appendixes list only a small part of experimental results for all the annuli under study. However, the graphs illustrating preliminary results obtained with the selected annulus contain all the experimental points. All the experimental data were used in correlating the design equations.

## 6.1 LOCAL HEAT TRANSFER FROM THE INNER TUBE OF AN ANNULUS HEATED FROM ONE SIDE

Heat transfer in an annulus with only the inner tube heated was investigated with test sections 1 and 2 at $4.0 \times 10^3 \leq \text{Re} \leq 6.0 \times 10^5$. The bulk of experimental data on determining the effect of physical properties variation was obtained with test section 2 at Re of $6 \times 10^4$ to $2.5 \times 10^5$, since the maximal accuracy was attained over this range of Re.

Figure 6.1 shows data on the distribution along the channel of the following variables: the temperature of the heated inner tube ($T_{w_1}$), the temperature of the outer tube ($T_{w_2}$) (heated by radiation from the inner tube), the temperature of the first thermal insulation shield ($T_{\text{sh}}$), the bulk temperature of the flow $T_f$, the

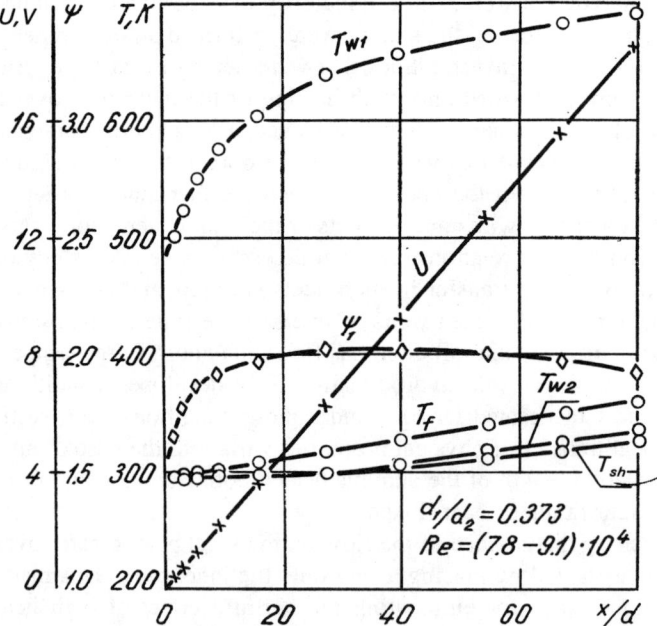

**FIG. 6.1** Variation of the principal quantities measured along an annulus with inner heated tube, $d_1/d_2 = 0.373$.

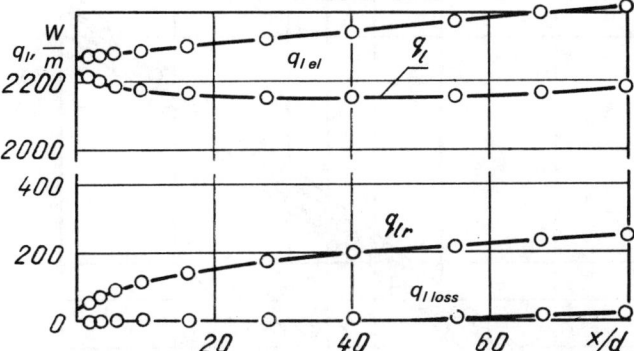

**FIG. 6.2** Variation of the heat flux components along an annulus with inner heated tube, $d_1/d_2 = 0.373$.

temperature factor $\psi$, and the voltage drop $U$. The rise in $\psi$ along the entrance region of the annulus is due to a steep rise in $T_{w_1}$, whereas the slight reduction in it at $x/d > 30$ is due to rising $T_f$ with the difference $(T_{w_1} - T_f)$ remaining constant. The temperature of the first reflecting shield of the vacuum thermal insulation is very close to the temperature of the outer tube, and hence the heat loss $q_{l\,\text{loss}}$ due to radiation of the outer tube is close to zero (Fig. 6.2). The convective component of the heat flux $q_l$—i.e., that part that is transmitted by convection from the wall of the inner tube to the cooling gas—is almost constant along the tube. It remains nearly unchanged since the slight rise in electrical heat release $q_{l\,\text{el}}$ due to the increase in electrical resistance of the tube is compensated by the rising radiant component $q_{lr}$. In this case the condition $q = $ constant is satisfied.

The distribution of the measured quantities described above satisfies the requirements formulated in designing the test section and confirms the feasibility of employing the previously presented general technique for data reduction.

The assumption that physical properties variation is nearly independent of Re, made in developing the experimental technique, is confirmed by results obtained with test section 1 for four different wall temperature levels (Fig. 6.3). The dependence of Nu on $\psi$ is the same over the entire range of Re. The relation

$$\text{Nu} = f(\text{Re}) \qquad (6.1)$$

can be approximated with sufficient accuracy for each temperature level by a straight line (in logarithmic coordinates), parallel to the power-law relation for a circular tube, for example, Eq. (1.29). Thus, it can be assumed that over the range of Re under study, the value of its power exponent $m$ is equal to 0.8, irrespective of the value of $\psi$

It is best and also most convenient to describe the effect of physical properties variation on heat transfer by the ratio $\text{Nu}/\text{Nu}_{\psi=1}$. However, it is very difficult to attain this ratio accurately, since it is impossible to calculate the scaling

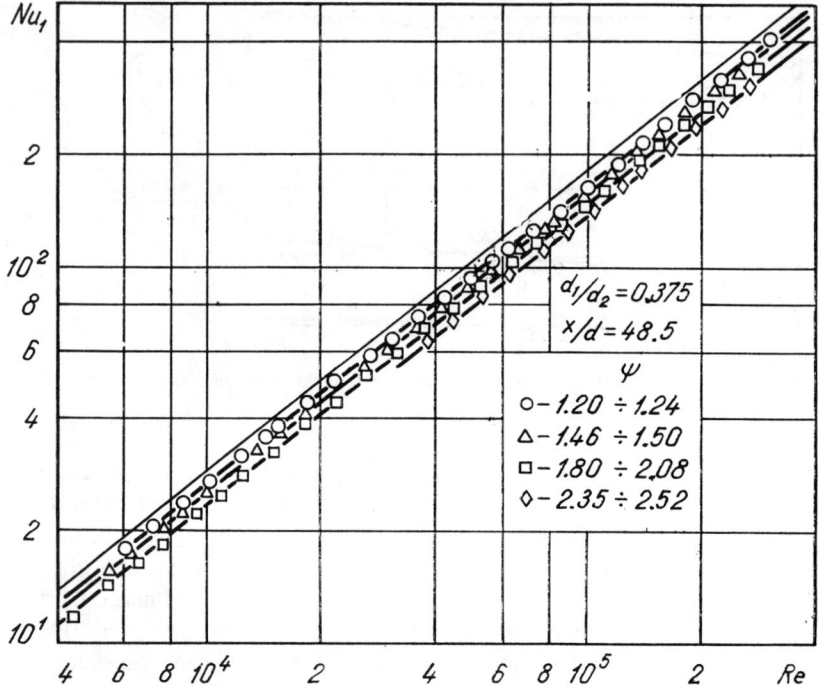

**FIG. 6.3** Heat transfer from the inner tube under fully developed heat transfer conditions at different values of the temperature factor. The solid curve represents the relation. $Nu = 0.0225 Re^{0.8} Pr^{0.6}$ [41].

quantity $Nu_{\psi=1}$ directly from experimental data. It is also impossible to utilize published recommendations for $Nu_{\psi=1}$ since in this case, they are of insufficient accuracy (particularly for the start of the heated length). Even a small error in determining $Nu_{\psi=1}$ is immediately reflected in the quantity describing the effect of physical properties variation on heat transfer. The problem is significantly simpler if experiments for determining the relation

$$Nu = f(Re, x/d) \tag{6.2}$$

are performed over one or several narrow ranges of Re. In this case, the value of $Nu_{\psi=1}$ can be calculated from the simplest power-law equation

$$Nu_{\psi=1} = c_l Re^{0.8} Pr^{0.6} \tag{6.3}$$

The value of $c_l$ for each $x/d$ is determined experimentally, by extrapolating experimental data, written in the form

$$\frac{Nu}{Re^{0.8} Pr^{0.6}} = f(\Psi). \tag{6.4}$$

at $\psi = 1$. As will be shown below, at $1 < \psi < 2$, Eq. (6.4) is a straight line in logarithmic coordinates. When a large volume of experimental data is available for $1.05 \leq \psi \leq 1.2$, this technique makes it possible to obtain rather exact values of $\mathrm{Nu}_{\psi=1}$ over the entire $x/d$ range for which the wall temperature is measured.

A series of runs was performed in test section 2, at $d_1/d_2 = 0.373$. Measurements over three narrow ranges of Re, removed from one another, showed that the effect of physical properties variation does not significantly depend on Re. In each set of tests the air flow rate and Re at the inlet were maintained constant and the wall temperature was varied. In determining $\mathrm{Nu}_{\psi=1}$, the experimental data of each set were represented as

$$\lg\left(\frac{\mathrm{Nu}}{\mathrm{Re}^{0.8}\,\mathrm{Pr}^{0.6}}\right) = f\left(\Psi, \frac{x}{d}\right) \tag{6.5}$$

at $\psi < 2$. The data were approximated by straight lines obtained using the method of least squares, and were extrapolated to $\psi = 1$. All these operations were performed according to a specially written program on a BESM-4M computer. The results for several values of $x/d$ are plotted in Fig. 6.4.

At a given mass flow rate of gas through the annulus, Re decreased somewhat along the duct due to rise in the bulk temperature. However, this does not affect the accuracy in determining the effect of $\psi$ on heat transfer.

The subsequent study of the effect of physical properties variation at different $d_1/d_2$ was performed mostly at $\mathrm{Re} > 7 \times 10^4$, where the accuracy of experimental data at high heat fluxes was at maximum. The values of $\mathrm{Nu}_{\psi=1}$ were determined similarly.

As shown by Petukhov and Royzen [15], the power exponent of Re at $d_1/d_2$

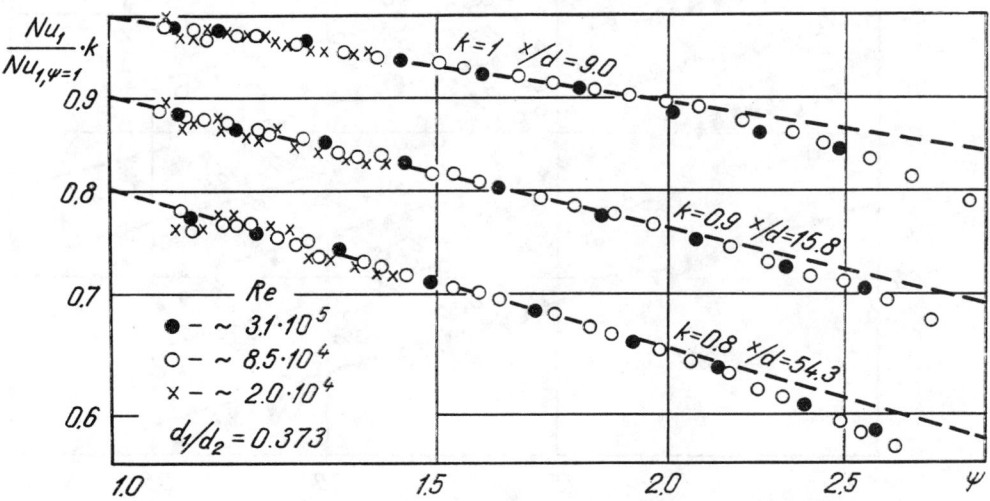

**FIG. 6.4** Effect of temperature factor on heat transfer in an internally heated annulus at different Re.

< 0.2 is somewhat smaller than 0.8; however, it is best not to correct for this, since the value of Re in experiments at $d_1/d_2 < 0.2$ changed only within 8%.

Over the entire range of $d_1/d_2$ under study, the effect of $\psi$ on heat transfer from the inner tube was qualitatively analogous to the case of heat transfer from a circular tube. The experimental results for $d_1/d_2 = 0.373$ and 0.205 plotted in Figs. 6.5 and 6.6 show that the effect of $\psi$ increases with distance along the annulus. In addition, at sufficiently high $\psi$ the relation

$$\frac{\mathrm{Nu}}{\mathrm{Nu}_{\Psi=1}} = f(\Psi) \qquad (6.6)$$

cannot be represented by a straight line in logarithmic coordinates. However, at $\psi \leq 2$ this nonlinearity can be neglected with sufficient accuracy. There the experimental data can be approximated by an expression

$$\frac{\mathrm{Nu}}{\mathrm{Nu}_{\Psi=1}} = \Psi^n \qquad (6.7)$$

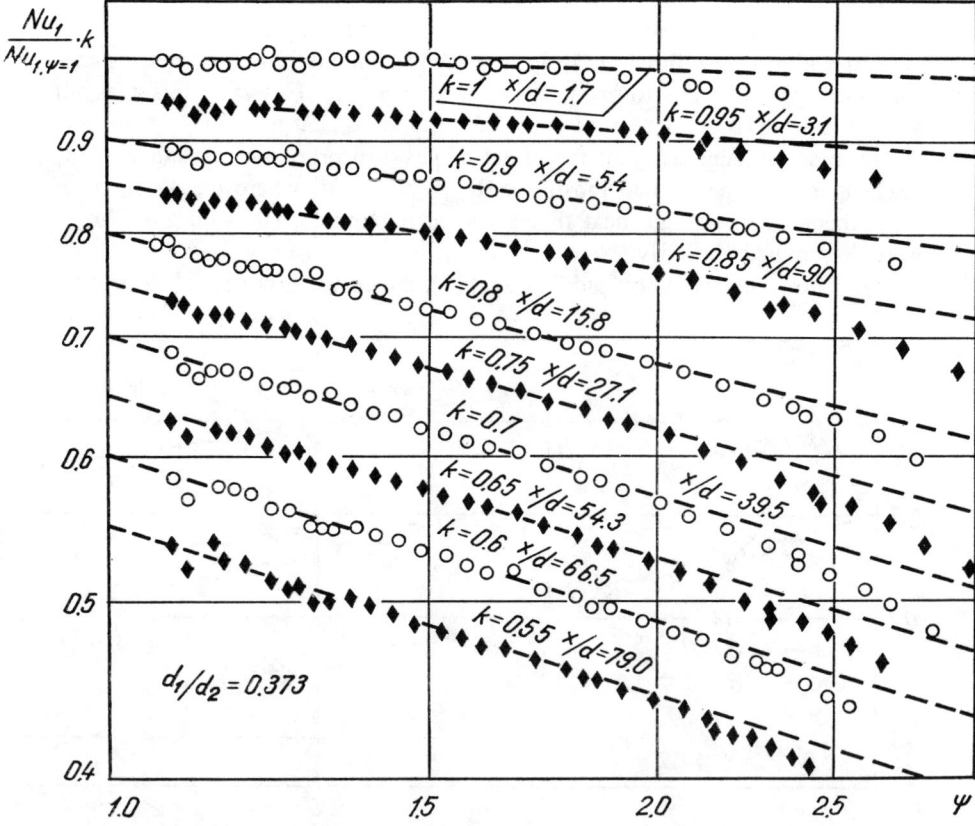

**FIG. 6.5** Effect of temperature factor on heat transfer in an internally heated annulus, $d_1/d_2 = 0.373$.

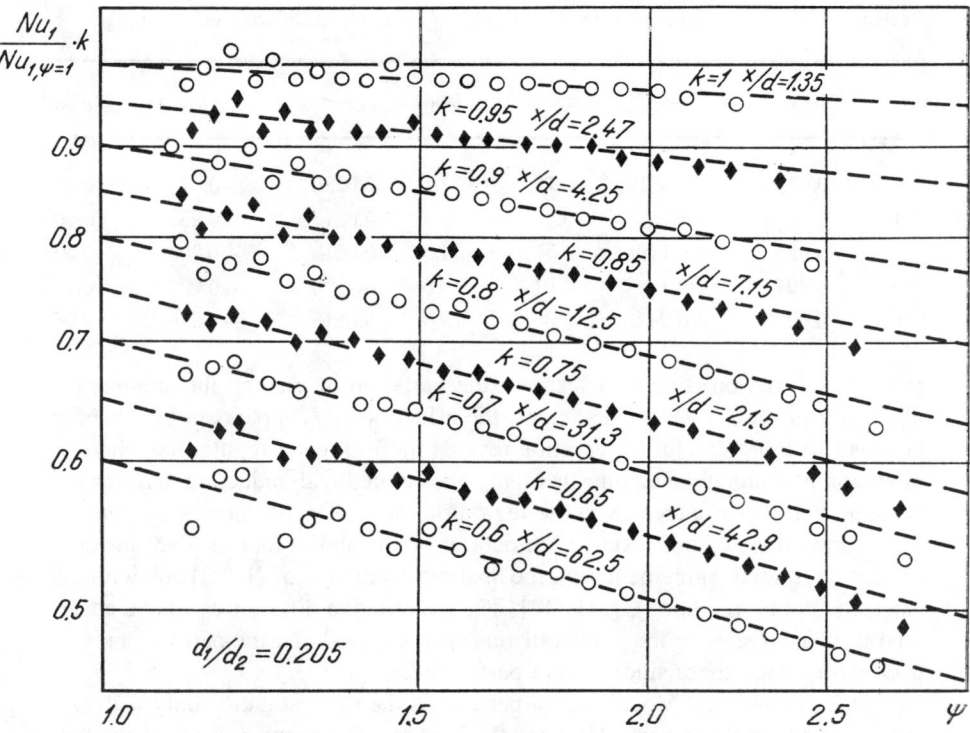

**FIG. 6.6** Effect of temperature factor on heat transfer from the inner tube in an internally heated annulus; $d_1/d_2 = 0.205$.

which is a straight line in logarithmic coordinates (dashed lines in Figs. 6.5 and 6.6) with a slope equal to $n$. In this case the power exponent $n$ is solely a function of $x/d$. The validity and accuracy of this approximation is illustrated in Table 6.1, which lists values of $c_l$, $n$, and the rms deviation $\sigma$ of data points from the correlation

$$\frac{\mathrm{Nu}}{c_l \, \mathrm{Re}^{0.8} \, \mathrm{Pr}^{0.6}} = \Psi^n \tag{6.8}$$

The value of $\sigma$ at all the $x/d$ does not exceed 1%, and deviation of $c_l$ from the mean in the thermally fully developed region ($x/d > 40$) is only $\pm 0.6\%$. Analogous results were also obtained for annuli with different values of $d_1/d_2$.

The temperatures of the annulus walls were measured by thermocouples attached to them. Due to thermal expansion of the tubes with rising $T_w$, the distance between the thermocouples increases, and the hydraulic diameter $d = d_2 - d_1$ in an internally heated annulus decreases. This results in some increase in the value of $x/d$, corresponding to the location of each thermocouple. The values of $x/d$ given in Figs. 6.4–6.6 are for low $T_w$.

Of main interest here is the power exponent $n$, since it corrects for physical

**TABLE 6.1** Values of $c_l$, $n$, and $\sigma$ for $\psi \leq 2$ and $d_1/d_2 = 0.373$ at Different $x/d$

| $x/d$ | $c_l$ | $n$ | $\sigma$, % | $x/d$ | $c_l$ | $n$ | $\sigma$, % |
|---|---|---|---|---|---|---|---|
| 1.7 | 0.0302 | −0.0019 | 0.71 | 27.1 | 0.0220 | −0.268 | 0.36 |
| 3.1 | 0.0275 | −0.063 | 0.51 | 39.5 | 0.0217 | −0.285 | 0.50 |
| 5.4 | 0.0256 | −0.118 | 0.51 | 54.3 | 0.0212 | −0.290 | 0.59 |
| 9.0 | 0.0243 | −0.169 | 0.51 | 66.5 | 0.0210 | −0.296 | 0.76 |
| 15.8 | 0.0236 | −0.243 | 0.36 | 79.0 | 0.0213 | −0.320 | 0.96 |

properties variation. Figure 6.7 shows the variation in $n$ along the annulus for different $d_1 d_2$ at $\psi \leq 2$. At first sight, the effect of $d_1/d_2$ appears rather chaotic; however, a more careful observation reveals the following regularities: the stabilization of $n$ along the annulus is highly accelerated with reduction in $d_1/d_2$; the absolute value of $n$ increases in the region of $x/d > 40$ with increasing $d_1/d_2$.

Comparison of our results with data of others shows that they are in satisfactory qualitative agreement with the analytic relation [Eq. (1.21)] and with the experimental correlation [Eq. (1.18)]. The quantitative differences can be attributed to inaccuracies in the results of refs. [8] and [28] and the narrow range of $d_1/d_2$ over which these studies were performed.

Dalle Donne and Meerwald [8] performed their investigation only with two annuli, in which $d_1/d_2$ were 0.503 and 0.725. Over this range of $d_1/d_2$, the power

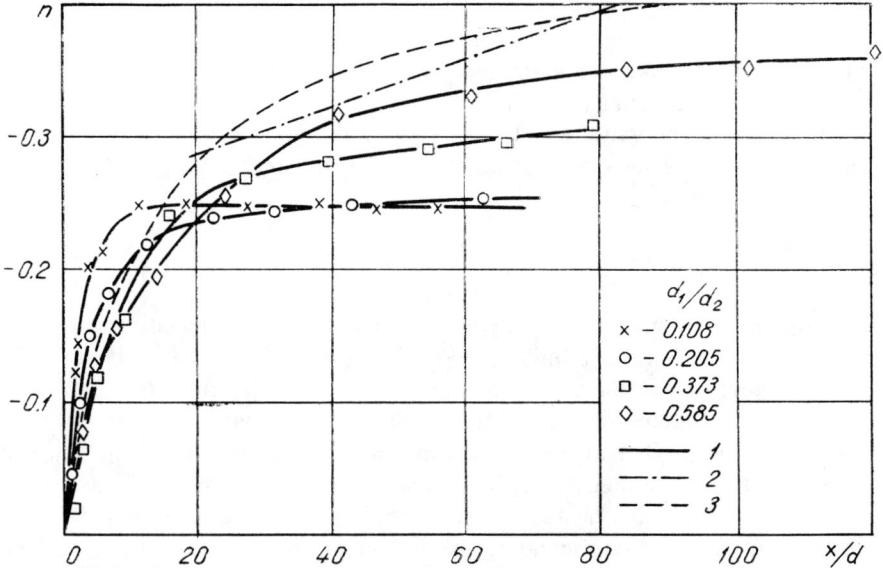

**FIG. 6.7** Variation in the power exponent $n$ along an internally heated annulus at $\psi \leq 2$. (1) Curves approximating our experimental results, (2) Eq. (1.18) [8], (3) Eq. (1.21) [28].

exponent $n$ does not depend much on this ratio, and hence Dalle Donne and Meerwald did not find any relationship between $n$ and $d_1/d_2$. For the same reason, the effect of the diameter ratio on $n$ was not observed by Galin and Yesin [28], whose calculations were performed for $d_1/d_2 = 0.375, 0.503$, and $0.725$.

The strong and complex dependence of the stabilization length of $n$ on $d_1/d_2$ suggests that in this case the development of the process along the duct cannot be described in terms of the parameter $x/d$. Subsequently, attempts at using the ratio $(x/d)(d_2/d_1)$ as a characteristic length were found to be rather fruitful in correlating the test data. In this case the quantity

$$dd_1/d_2 = d_1(d_2 - d_1)/d_2 \tag{6.9}$$

tends to $d_1$ as $d_1$ is reduced. From the physical point of view, this method is validated also by the fact that in the entrance region, at sufficiently high $d_2/d_1$, $d_1$ is a better parameter than $d = d_2 - d_1 \approx d_2$ for describing the development of processes on the inner tube surface.

A plot of the power exponent $n$ versus $(x/d)(d_2/d_1)$ is shown in Fig. 6.8,

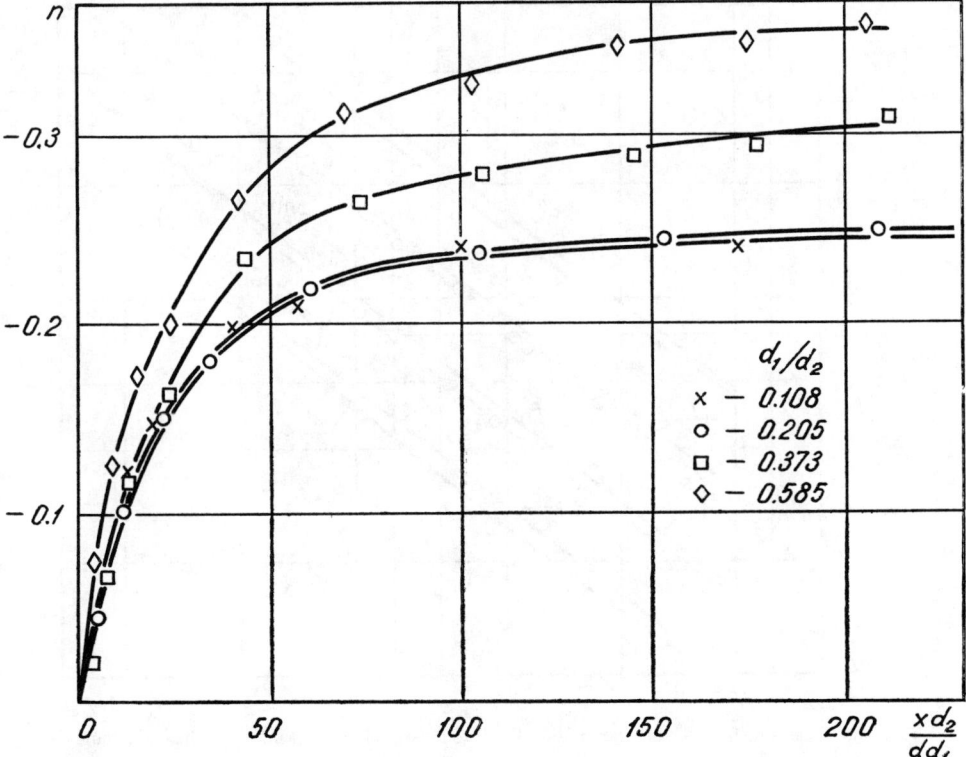

**FIG. 6.8** The power exponent $n$ versus $(x/d)(d_2/d_1)$ for an internally heated annulus at $\psi \leq 2$.

from which it is seen that for all the $d_1/d_2$, stabilization of $n$ is reached at $(x/d)(d_2/d_1) \approx 200$.

A final correlation of experimental data on the effect of physical properties variation on heat transfer from the inner tube (including data at $\psi \geq 2$) will be performed after analysis of results on heat transfer from the outer tube and of an annulus heated from both sides, since these problems are interrelated.

To find correlations suitable for our test data on heat transfer in annuli at constant physical properties, we performed several additional sets of runs at $\psi$ between 1.18 and 1.25 and different Re. The results on heat transfer from the inner tube, according to data in Figs. 6.5 and 6.6, were extrapolated to $\psi = 1$. The results obtained in the thermally fully developed region (where Nu is independent of $x/d$) were compared only with the Petukhov-Royzen relation [Eq. (1.7)] and the Bobkov-Ibragimov-Savanin relation [Eqs. (1.12) and (1.13)], which are presently the most accurate. The value of $Nu_{p\infty}$ in Eq. (1.7) was determined from Eq. (1.29). As seen in Fig. 6.9, which presents the results of this comparison, both of the above relations yield very similar results. Our ex-

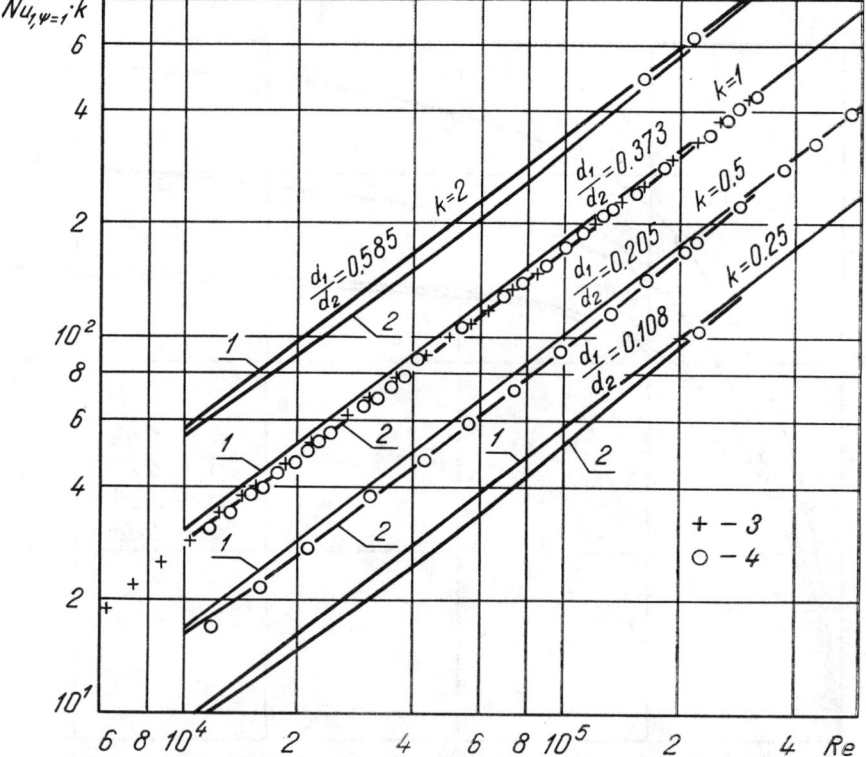

**FIG. 6.9** Heat transfer over the thermally fully developed region of an internally heated annulus. (1) Curves from Eq. (1.7); (2) curves from Eqs. (1.12) and (1.13); (3) and (4) data obtained with test sections 1 and 2, respectively.

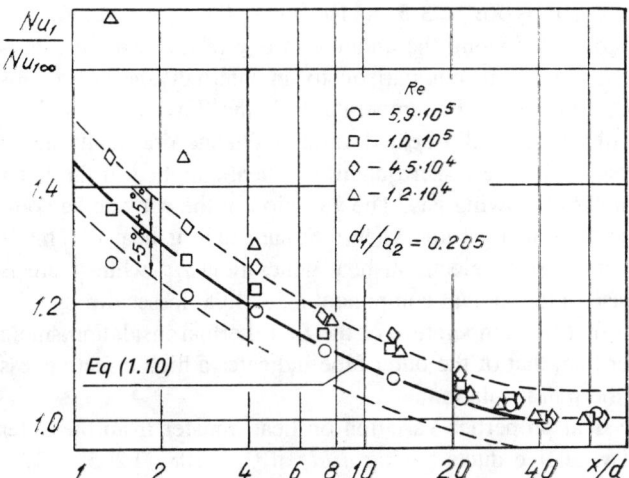

**FIG. 6.10** Heat transfer in the thermal entrance region of an internally heated annulus.

perimental data are virtually identical with the results of calculations from Eqs. (1.12) and (1.13).

Equation (1.10) is the only one currently available for calculating heat transfer at the thermal entrance region, and it was compared to our data. In this region the agreement is somewhat less satisfactory than in the fully developed region, particularly at small $x/d$ and $d_1/d_2$ (Fig. 6.10). Equation (1.10) makes no allowance for the effect of Re on the length of the thermal entrance region, whereas our data point to a highly significant effect of Re.

The adiabatic temperature of the outer tube, measured by us in the thermally fully developed region, was 10–15% higher than that calculated from Eq. (1.9).

## 6.2 LOCAL HEAT TRANSFER IN AN EXTERNALLY HEATED ANNULUS

The experiments for determining heat transfer in an externally heated annulus had the same purpose as when the inner tube was heated, that is, to determine the relationship between variability of physical properties and heat transfer, and hence the experimental arrangements were similar in both cases.

Local heat transfer from the outer tube was investigated with test section 2 at Re of $7 \times 10^3$ to $5 \times 10^5$. As in the case of an internally heated annulus, physical properties variation was not found to depend on Re. To ensure maximal accuracy of results and to obtain the highest possible values of $\psi$, most experiments for determining the relationship

$$\mathrm{Nu}_2/\mathrm{Nu}_{2,\,\Psi=1} = f\left(\Psi,\, \frac{x}{d},\, \frac{d_1}{d_2}\right) \tag{6.10}$$

were performed at $6 \times 10^4 \leq \text{Re} \leq 3.5 \times 10^5$ and $1.03 < \psi < 2.36$. The variation of various parameters along the annulus in one of the experiments is shown in Figs. 6.11 and 6.12. In comparison to an internally heated anulus under similar operating conditions, the temperatures of the flow, walls, and the first reflecting shield of an externally heated annulus change drastically along the annulus. This is caused by the larger quantity of heat supplied in the latter care to the same unit mass of flowing gas. The variation of the convective component of the heat flux $q_l$ is also greater when the outer tube is heated. This is due to the significant increase in electrical heat generation $q_{l\,el}$ while there is only a moderate increase in the radiant component $q_{l r}$ and the losses through the shields $q_{l\,loss}$. The fact that the temperature of the first thermal insulation shield is only 35–45°C lower than that of the outer tube indicates a high effectiveness of the vacuum shield thermal insulation.

The effect of physical properties variation on heat transfer from the outer tube was investigated with five ducts having $d_1/d_2$ of 0, 0.108, 0.205, 0.373, and 0.585. Qualitatively, the effect of $\psi$ on the ratio $\text{Nu}_2/\text{Nu}_{2,\psi=1}$ (Figs. 6.13–6.15) is analogous to that examined in the preceding section. As in an internally heated annulus, Re does not affect the dependence of heat transfer on $\psi$. This made it possible to use the same technique for determining the value of Nu at constant physical properties.

The results for each $d_1/d_2$ and each $x/d$ at $\psi \leq 2$ were approximated by Eq. (6.7). The variation of $n$ with $x/d$ at different $d_1/d_2$ is shown in Fig. 6.16; it is seen that over the entire $x/d$ range, the value of $n$ tends, with decreasing $d_1/d_2$, to its value for a circular tube. Comparison of these data with results obtained for an internally heated annulus (Fig. 6.7) shows that the behavior of $n$ as a function

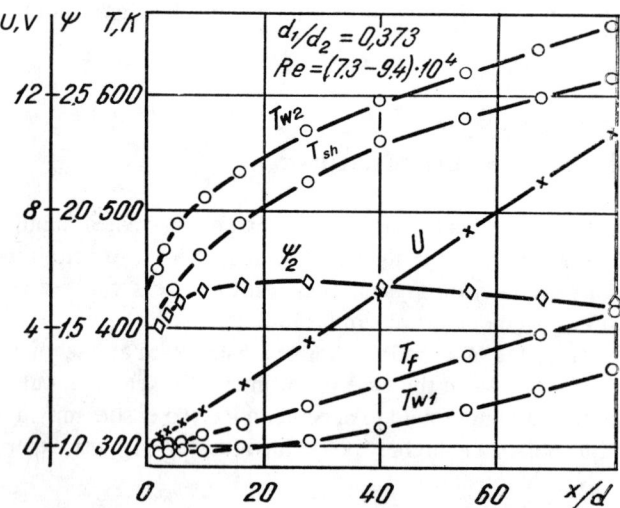

**FIG. 6.11** Variation of temperature, temperature factor, and voltage along an externally heated annulus.

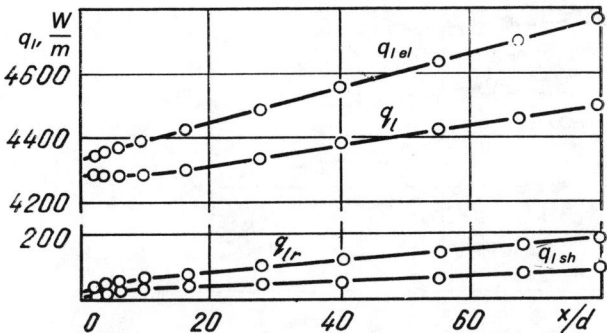

**FIG. 6.12** Variation of heat flux components along an externally heated annulus, $d_1/d_2 = 0.373$.

of $\psi$ becomes similar in both cases with increasing $d_1/d_2$. This is entirely valid, since at sufficiently high $d_1/d_2$ the annulus becomes equivalent to parallel flat plates, and geometric effects of the inner and outer surfaces almost disappear.

In this case the stabilization of $n$ along the annulus is much slower. One hundred hydraulic diameters is insufficient for attaining a constant value of $n$, and thus thermally fully developed conditions. Unlike the preceding case, in an externally heated annulus, the ratio $x/d$ is a good parameter for describing the effect of length on heat transfer.

Some experiments were performed over a wide range of Re, at $\psi$ from 1.15 to 1.20, in order to compare results on heat transfer from the outer tube (with the inner tube at adiabatic temperature) with relations suggested by other investigators. Results in the thermally fully developed region were extrapolated to $\psi = 1$ using the power exponents $n$, which was found from Figs. 6.13–6.15. These values are in satisfactory agreement with the Petukhov-Royzen relation [Eq. (1.8)] and the Bobkov-Ibragimov-Savanin relation [Eqs. (1.12) and (1.14)], which give virtually identical results for an externally heated annulus (Fig.

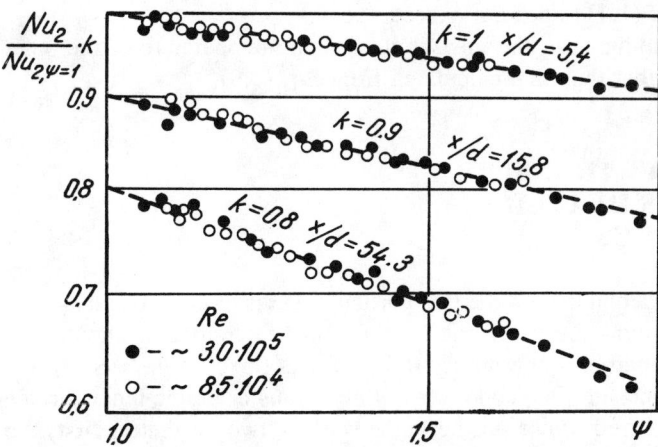

**FIG. 6.13** Effect of $\psi$ on heat transfer from the outer tube at different Re, $d_1/d_2 = 0.373$.

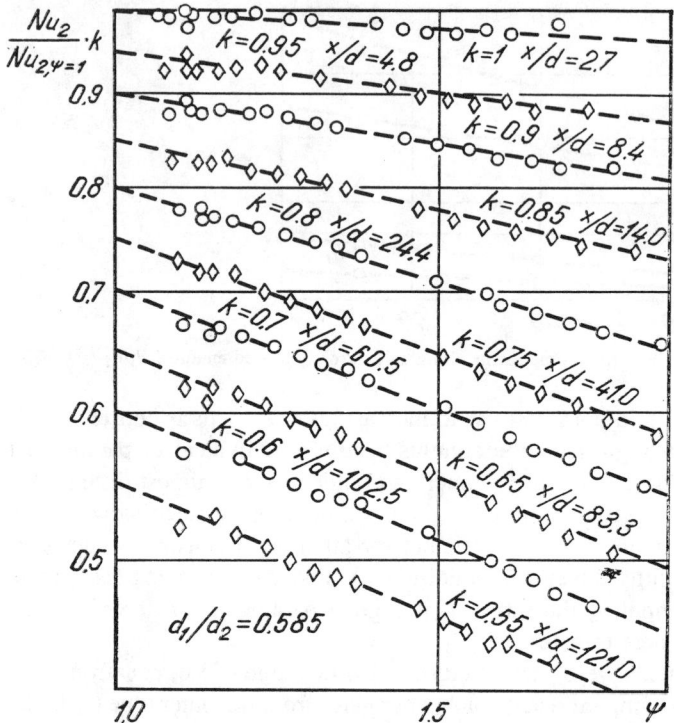

**FIG. 6.14** Effect of temperature factor on heat transfer in an externally heated annulus, $d_1/d_2 = 0.585$.

6.17). However, in the entrance region (Fig. 6.18) the experiments indicate significant depends of the degree of thermal development on Re. This effect is not incorporated in the Petukhov-Royzen equation, Eq. (1.11). At $x/d > 3$ the divergence from Eq. (1.11) does not exceed $\pm 5\%$.

As in the preceding case, the measured adiabatic temperature of the inner tube, is 10–15% higher than that calculated from Eq. (1.9).

## 6.3 LOCAL HEAT TRANSFER IN AN ANNULUS HEATED FROM BOTH SIDES

Our experimental data obtained in the two particular cases of $q_1 = 0$ and $q_2 = 0$ are helpful in determining the governing relations for heat transfer in annuli with variable physical properties and arbitrary ratios of heat fluxes at the walls $q_1/q_2$. When heated from one side, the conditions of heat transfer on the inner surface of the annulus differ from those on the outer wall by two attributes: first, the sign and magnitude of the radius of the curvature of the surface, and second, the ratio of the heated to the nonheated perimeter. At constant physical properties the flow pattern does not depend on the thermal processes. However, in the case of variable physical properties the thermal processes have a significant effect on the flow pattern, a change which influences the heat transfer pattern. Hence it is quite possible that the ratio of heated and nonheated perimeters has an effect on $Nu/Nu_{\psi=1}$.

The results of our study of heat transfer with only internally or only externally heated annuli show that $Nu_i/Nu_{i,\,\psi=1}$ depends significantly on $d_1/d_2$; the effect of physical properties variation increases with the ratio of the heated to the nonheated perimeter. The only exception to this rule is the thermal entrance region in an internally heated annulus (Fig. 6.7).

Assuming that at moderate heating rates the dependence of $Nu_i/Nu_{i,\,\psi=1}$ on $d_1/d_2$ is due only to geometric factors, then for a one-sided heating the effect of physical properties variation on heat transfer is independent of the boundary conditions at the unheated wall. When an arbitrary ratio of wall heat fluxes is employed, it is possible to make allowance for the effect of physical properties variation on heat transfer in annuli $Nu_{ij}/Nu_{ij,\,\psi=1}$ using equations obtained when

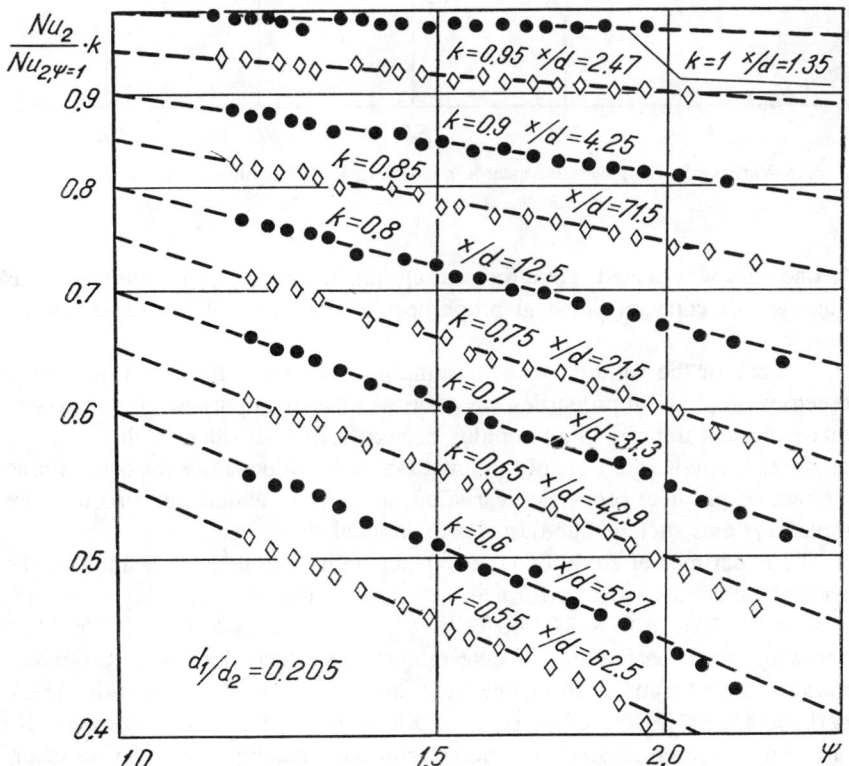

**FIG. 6.15** Effect of temperature factor on heat transfer in an externally heated annulus, $d_1/d_2 = 0.205$.

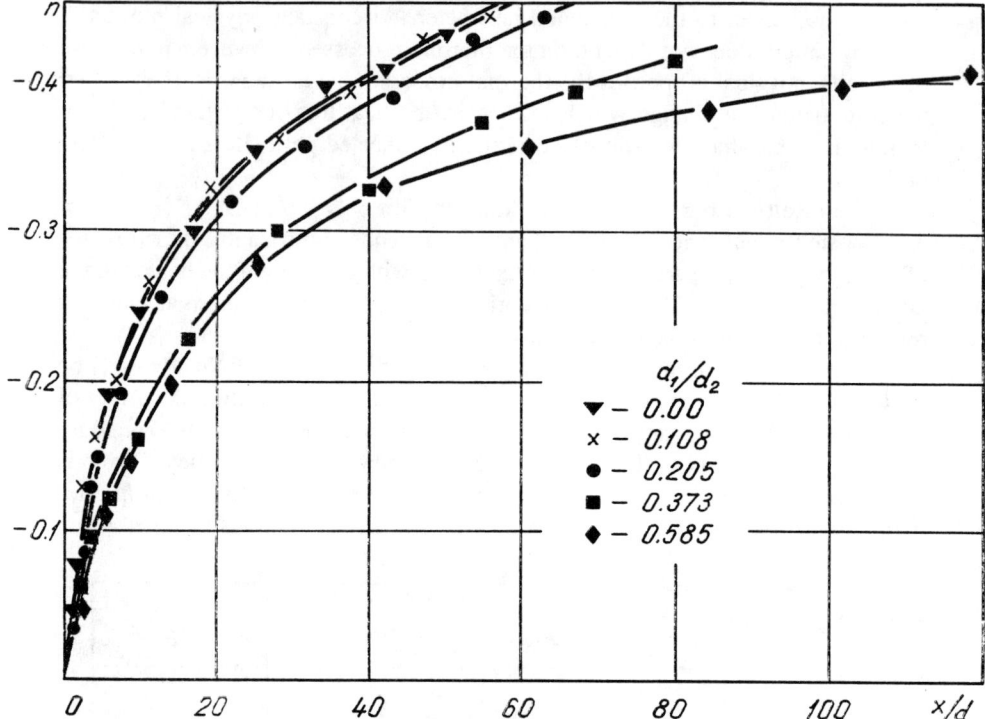

**FIG. 6.16** Variation of the power exponent $n$ in an externally heated annulus at $\psi \leq 2$.

only one wall was heated. Here the scaling quantity $Nu_{ij,\ \psi=1}$ is determined (as in the case of constant physical properties) on the basis of the superposition principle.

A check of the validity of the assumptions regarding the effect of surface geometries on physical properties variation was the main purpose of our experiments with heat transfer in an annulus heated from both sides. If this effect is negligible, it is possible to apply the method of local modeling for determining the effect of physical properties variation not only in annuli but also in more complex systems such as bundles of rods in axial flow.

The experimental study of heat transfer with simultaneous heating of the inner and outer tube was performed with test section 2 with $d_1/d_2 = 0.585$, 0.373, and 0.205, at $8 \times 10^3 \leq Re \leq 6 \times 10^5$ and $1.04 \leq \psi \leq 2.5$. Most experiments were performed for determining the effect of physical properties variation on heat transfer from internally and externally heated annuli. These experiments were performed at Re from $5.8 \times 10^4$ to $3.2 \times 10^5$; however, Re of a given measurements set was nearly constant, making it possible to determine $Nu_{ij,\ \psi=1}$ with sufficient accuracy using the same technique as when only one wall was heated.

At the end of the heated length (in the location of the last thermocouples) the temperature of both walls was maintained at equal value. This was dictated by the desire to obtain as close as possible values of $\psi$ on both walls. In addition, it simplified selection and adjustment of the operating conditions, since in this case the variation of ratio $q_1/q_2$ for any given value of $d_1/d_2$ is ensured automatically to be within $\pm 2\%$. The characteristic distributions of temperature, temperature factor, voltage, and heat fluxes along the annulus are shown in Figs. 6.19 and 6.20. As a result of the relatively large rise in flow temperature (the entire wetted perimeter was heated), the attained maximum values of $\psi$, as compared with heating from one side, shifted in the direction of smaller $x/d$. The radiative component of the heat flux is virtually absent, since the temperature difference between the walls is small. The variation in the temperature of the tube walls along the annulus is most remarkable. At $x/d > 20$, the temperature of the outer tube rises more rapidly than that of the inner one. At constant $q_1/q_2$ over this region, such a temperature variation is possible only in one case—when physical properties variation affects the heat transfer from the outer tube more than that from the inner one.

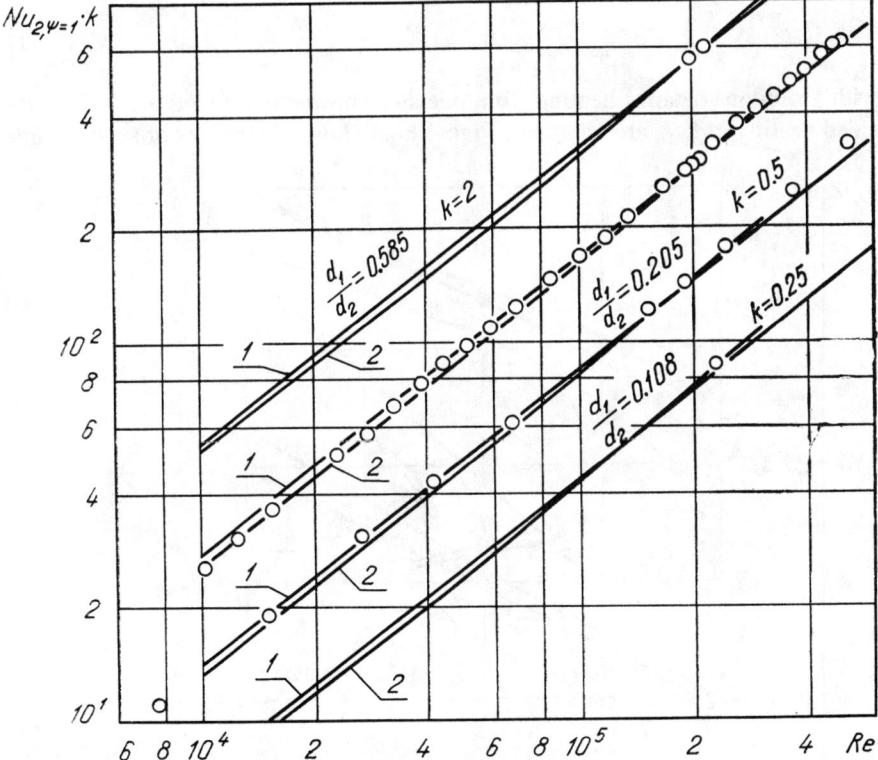

**FIG. 6.17** Heat transfer over the thermally fully developed region in an externally heated annulus. (1) Eq. (1.8), (2) Eqs. (1.12), and (1.14), the points represent our experimental data.

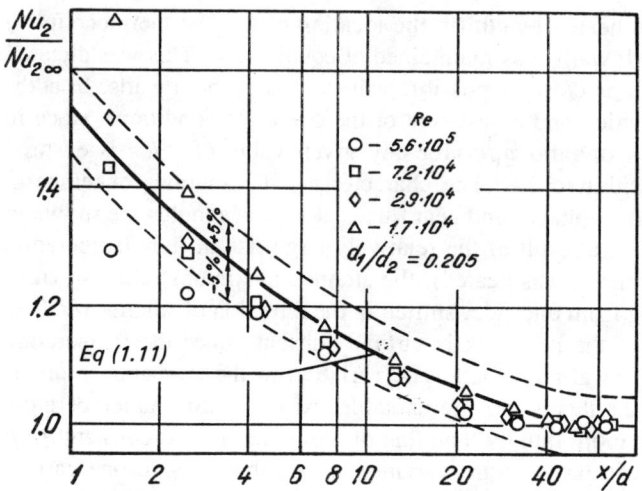

**FIG. 6.18** Heat transfer in thermal entrance region of an externally heated annulus.

Comparison of results obtained in the form

$$\frac{\mathrm{Nu}_{ij}}{\mathrm{Nu}_{ij,\,\Psi=1}} = f_{ij}\left(\Psi_i,\,\frac{x_i}{d},\,\frac{d_1}{d_2}\right) \tag{6.11}$$

with analogous data for heating from one side showed that functions $f_i$ (for one-sided heating) and $f_{ij}$ are identical. Figure 6.21 shows the results of such a com-

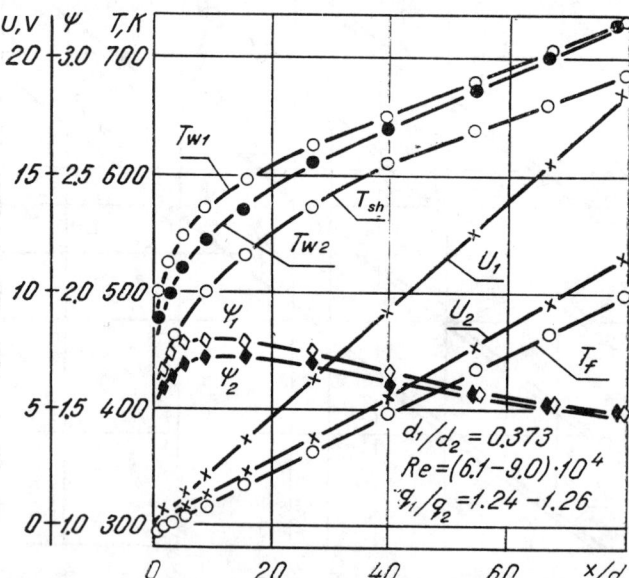

**FIG. 6.19** Variation of temperature, temperature factor, and voltage along an annulus heated from both sides, $d_1/d_2 = 0.373$.

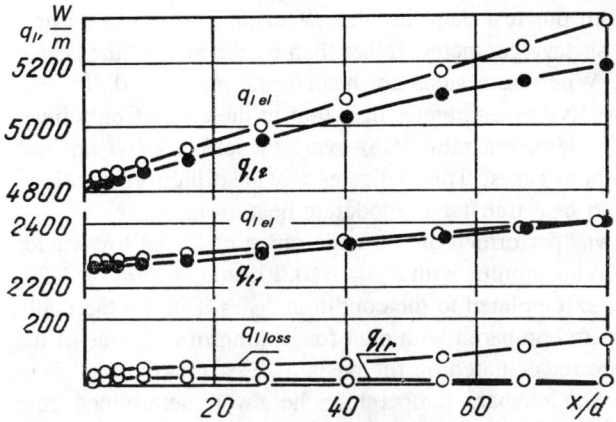

**FIG. 6.20** Variation of heat flux components and heat losses along an annulus heated from both sides.

parison for only one annulus, with $d_1/d_2 = 0.205$. The situation observed in the remaining annuli is basically the same, the effect of physical properties variation on heat transfer from each of the walls is independent of whether it is one or both walls that are heated.

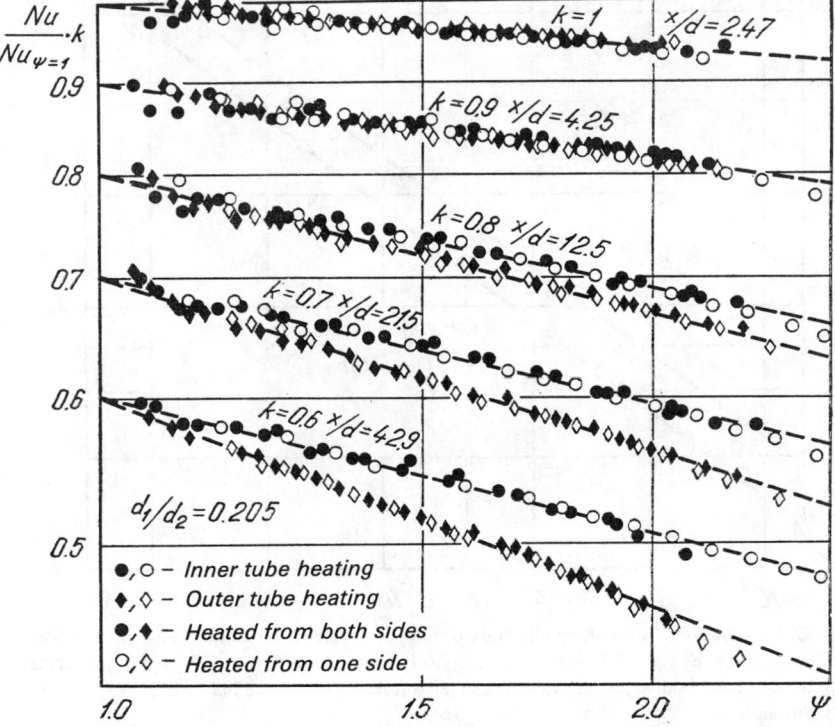

**FIG. 6.21** Comparison of experimental data obtained in annuli heated from one and both sides.

**128** HEAT TRANSFER OF GAS COOLANTS

It was concluded from our test data that the effect of $\psi$ at moderate heat fluxes is governed by the surface geometry, rather than by the ratio of the heated to nonheated perimeters. When both sides are heated and $d_1/d_2 = 0.205$, the heated perimeter is almost five times greater than that in the case of only inner tube heating at the same diameters ratio. However, the effect of $\psi$ on heat transfer was the same in both cases. This indicates that heat-induced acceleration of flow does not affect heat transfer in moderate heating rates.

A small set of runs was performed over a wide range of Re, at low values of $\psi$ ($\psi_{ij} = 1.06-1.26$), in an annulus with $d_1/d_2 = 0.205$ and at $q_1/q_2 = 1.29$. The results of these tests, extrapolated to the condition $\psi_{ij} = 1$ in the thermally fully developed region, were compared with data for heating of only one of the walls. The latter data were recalculated on the basis of Eqs. (1.4) and (1.5) to the case of $q_1/q_2 = 1.29$. The adiabatic temperatures here were determined from Eq. (1.9). As seen in Fig. 6.22, the results of direct measurement, and the values calculated from results on heating of one side only are virtually identical. The location of the experimental data points relative to the Petukhov-Royzen correlating curves (for $q_1/q_2 = 1.29$) is the same for heating of either one or

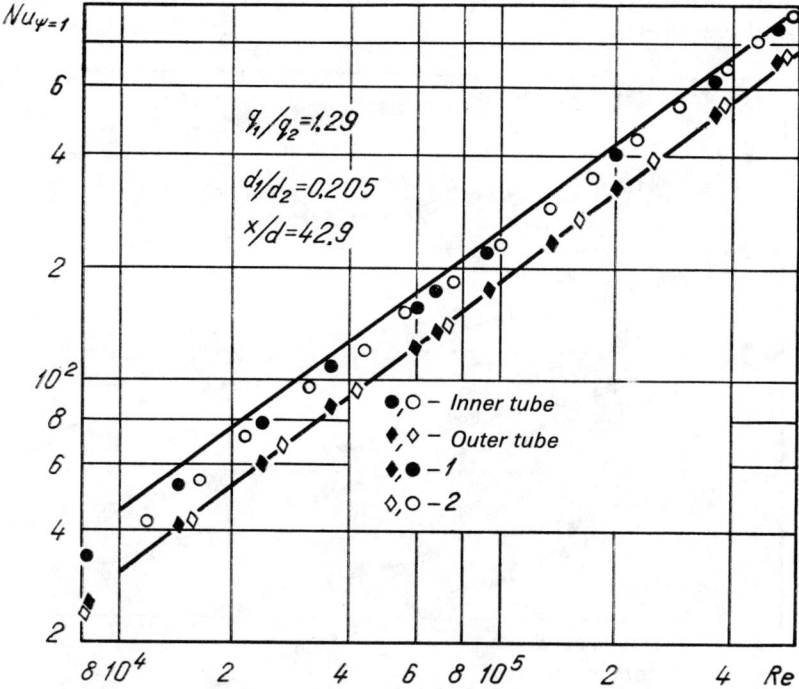

**FIG. 6.22** Heat transfer in the thermally fully developed region of an annulus heated from both sides. The solid curves were calculated according to Petukhov and Royzen [17]. (1) Experimental data for heating from both sides, (2) values calculated from Eqs. (1.4) and (1.5) using experimental data of $Nu_1$ and $Nu_2$.

both sides (see also Figs. 6.9 and 6.17). It is clearly seen from Fig. 6.22 that heat transfer from the inner tube is higher than from the outer one.

The above results allow the conclusion that in the case of moderate heating rates the effect of the temperature factor on heat transfer from one surface is virtually independent of the thermal conditions on the other.

## 6.4 CORRELATION OF RESULTS ON HEAT TRANSFER IN ANNULI

In certain cases it is convenient to assess the effect of physical properties variation on heat transfer by means of the temperature factor $\psi = T_w/T_f$. However, relations based on $\psi$ have the significant shortcoming of being inconvenient for use at boundary conditions of the second kind.† In most practical cases the heat flux is specified and it is required to find the wall temperature $T_w$. In spite of the fact that most measurements of heat transfer to the flow were performed with boundary conditions of the second kind, up to now experimental data are usually correlated in the form

$$\frac{\mathrm{Nu}}{\mathrm{Nu}_{\Psi=1}} = f\left(\frac{x}{d}, \Psi, n_\lambda, n_\mu, n_c\right) \qquad (6.12)$$

This is done because $T_w$ can be measured, which significantly simplifies the correlation. In addition, $\psi$ itself has a clear physical interpretation as a parameter of the variation in physical properties across the thickness of the boundary layer.

However, as shown clearly by Kurganov and Petuhov [41], design correlations obtained on the basis of Eq. (6.12) are not suitable to actual use. It is also unlikely that such relations could be improved for use with a wide range of gases and at high heating rates.

Let the right-hand side of Eq. (6.12) be written as

$$\frac{\mathrm{Nu}}{\mathrm{Nu}_{\Psi=1}} = \frac{\frac{q}{T_w - T_f} \cdot \frac{d}{\lambda}}{\mathrm{Nu}_{\Psi=1}} = \frac{\frac{1}{(\Psi-1)} \frac{qd}{\lambda T_f}}{\mathrm{Nu}_{\Psi=1}} = \frac{Q^+}{(\Psi-1)\mathrm{Nu}_{\Psi=1}} \qquad (6.13)$$

where $qd/\lambda T_f = Q^+$ is the nondimensional heat flux parameter. If $Q^+/\mathrm{Nu}_{\psi=1}$ is designated by $K_f$, we can write Eq. (6.12) as

$$\frac{\mathrm{Nu}}{\mathrm{Nu}_{\Psi=1}} = \frac{K_f}{\Psi - 1} = f\left(\frac{x}{d}, \Psi, n_\lambda, n_\mu, n_c\right) \qquad (6.14)$$

or

$$K_f = (\Psi - 1) \cdot f\left(\frac{x}{d}, \Psi, n_\lambda, n_\mu, n_c\right) \qquad (6.15)$$

†These are $q = f(x, y, z, t)$, where $x$, $y$, and $z$ are the Cartesian coordinates and $t$ is the time.

It can be easily shown by substituting $\mathrm{Nu}_{\psi=1} = qd/(T_w - T_f)\lambda$ into (6.13) that $K_f$ can be written as

$$K_f = \frac{(T_w - T_f)_{\Psi \to 1}}{T_f} = (\Psi - 1)_{\Psi \to 1}. \tag{6.16}$$

Here $K_f$ plays the role of the relative temperature difference at constant physical properties of the gas, corresponding to $T_f$. Following Kurganov and Petukhov [41], who noted that a unique relationship exists between the sought change in relative temperature over the flow cross section ($\psi - 1$) and its nominal value $K_f$, we can solve Eq. (6.15) for ($\psi - 1$). Then the latter difference, which is an expression for the wall temperature, can be represented in the form

$$\Psi - 1 = \varphi_1\left(\frac{x}{d},\ K_f,\ n_\lambda,\ n_\mu,\ n_c\right) \tag{6.17}$$

or

$$\frac{\mathrm{Nu}}{\mathrm{Nu}_{\Psi=1}} = \frac{K_f}{(\Psi - 1)} = \varphi_2\left(\frac{x}{d},\ K_f,\ n_\lambda,\ n_\mu,\ n_c\right). \tag{6.18}$$

The right-hand sides of Eqs. (6.17) and (6.18) contain only known quantities, which makes it possible to obtain explicit formulas for determining the wall temperature at a given heat flux.

Since we wished to use the large experience accumulated by other investigators in correlating results on heat transfer in tubes, we decided to use available relations from this branch as a basis for obtaining relationships for calculating heat transfer in annuli.

Our results on heat transfer from a single circular tube (without an internal insert) were compared with results on heat transfer from the outer tube of an annulus using the form

$$\frac{1 - \mathrm{Nu}_2/\mathrm{Nu}_{2,\ \psi=1}}{1 - \mathrm{Nu}_p/\mathrm{Nu}_{p,\ \psi=1}} = \omega_2 \tag{6.19}$$

It was found that over the range of flow and geometric variables under study and at the same values of $K_f$ and $x/d$, $\omega_2$ depends significantly only on $d_1/d_2$. An analogous picture is also obtained for heat transfer from the inner tube if $(x/d)(d_2/d_1)$ rather than $x/d$ is used as the length parameter. Consequently, it can be assumed with a sufficient accuracy that the quantities $\omega_2$ and

$$\frac{1 - \mathrm{Nu}_1/\mathrm{Nu}_{1,\ \psi=1}}{1 - \mathrm{Nu}_p/\mathrm{Nu}_{p,\ \psi=1}} = \omega_1 \tag{6.20}$$

are functions only of the ratio $d_1/d_2$. Based on our experimental data, we found the following empirical expressions for these functions:

$$\omega_2 = 1 - 0.262 \exp\left[-0.022 \left(\frac{d_2}{d_1}\right)^{2.74}\right] \quad (6.21)$$

$$\omega_1 = 0.44 + 0.6 \exp\left(-0.58 \frac{d_2}{d_1}\right) \quad (6.22)$$

Equations (6.21) and (6.22) satisfy the limiting transitions: if $d_1/d_2 = 0$ (circular tube), then $\omega_2 = 1$, and if $d_1/d_2 = 1$ (flat slot), then $\omega_1 = \omega_2 = 0.744$. Equations (6.21) and (6.22) are compared with experimental results in Fig. 6.23.

The expression for the ratio $\mathrm{Nu}/\mathrm{Nu}_{\psi=1}$ that was obtained by Kurganov and Petukhov [41] is given in Eq. (1.31). This relation makes allowance for properties of individual gases within the coolant. Data on the effect of physical properties variation of other gases (except for air) on heat transfer in annuli are unavailable. However, with reference to the fact that a tube can be regarded as some version of an annulus, it can be assumed that the effect of variation of physical properties ($\lambda$, $\mu$, $c$) will be the same in an annulus as in tubes. Formulation of equations for annuli on the basis of Eq. (1.31) allows one to expect satisfactory results in extrapolating experimental data to a range of variables for which no experiments were performed.

On the basis of these considerations we obtained correlations for heat transfer in annuli with variable physical properties. In the case of one-sided heating we have

$$\frac{\mathrm{Nu}_i}{\mathrm{Nu}_{i,\,\psi=1}} = 1 - \{1 - \exp[-K_{fi}(a\varphi_i + n_\mu \Phi_i K_{fi})]\}\omega_i \quad (6.23)$$

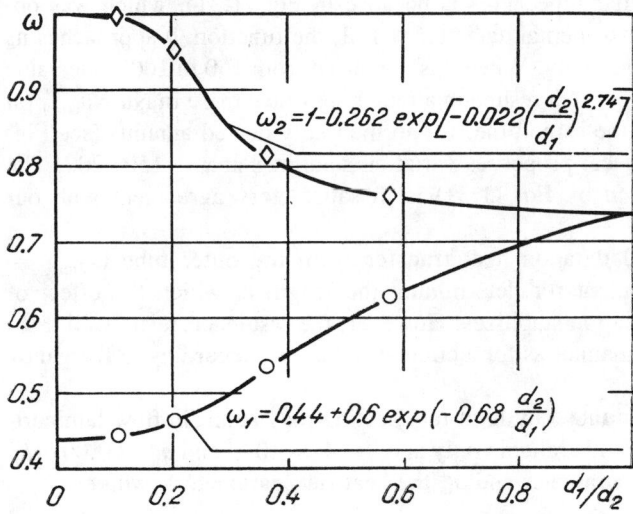

FIG. 6.23 Plot of $\omega$ versus $d_1/d_2$.

where $i = 1, 2$,

$$a = -0.53\, n_\rho - \frac{1}{3} n_\lambda - \frac{1}{4} n_c$$

$$\varphi_i = 1 - \exp(-0.1\, \tilde{x}_i)$$

$$\Phi_1 = 0.37\,[1 - \exp(-1.32 \cdot 10^{-3}\, \tilde{x}_1^{1.66})]$$

$$\Phi_2 = \frac{1.25\,(0.01\, \tilde{x}_2)^2}{1 + (0.01\, \tilde{x}_2)^2}$$

$$\tilde{x}_1 = \frac{x_1\, d_2}{d\, d_1}, \quad \tilde{x}_2 = \frac{x_2}{d}$$

$$K_{fi} = \frac{q_i\, d}{\lambda T_f\, \mathrm{Nu}_{i,\, \Psi = 1}}$$

The values of coefficients $a$ and $n$ should be taken from Table 1.1.

Equation (6.23), written as

$$\psi_i = 1 + \frac{K_{fi}}{1 - \{1 - \exp[-K_{fi}(a\varphi_i + n_\mu\, \Phi_i\, K_{fi})]\}\, \omega_i}, \quad (6.24)$$

makes it possible to immediately calculate the local wall temperature if the values of $q$, $\overline{\rho u}$, $T_f$, $d_1$, and $d_2$ are known.

Experimental results on annuli heated from both sides can also be correlated with Eqs. (6.23) and (6.24), provided that $\mathrm{Nu}_i$, $\mathrm{Nu}_{i,\psi=1}$, and $K_{f,i}$ are replaced, respectively, by $\mathrm{Nu}_{ij}$, $\mathrm{Nu}_{ij,\psi=1}$, and $K_{f,ij}$. Here $j = 2$ when $i = 1$, and $j = 1$ when $i = 2$.

The parameter $\Phi_1$ of Eqs. (6.23) and (6.24) is expressed differently for annuli than for a circular pipe. This is because in Eq. (1.31), which was obtained mostly from experimental data at $\tilde{x} \leq 133$, the function $\Phi$ approaches its asymptotic value rather slowly. When $\tilde{x}$ is increased from 150 to 1000, the value of $\Phi$ increases by 40%, which results in a rather significant rise in $\mathrm{Nu}/\mathrm{Nu}_{\psi=1}$ (at $K_f = 1$ it is 30%). At the same time, for an internally heated annulus (see Fig. 6.8), the effect of physical properties variation stabilizes at $\tilde{x} > 150\text{–}200$. The expression for $\Phi_1$ given by Eq. (1.31) is in satisfactory agreement with our results only at $\tilde{x}_1 < 80$.

Our experimental data on heat transfer from the outer tube ($\tilde{x}_{2\max} = 55\text{–}120$) are not sufficient for determining the length at which the effect of physical properties variation stabilizes. However, the results correlate well if $\Phi_2$ is defined in the same manner as for a circular pipe, i.e., according to Kurganov and Petukhov [41].

The acceleration induced by high-rate heating may result in flow laminarization. Hence, Eq. (1.31) is reliable only at $K' \leq 4 \times 10^{-7}$ and $q_\mathrm{in}^+ \leq 0.007$ ($K'$ being the acceleration parameter and $q_\mathrm{in}^+$ the heat flux parameter), where

$$q_\mathrm{in}^+ = \frac{q_w}{\overline{\rho u}\, c_p\, T_\mathrm{in}} = \frac{1}{4} \frac{1}{T_\mathrm{in}} \left( \frac{\partial T_\mathrm{in}}{\partial\,(x/d)} \right) \quad (6.25)$$

$$K' = \frac{4q_f^+}{\text{Re}} \qquad (6.26)$$

$$q_f^+ = \frac{q_w}{\overline{\rho u}\, c_p\, T_f} = \frac{1}{4}\left(\frac{1}{T_f}\, \frac{\partial T_f}{\partial (x/d)}\right) \qquad (6.27)$$

Parameters $q_{\text{in}}^+$ and $q_f^+$ are proportional to the relative rate of change in flow temperature along the tube. In the case of an ideal gas with negligible acceleration due to friction-induced axial pressure drop, they are also proportional to the relative flow acceleration, since

$$\frac{1}{T}\, \frac{\partial T}{\partial (x/d)} \simeq \frac{1}{u}\, \frac{\partial u}{\partial (x/d)} \qquad (6.28)$$

Note that under these conditions the parameter $K'$ is identical to the acceleration parameter in external boundary layers

$$K = \frac{\nu}{u_\infty^2}\left(\frac{du_\infty}{dx}\right) \qquad (6.29)$$

suggested by Moretti and Kays [105].

In the general case the heat fluxes on both walls of the annulus are not equal to one another. In order that the aforementioned axial effects be uniquely defined by parameters $q_{\text{in}}^+$ and $q_f^+$ at any $d_1/d_2$ and $q_1/q_2$, one must use an expression for the reduced heat flux. Then

$$q_{\text{in}}^+ = \frac{q_1 d_1 + q_2 d_2}{(d_1 + d_2)\,\overline{\rho u}\, c_p\, T_{\text{in}}}$$

$$q_f^+ = \frac{q_1 d_1 + q_2 d_2}{(d_1 + d_2)\,\overline{\rho u}\, c_p\, T_f} \qquad (6.30)$$

In the series of runs whose data served for constructing Eqs. (6.23) and (6.24), the values of $q_{\text{in}}^+$ did not exceed 0.004, and the values of $K'$ were smaller than $10^{-7}$. Several runs were performed at $K'$ between $3 \times 10^{-7}$ and $5.3 \times 10^{-7}$. In these experiments we noted a sharp reduction in heat transfer at $x/d > 15$, both with one-sided and two-sided heating. The deviation of test points from Eq. (6.23) was from 20 to 40%.

It follows from these studies and from the results obtained by Coon and Perkins [106] and by Bankston [107] on flow laminarization in heated tubes that Eqs. (6.23) and (6.24) are suitable at $q_{\text{in}}^+ \leq 0.007$ and $K' \leq 10^{-7}$.

Using measured values of the air flow rate, the inlet temperature, and the convective components of heat fluxes, the local wall temperatures were calculated from Eq. (6.24). The Nusselt numbers for a constant-property flow were calculated from Eq. (6.3), and the values of $c_l$ were determined by the technique described in Section 6.1. The rms deviation of the calculated local temperatures of the heated walls from the measured values comprised 1.2–2.3% of the temperature difference for the outer tube, and 1–3.2% for the inner tube. This was the case at low as well as high values of $\psi$, which indicates that Eqs. (6.23) and (6.24) are sufficiently accurate. The results of calculations for $\psi$ from 1.7 to 2.7 are plotted in Figs. 6.24 and 6.25.

It follows from the above that in annuli the heat transfer to turbulent flows of gases with variable physical properties should be calculated from Eqs. (6.23) and (6.24); the values of $Nu_{\psi=1}$ should be determined from Eqs. (1.7)–(1.11). If calculations are performed in a region sufficiently far from the start of heating, then $Nu_{\psi=1}$ can also be calculated from Eq. (1.12). For the boundary condition $q = $ constant Eqs. (6.23) and (6.24) are valid over a wide range of flow variables and with any ratio of $d_1/d_2$, including plane channels.

## 6.5 HEAT TRANSFER IN AN ANNULUS WITH AN OBSTACLE

It frequently happens that in heat exchangers comprised of ducts with complex geometry (annuli, bundles of fuel rods in axial flow, etc.), it is necessary, due to

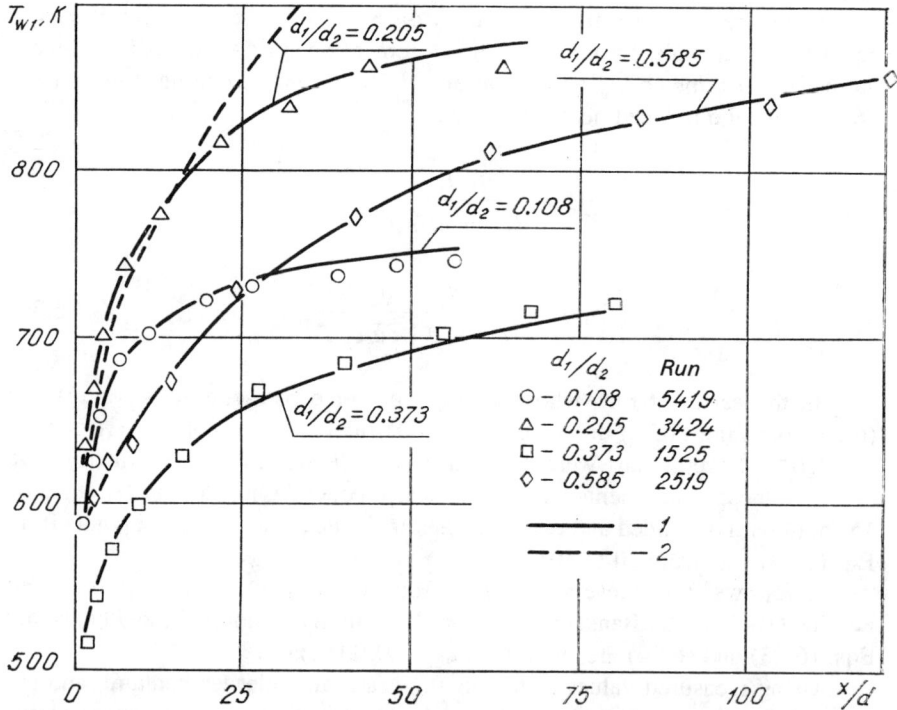

**FIG. 6.24** Comparison of measured and calculated wall temperatures of the inner tube in an internally heated annulus. (1) Calculated from Eq. (6.24); (2) calculated with the effect of $\psi$ on heat transfer corrected for by Eq. (1.21).

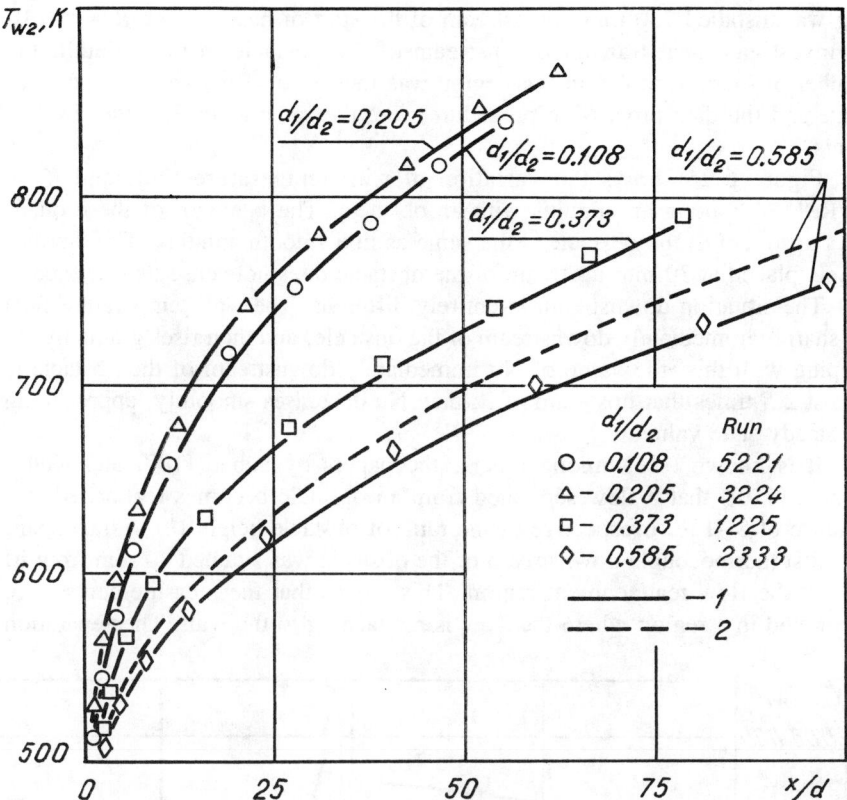

**FIG. 6.25** Comparison of measured and calculated temperatures of the outer wall of an externally heated annulus. (1) Calculated from Eq. (6.24), (2) calculated from Eq. (1.32).

structural requirements, to place various spacing devices that partially cover the flow passages in certain places. Such devices result in significant disturbance of the flow, which, in its turn, augments the heat transfer rate [108]. The disturbing effect of spacing grids is local, and the flow gradually stabilizes at sufficient distance from the obstacle.

Gas-cooled reactors with rod-type fuel-element assemblies always have densely placed spacing grids. It is hence important both to know their effect on heat transfer at constant physical properties of the coolant, and to be able to estimate their effect in the case of significant temperature variation across the flow.

An annulus with a thin transverse fin on the inner tube is suitable for simulating the perturbing effect of spacing grids. Such a fin causes flow separation over the inner tube. The flow structure in this region should be similar to a flow downstream of a grid. To investigate heat transfer in a highly disturbed flow of a gas with variable physical properties, we placed a thin ring (Fig. 2.12) with height $h = (d_r - d_1)/2 = 4.25$ mm in an annulus having $d_1/d_2 = 0.373$. The ring was installed 730 mm downstream of the start of heating (at $x/d = 41.3$). To investigate heat transfer downstream of the obstacle in more detail, the number of thermocouples in this region was increased. The experimental technique and the data processing procedure were the same as in the other experiments.

Figure 6.26 shows the variation of wall temperature, Nu, and $K = \text{Nu}/\text{Re}^{0.8}\text{Pr}^{0.6}$ along an annulus with an obstacle. The behavior of these quantities upstream of the obstacle is the same as in a smooth annulus. The thermocouple placed at 30 mm upstream of the obstacle does not sense its presence.

The situation downstream is entirely different. The wall temperature falls off sharply immediately downstream of the obstacle, and then rises gradually. In keeping with this, the value of Nu immediately downstream of the obstacle is almost 2.5 times that upstream of it; then Nu decreases smoothly, approaching the steady-state value.

It is known (see, among others, the papers by Seban [109] and Ktalkherman [110]) that a flow separated from an obstacle becomes reattached at a distance of 8–12$h$, irrespective of the ratio of obstacle height ($h$) to duct span. The first thermocouple downstream of the obstacle was situated 40 mm from it, i.e., in the flow reattachment region. This means that the measurements were performed in a region where the flow is reattaching to the wall. The separation

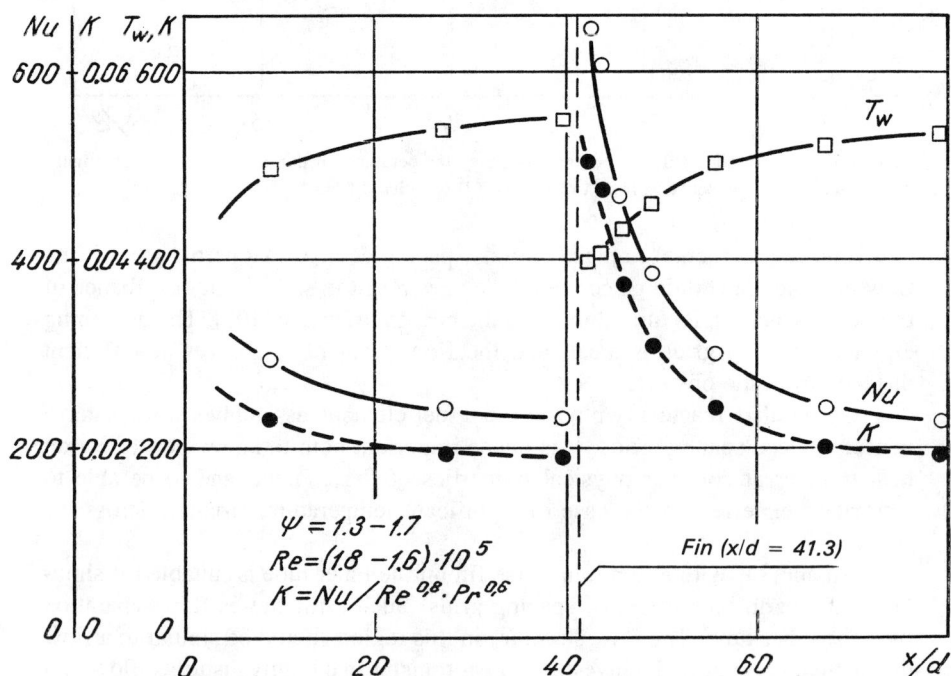

**FIG. 6.26** Wall temperature, Nu, and $K$ variations along an annulus with an obstacle.

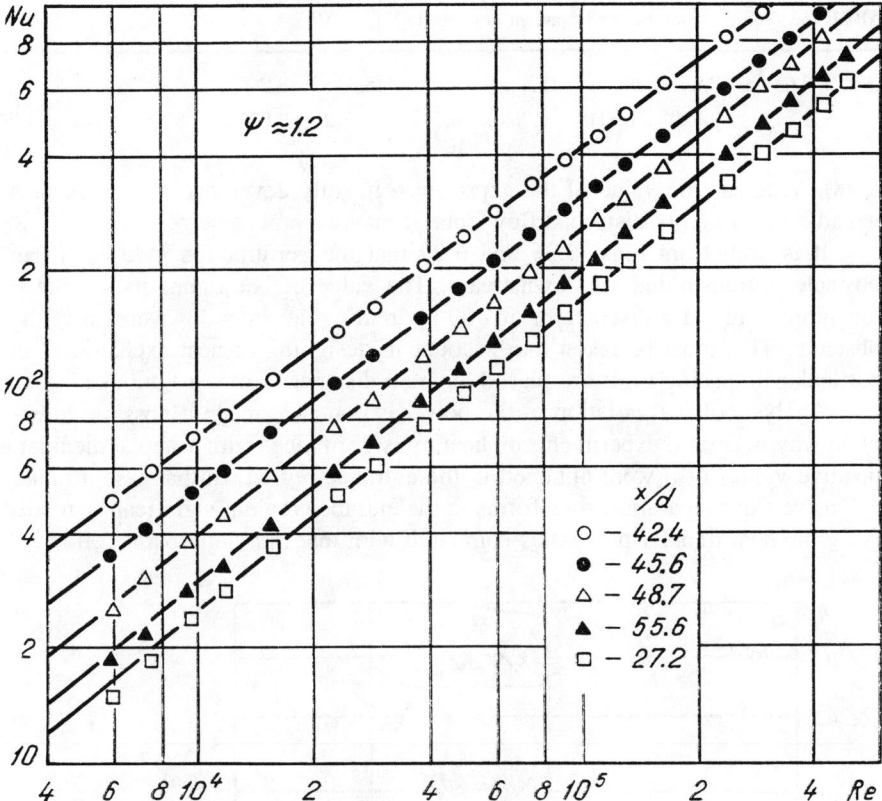

**FIG. 6.27** Heat transfer in an annulus with an obstacle.

region was not investigated by us, since its length is insignificant as compared with the length of the entire annulus.

Results of heat transfer measurements for certain $x/d$ are shown in Fig. 6.27. It is seen that the power exponent $m$ of Re is close to 0.8 at almost all the values of $x/d$, decreasing somewhat only near the obstacle, but even there it is not below 0.74.

To determine the power exponent of $\psi$, the experimental results for each $x/d$ were represented in the form

$$\frac{\mathrm{Nu}}{\mathrm{Re}^m \, \mathrm{Pr}^{0.6}} = f(\Psi) \qquad (6.31)$$

where $m$ was determined for each $x/d$ directly from experimental data (Table 6.2).

Such a presentation showed graphically that flow disturbances induced by a transverse baffle greatly affect the value of the heat-transfer coefficient as well as the manner in which the temperature factor influences the heat transfer (Fig.

**TABLE 6.2** Values of the Power Exponent $m$ of Re in Eq. (6.31)

| $x/d$ | 9.0   | 27.2  | 39.7  | 42.4  | 43.6  | 45.5  | 48.7  | 55.6  | 66.8  | 79.2  |
|-------|-------|-------|-------|-------|-------|-------|-------|-------|-------|-------|
| $m$   | 0.811 | 0.820 | 0.820 | 0.754 | 0.745 | 0.762 | 0.807 | 0.833 | 0.828 | 0.827 |

6.28). Whereas the value of the exponent $n$ in fully developed flow is always negative, in a highly disturbed flow zone it may even become positive.

It is seen from Figs. 6.26 and 6.29 that the perturbation induced by an obstacle is transmitted far downstream. The value of Nu attains its stabilized magnitude only at a distance of 50–60 hydraulic diameters downstream of the obstacle. This must be taken into account in designing of heat exchangers, in particular if spacing grids are placed at small distances from one another.

Analysis of the variation in the power exponent $n$ made it possible to explain why in certain experiments on heat transfer in tubes with a step at the inlet, positive values of $n$ were obtained in the entrance region. In that case a rather extensive flow separation zone forms in the entrance region, significantly modifying the heat transfer process as compared with that in a nonseparated flow.

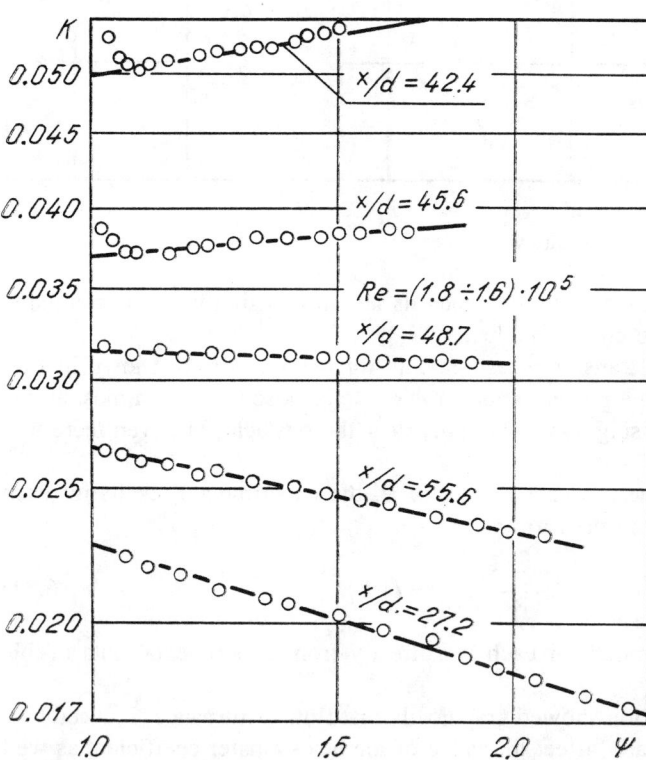

**FIG. 6.28** Effect of the temperature factor on heat transfer in an annulus with an obstacle.

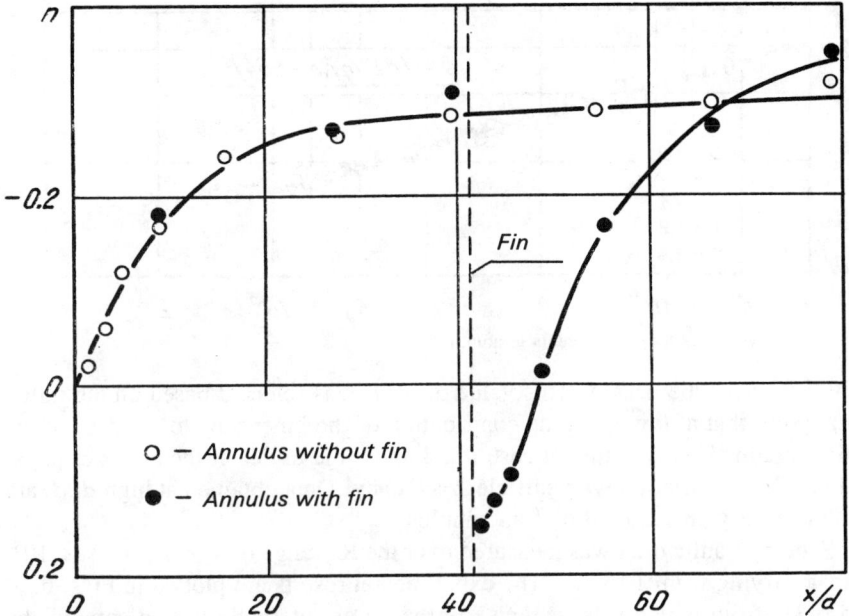

**FIG. 6.29** Variation of the power exponent $n$ along an annulus with an obstacle.

## 6.6 HYDRAULIC DRAG AT LARGE TEMPERATURE DIFFERENCES

The hydraulic drag in duct flows of gases with variable properties has been investigated much less than the heat transfer. Only a few studies were performed with circular tubes ([31, 111, 112]), in which it was shown that the drag is less sensitive to changes in physical properties than the heat transfer. At present, hydraulic drag is usually calculated from the expression

$$\frac{\xi}{\xi_{\Psi=1}} = \Psi^{-0,1} \tag{6.32}$$

where $\xi_{\psi=1}$ is the friction drag coefficient at constant properties. Measurements performed by Lowdermilk et al. [113] with rectangular and triangular ducts showed that the drag is almost independent of ψ. Analogous results were obtained for internally heated annuli [8] with $d_1/d_2$ = 0.503 and 0.725. An annulus is very convenient for such studies since static-pressure measurements can be performed on the nonheated wall, while the opposite wall is heated. This makes it possible to significantly improve the experimental accuracy.

The friction coefficients were determined with test section 1, using the technique described in Chapter 2. The experiments were performed in hydrodynamic fully developed flows in a single annulus with $d_1/d_2$ = 0.375 and with

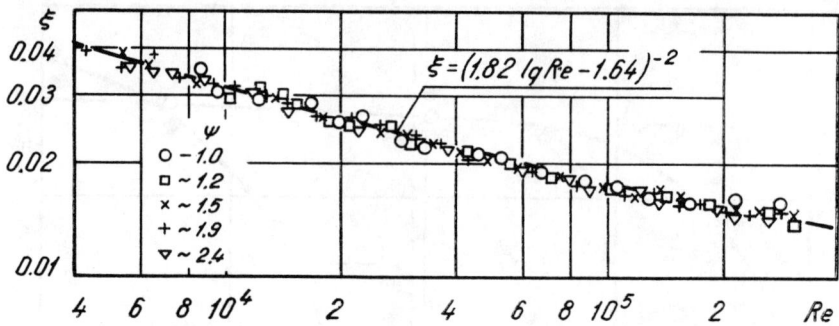

**FIG. 6.30** Hydraulic drag coefficients in annuli.

only the inner tube heated. This value of $d_1/d_2$ was selected based on the initial assumption that at low $d_1/d_2$ the contribution of the inner tube to the drag of the entire annulus is small. In that case it is impossible to detect the effect of physical properties variation with sufficient accuracy. Data obtained at high $d_1/d_2$ are no longer very characteristic of an annulus.

The hydraulic drag was measured over the Re range of $4 \times 10^3$ to $3 \times 10^5$, with $\psi$ varying from 1 to 2.5. The experimental results are plotted in Figs. 6.30 and 6.31, from which it is seen clearly that in an internally heated annulus the drag is independent of $\psi$ over the range of flow variables under study. The friction coefficient can be calculated from isothermal flow relations such as the Filonenko formula [114]:

$$\xi = (1.82 \lg \text{Re} - 1.64)^{-2} \qquad (6.33)$$

This expression is very similar to the familiar relation

$$\frac{1}{\sqrt{\xi}} = 2.0 \lg (\text{Re} \sqrt{\xi}) - 0.8 \qquad (6.34)$$

describing the universal Prandtl frictional resistance law for smooth tubes. However, Eq. (6.33) is more suitable for calculations, since $\xi$ is expressed explicitly.

Our experimental results are in satisfactory agreement with data of Quármby [115], obtained under adiabatic conditions in annuli with $d_1/d_2 = 0.348$, $0.174$, and $0.106$. The results of [115] show that frictional drag can be calculated from formulas for circular tubes with the reference dimension represented by the hydraulic diameter.

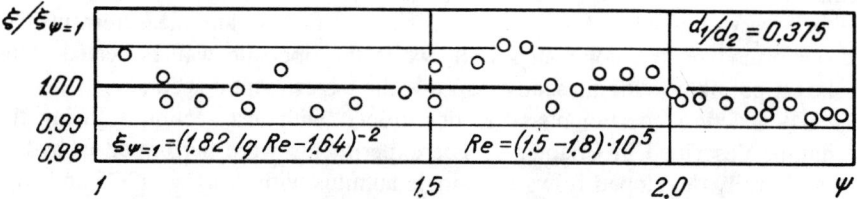

**FIG. 6.31** Hydraulic drag coefficient versus temperature factor.

CHAPTER
# SEVEN

## HEAT TRANSFER AND HYDRAULIC DRAG IN AN ANNULUS WITH A HELICALLY TWISTED INNER TUBE

Heated surfaces with complex geometries are frequently used for increasing the effectiveness and reliability of devices with high energy transfer rates.

Study of the flow pattern and heat transfer in systems with complex geometries makes it possible to design more effective and reliable heat exchangers. Of certain interest in this respect are helically shaped tubes and rods with cross sections in the shape of an ellipse, oval, cross, and other. The improved effectiveness provided by these shapes was confirmed by studies of heat transfer and hydraulic drag in tubes with such geometries, performed by Paegle et al. [116], Kidd [117], and Lawson et al. [118]. Helically shaped tubes have the advantage of a duct formed by a circular tube and an auger-shaped turbulence promoter whose entire wetted surface participates in the transfer of heat. The use in nuclear reactors of fuel elements in the form of specially shaped rods [119] makes it possible to increase the heat-release area.

It is best to investigate heat transfer from such surfaces in an annulus, where the inner tube is replaced by such a surface.

## 7.1 HEAT TRANSFER FROM A HELICALLY SHAPED INNER TUBE OF AN ANNULUS

Test section 1 (Fig. 2.7) was used for investigating the heat transfer from four helically twisted tubes with a three-lobe type cross section, and five helically twisted tubes with an oval-shape cross section, all having different twist pitches. Tests were conducted over the ranges $5 \times 10^3 \leq Re \leq 4 \times 10^4$ and $1.05 \leq \psi \leq 2.9$. Each such tube was placed coaxially into a circular tube to form an annulus. The geometries of the tubes and the annulus are described by Table 2.1.

For comparison of results, the same test section was used to investigate heat transfer from a circular inner tube ($d_1/d_2 = 0.375$) with the same perimeter as the helical tubes.

The wall temperature of the helical tubes was measured in two points around the periphery—at minimum distance from the tube axis (trough of the three-lope profile, minor axis of the oval), and at maximum distance from the axis (crest of the three-lobe profile, major axis of the oval). The values of Nu were calculated from the local wall temperature and the mean-perimeter heat flux, that is, without correction for heat dissipation around the perimeter of the tube. At maximum Re, the heat transfer at the crest was 7–10% higher than in the trough, and 13–18% higher at the major axis than at the minor axis (Fig. 7.1). At minimum Re this difference was 2–4% and 4–6%, respectively. The nonuniformity of heat transfer rate over the tube perimeter increased somewhat with their twist pitch. The values of Nu used for comparison and correlation of test data were averaged over the perimeter.

A series of runs was conducted for determining the heat transfer from the three-lobe tube with minimum twisting pitch of $s/d_{dc} = 13.2$. These tests were performed at different values of $\psi$, over the Re range under study. They showed that $\psi$ does not exert a significant effect on the manner in which Nu depends on Re (Fig. 7.2). This means that experiments for determining the effect of phys-

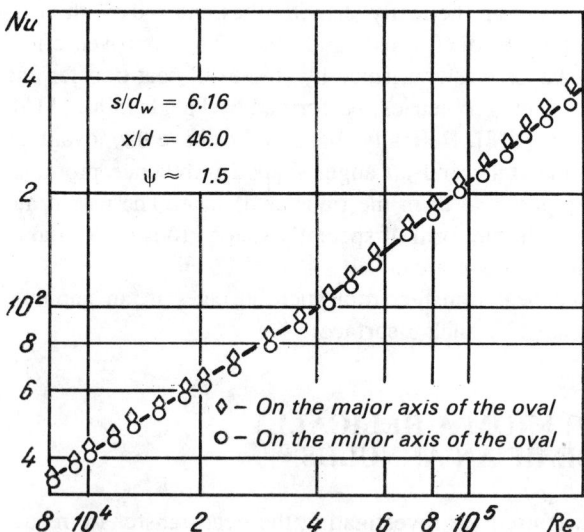

**FIG. 7.1** Plot of Nu obtained from the wall temperatures at the major and minor axes of an oval.

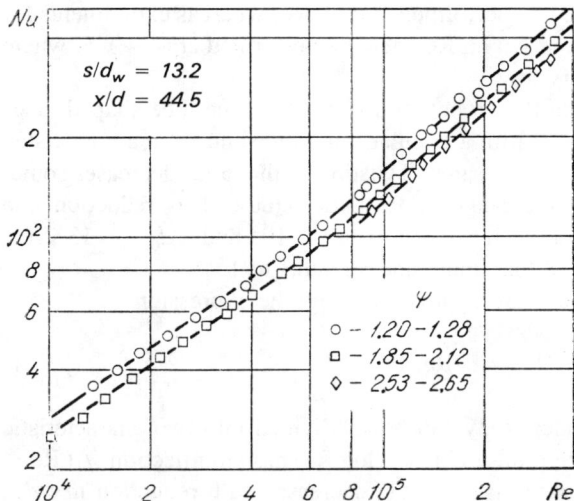

**FIG. 7.2** Plot of Nu versus Re at different values of the temperature factor for a three-lobe shaped inner tube.

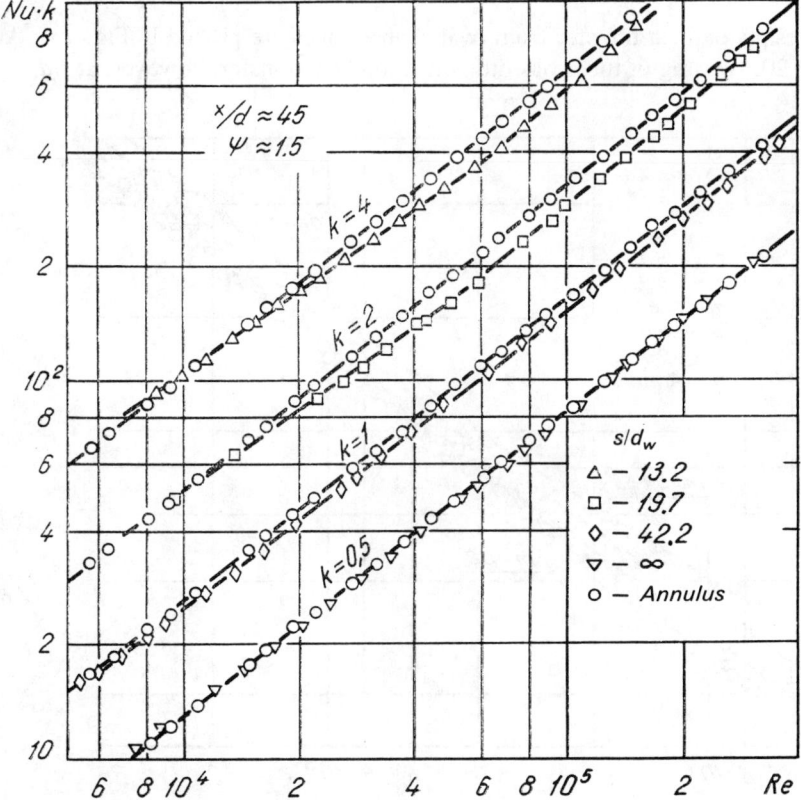

**FIG. 7.3** Heat transfer from a three-lobe shaped inner tube of an annulus.

ical properties variation could be performed at high Re, whereas experiments for determining the dependence of Nu on Re could be performed at $\psi \approx 1.5$, where the experimental error is very low.

Figure 7.3 shows the results on heat transfer from three-lobe shaped tubes at $x/d = 45$. The heat transfer from a circular inner tube and from a nontwisted tube with a three-lobe profile is the same. Twisting of the tubes decreases somewhat the heat transfer over the range of Re under study. This reduction is a function of Re and the twist pitch. At $Re = 6 \times 10^4$ and $s/d_{dc} = 13.2$ the reduction in heat transfer is at maximum and amounts to 15%.

If the experimental results are approximated by the expression

$$\mathrm{Nu} = c_l \, \mathrm{Re}^m \, \mathrm{Pr}^{0.6} \psi^n \tag{7.1}$$

then the entire Re range under study can be subdivided into two characteristic zones. At $Re \approx 6 \times 10^4$, the plot of $\mathrm{Nu} = f(\log \mathrm{Re})$ has an inflection. At $Re < 6 \times 10^4$ the power exponent $m$ in Eq. (7.1) decreases with reduction in $s/d_w$, whereas at $Re > 6 \times 10^4$ it increases. In the case of a nontwisted tube ($s/d_w = \infty$) and of a circular inner tube, the value of $m$ is close to 0.8 over the entire Re range.

Results on heat transfer from oval-shaped tubes are plotted in Fig. 7.4. At $s/d_w > 20$, twisting of tubes has little effect on heat transfer; however, at $s/d_w <$

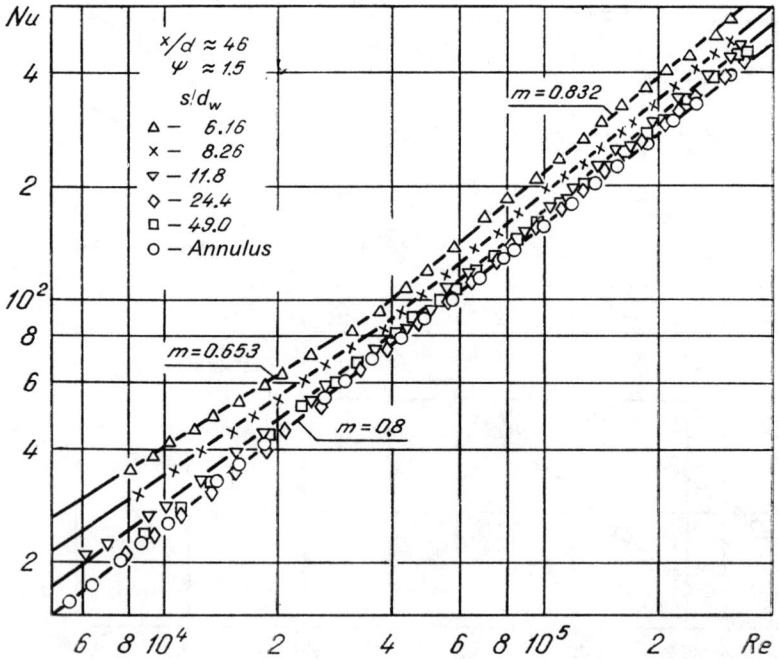

**FIG. 7.4** Heat transfer from an oval-shaped inner tube of an annulus.

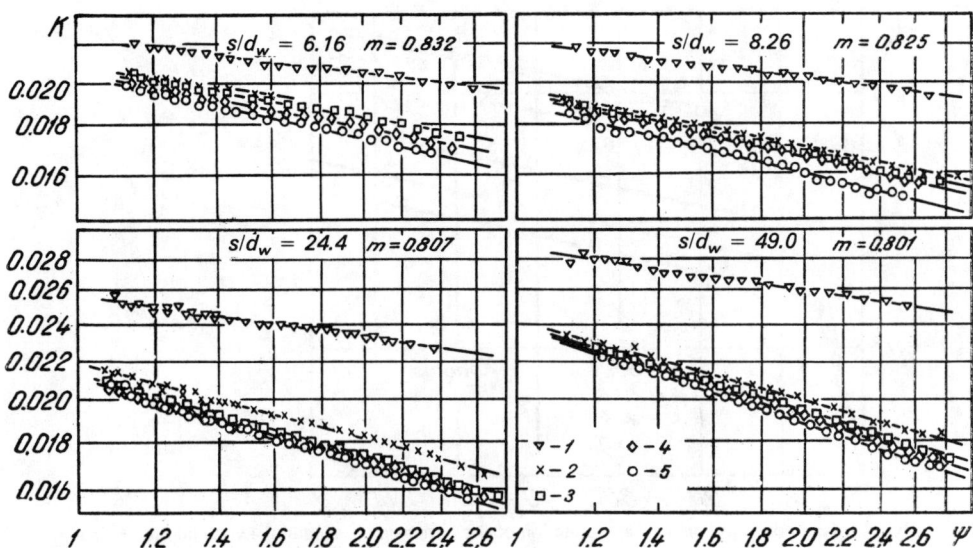

**FIG. 7.5** Effect of the temperature factor on heat transfer from an oval-shaped tube with different twisting pitches.

10 it causes heat transfer to increase rather significantly. As in the case of three-lobe shaped tubes, the range of Re under study can be subdivided into two zones with different values of $m$. These zones interface at Re = $4 \times 10^4$. The effect of $s/d_{dc}$ on the value of $m$ for twisted oval-shaped tubes is satisfactorily expressed at Re < $4 \times 10^4$ by

$$m = 0.8(1 - 0.35\, e^{-0.105\, s/d_w}) \tag{7.2}$$

and at Re > $4 \times 10^4$ by

$$m = 0.8(1 + 0.06\, e^{-0.074\, s/d_w}) \tag{7.3}$$

When $s/d_w$ is increased, $m$ tends to 0.8.

The bulk of the experiments for determining the effect of physical properties variation on heat transfer were performed at Re ≈ $1.5 \times 10^5$. In this case, as in annuli with circular tubes, the effect of $\psi$ increases along the annulus (Fig. 7.5). The effect of $\psi$ on heat transfer lessens with reduction in $s/d_w$.

To simplify the correlations, the effect of physical properties variation was incorporated by using the term $\psi^n$ in Eq. (7.1), where the power exponent $n$ was assumed to be independent of the temperature factor over the entire $\psi$ range under study. The variations of $n$ along an annulus for oval-shaped tubes with different twisting pitches and for a circular inner tube are shown in Fig. 7.6.

It is seen by comparing Figs. 7.6 and 6.7 that for a circular inner tube with almost the same value of $d_1/d_2$, the values of $n$ in Fig. 7.6 are somewhat higher.

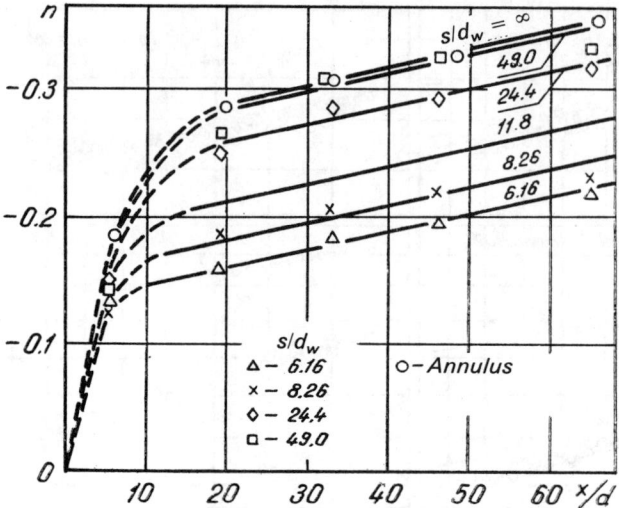

**FIG. 7.6** The power exponent $n$ as a function of the dimensionless annulus length and the twisting pitch of an oval-shaped tube.

This is because the curves constructed in Fig. 6.7 were plotted from results at $\psi \leq 2$, whereas in this chapter the results were obtained experimentally at $\psi$ of 1.05 to 2.6. It is well known that the absolute values of $n$ increase somewhat with increasing $\psi$ at sufficiently high values of the latter (see Fig. 6.5).

At $20 \leq x/d < 65$, the experimental data for helically twisted tubes with oval and three-lobe cross-section shapes are satisfactorily described by Eq. (7.1) with the power exponent

$$n = -[0.26(1 - 0.8\, e^{-0.08\, s/d_w})] + 0.0014\, x/d \qquad (7.4)$$

As $s/d_w$ is increased, $n$ tends to a value characteristic of a circular inner tube.

The results on heat transfer from helically twisted tubes with oval cross section were correlated by Eq. (7.1), in which $c_l$ is a function of the degree of heat transfer stabilization and of $s/d_w$. At $Re > 4 \times 10^4$, stabilization of highly twisted heat transfer tubes accelerates with rising Re (Fig. 7.7). The length of the thermal entrance region of an annulus with a circular inner tube was 30–35 diameters, whereas for a twisted tube it was from 10 to 35 diameters (the lower limit is for a tube with an oval-shaped cross section having the minimum $s/d_w$ at $Re > 2 \times 10^5$). The volume of experimental data was too limited for a more detailed investigation of heat transfer from the thermal entrance region.

Table 7.1 shows values of the coefficient $c_l$ and the power exponent $m$ in the thermally fully developed region of tubes with oval-shaped cross section. The rms deviation of measured Nu values from those calculated using Eqs. (7.1)–(7.4) and Table 7.1 (tubes with an oval-shaped cross section) is 2.3%.

The gain in heat transfer due to twisting of tubes can be expressed by a ratio

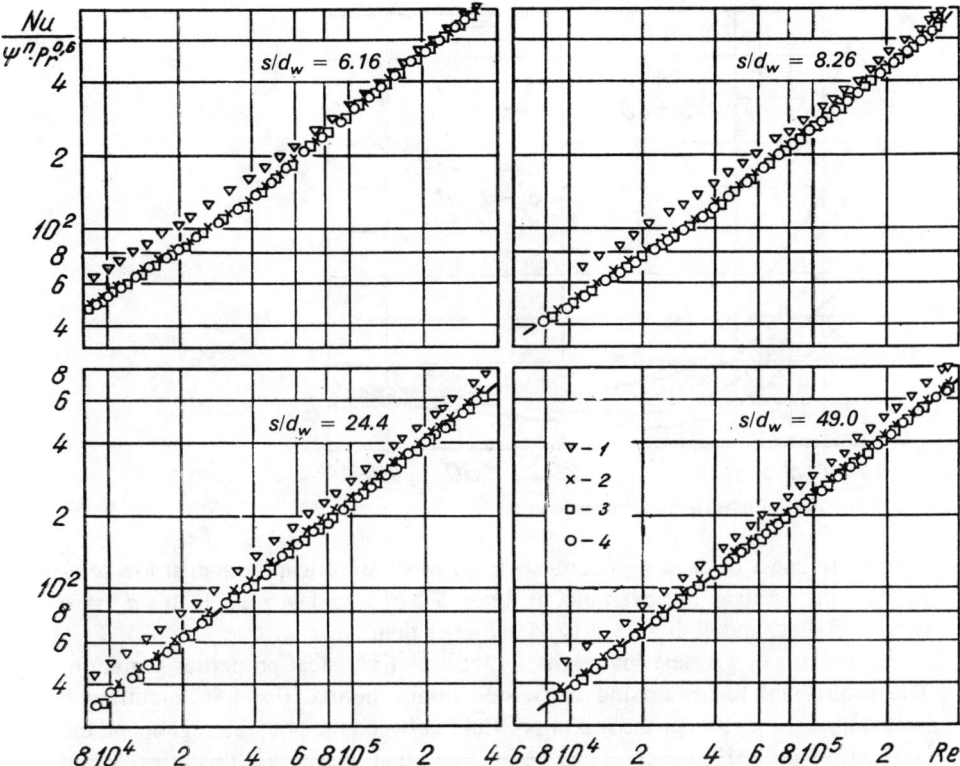

**FIG. 7.7** Heat transfer from tubes with an oval-shaped cross section at different $x/d$.

of Nu in an annulus with a twisted and a circular inner tube, i.e., $Nu/Nu_1$. In the thermally fully developed region, the effect of the twisting pitch of tubes with an oval-shaped cross section is rather complex (Fig. 7.8). At the Re where Nu = $f$(Re) has an inflection, gradual twisting ($s/d_w > 12$) reduces the heat transfer. However, at $s/d_w < 12$ the heat transfer increases with decreasing $s/d_w$ over the entire range of Re. It is seen from Fig. 7.8 that at high temperature differences

**TABLE 7.1** Values of $c_l$ and $m$ for Helically Twisted Tubes of Oval-Shaped Cross Section in the Thermally Fully Developed Region

| $s/d_w$ | Re < 4·10⁴ | | Re > 4·10⁴ | |
|---|---|---|---|---|
| | $c_l$ | $m$ | $c_l$ | $m$ |
| 6.16 | 0.1332 | 0.653 | 0.0202 | 0.832 |
| 8.26 | 0.0856 | 0.682 | 0.0188 | 0.825 |
| 11.8 | 0.0509 | 0.719 | 0.0178 | 0.820 |
| 24.4 | 0.0290 | 0.778 | 0.0207 | 0.807 |
| 49.0 | 0.0239 | 0.798 | 0.0237 | 0.801 |

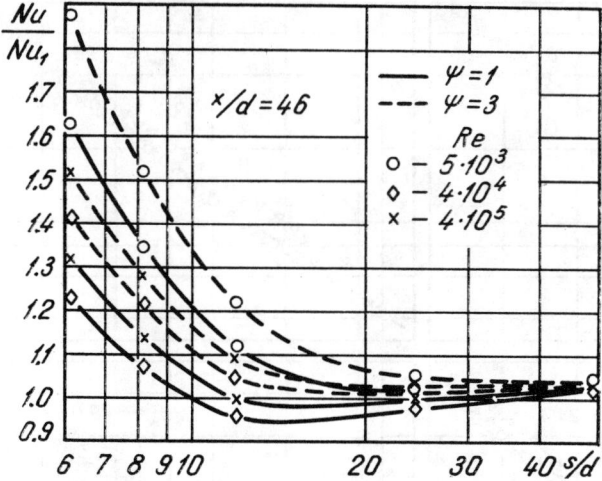

**FIG. 7.8** Effect of tube twisting on heat transfer.

the effectiveness of twisting manifests itself more significantly than at low temperature differences. For example, at Re = $5 \times 10^3$ and $\psi \approx 1$, Nu at $s/d_w = 6$ is 60% higher, and at $\psi = 3$ it is 85% higher than $Nu_1$.

The item of greatest interest is the effect of physical properties variation. The centrifugal forces arising in twisted nonisothermal flows in annuli with heated inner tubes hinder the mixing of fluid between the hot wall region and the cooler flow core. However, it can be assumed that in our case this effect is not decisive, since the reduction in heat transfer due to the temperature factor is smaller for twisted tubes than for nontwisted tubes. Analysis of studies on heat transfer in annuli with rotating inner tube, performed by Shchukin [120], also indicates that the direction of the heat flux has no effect on the heat transfer coefficient under conditions somewhat similar to ours. It can apparently be assumed that the decisive effect on heat transfer (as in the case of rotating tubes) is caused by secondary flows and separated flow regions at low $s/d_w$. The slight reduction in heat transfer for twisted tubes with three-lobe cross-section shape

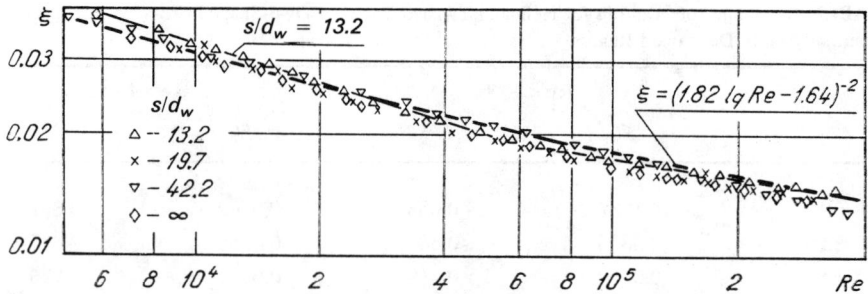

**FIG. 7.9** Hydraulic drag of an annulus with a three-lobe shaped inner tube.

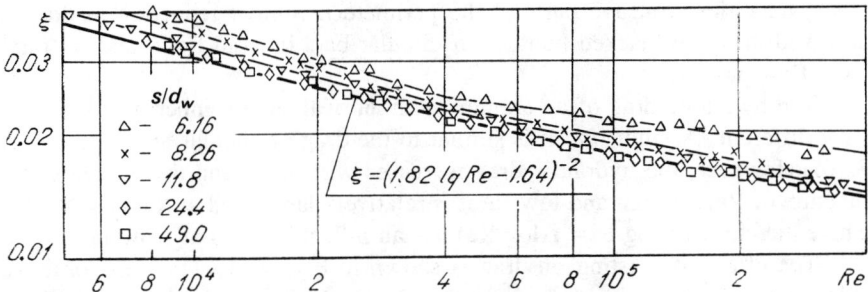

**FIG. 7.10** Hydraulic drag of an annulus with an oval-shaped inner tube.

can be attributed to the presence of concave perimeter elements, which are responsible for formation of zones with reduced velocities.

Results of studies of heat transfer in annuli with helically twisted tubes show that physical properties variation depends significantly on surface geometry. They also confirm the assumption of Kalinin et al. [121] that the influence of flow temperature variation on smooth ducts and in ducts with artificial turbulence promoters is different.

## 7.2 HYDRAULIC DRAG OF AN ANNULUS WITH A SHAPED, HELICALLY TWISTED INNER TUBE

In addition to studying heat transfer from helically twisted tubes of various cross-section shapes, we used the same test section for measuring their hydraulic drag. The results obtained in an annulus with a three-lobe shaped inner tube are plotted in Fig. 7.9, which also shows a line obtained from Eq. (6.33). As in the case of heat transfer, twisting of the tube has little effect on the drag and an inflection in the curve of $\log \xi = f(\log Re)$ is observed at $Re \approx 6 \times 10^4$. At sufficiently high values of $s/d_w$ and at high Re, the drag is even somewhat lower than in a smooth annulus. This can be attributed to the presence of regions with

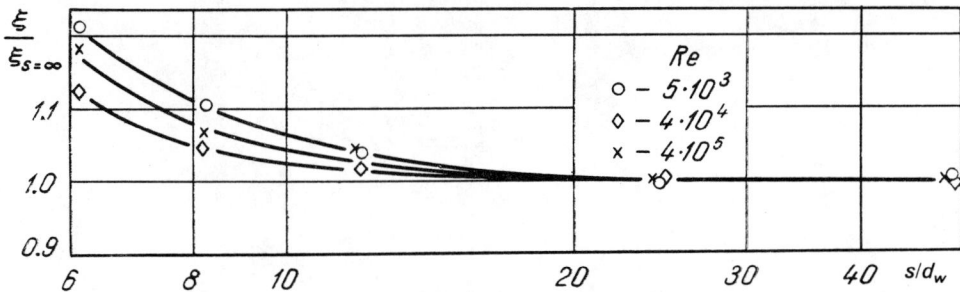

**FIG. 7.11** Effect of twisting pitch on the hydraulic drag coefficient.

lower velocities (concave parts of the perimeter). Similar reduction in drag is observed in tightly packed bundles of circular bars in axial flow, and in triangular flow passages.

The hydraulic drag of an annulus with an oval-shaped inner tube is somewhat different. At high $s/d_w$ it is similar to the drag of circular annuli (see Fig. 7.10). However, the hydraulic drag increases with $s/d_w$, and this increase is a function of Re. At high and low Re it is relatively large, and at Re $= 4 \times 10^4$, where the curve of log $\xi = f(\log Re)$ has an inflection, it is at minimum.

The effect of twisting on drag is shown in Fig. 7.11, where the ordinate axis represents the ratio of $\xi$ for an annulus with a twisted tube to that for a shaped, but nontwisted tube. These data are in satisfactory agreement with results on heat transfer.

In conclusion, the use of heated surfaces with the above geometries is advantageous from the point of view of heat transfer augmentation only when the relative twisting pitch $s/d_w$ is smaller than 10. At high $s/d_w$, both the heat transfer and the hydraulic drag are virtually the same as their counterparts for circular annuli.

CHAPTER
# EIGHT
## CONCLUSIONS AND PRACTICAL RECOMMENDATIONS

In the preceding chapters we analyzed the results of extensive experimental and analytic studies of heat transfer, thermal, and hydrodynamic variables in gas-cooled annuli and in axial flow over cylinders. Our data were compared with those obtained by others. This chapter contains general conclusions and certain practical recommendations for calculating heat transfer and hydraulic drag in flows of heated gases in annuli. Practical recommendations regarding certain flow variables of a turbulent boundary layer on a cylinder in axial flow are also presented.

## 8.1 FEATURES OF THE BOUNDARY LAYER ON A CYLINDER IN AXIAL FLOW

A great deal of attention was devoted in this book to the study of the turbulent boundary layer on a cylinder in axial flow. The results of this study were used for calculating the skin friction and for analyzing experimental data on heat transfer in the entrance region of an annulus.

The cylinder boundary layer develops in a somewhat different manner than that of a plate, which in the general case can be treated as a cylinder with an infinite radius.

In a fluid at rest, the flux of any physical quantity (such as heat) generated at a surface is constant at a fixed distance from that surface, irrespective of the mechanism of transport of this quantity. For a cylinder, such a flux decreases in proportion to the distance from the center of the cylinder. When a translational fluid motion is superimposed on the above situation, the total transport changes. Here the gradient of any quantity (for example, the longitudinal component of velocity, temperature, concentration) will be greater on the surface of a cylinder than on a plate. The fluxes of momentum, heat, and mass induced by these gradients will be correspondingly large.

If the principal variation of the physical quantity under study occurs in the very thin boundary layer adjoining the cylinder, and if the thickness of this layer is significantly smaller than the radius of the cylinder, then the effect of cylindrical geometry virtually vanishes. In that case the flow over the cylinder becomes identical with that over a flat plate. However, if the boundary layer thickness is of the same (or larger) order of magnitude as the cylinder radius, the effect of cylindrical geometry is significant.

The features of a cylinder boundary layer in axial flow can be most clearly determined by comparing measured or calculated velocity and temperature boundary layer profiles of a cylinder with analogous profiles of a plate.

The turbulent boundary layer of a flat plate is usually subdivided into two parts: the wall or inner region, and the outer region. Very near the wall a very thin viscous sublayer is formed. The flow within the sublayer is controlled by molecular viscosity. The viscous sublayer is followed by a transition region, in which the flow is controlled by the combined effect of turbulent fluctuations and molecular viscosity. Then the flow takes on a developed turbulent nature, and the velocity distribution becomes logarithmic. The wall region occupies about 0.2% of the thickness of the entire boundary layer, and the flow in it is described by the so-called law of the wall, $u^+ = f(y^+)$. The remaining part of the boundary layer is made up of the outer region, which is described by the so-called velocity-defect law $u_\infty^+ - u^+ = f(y/\delta)$.

Measurements and calculations of velocity profiles in the boundary layer confirm the above conclusions on the effect of cylindrical geometry (Fig. 8.1). With reduction in the cylinder radius, the velocity profiles increasingly deviate from analogous profiles for a flat plate. The effect of curvature is strongest in the outer part of the boundary layer and is almost negligible in the sublayer and in the transition region. The logarithmic part of the profile reduces sharply with reduction in $r_0^+$ and vanishes almost entirely at sufficiently small values of this quantity. The transition region is immediately followed by the outer part of the boundary layer.

The velocity deficiency is smaller in a cylinder boundary layer than in that of a plate (Fig. 8.2). This divergence from a flat plate velocity profile increases as the cylinder diameter becomes smaller and as the boundary layer becomes thicker.

Analogous conclusions can be drawn also for the thermal boundary layer. Although such experiments were performed only with air, for which the Pr is close to 1, certain conclusions can be drawn also for other fluids. At high Pr most of the temperature difference is across the thin laminar sublayer. Conse-

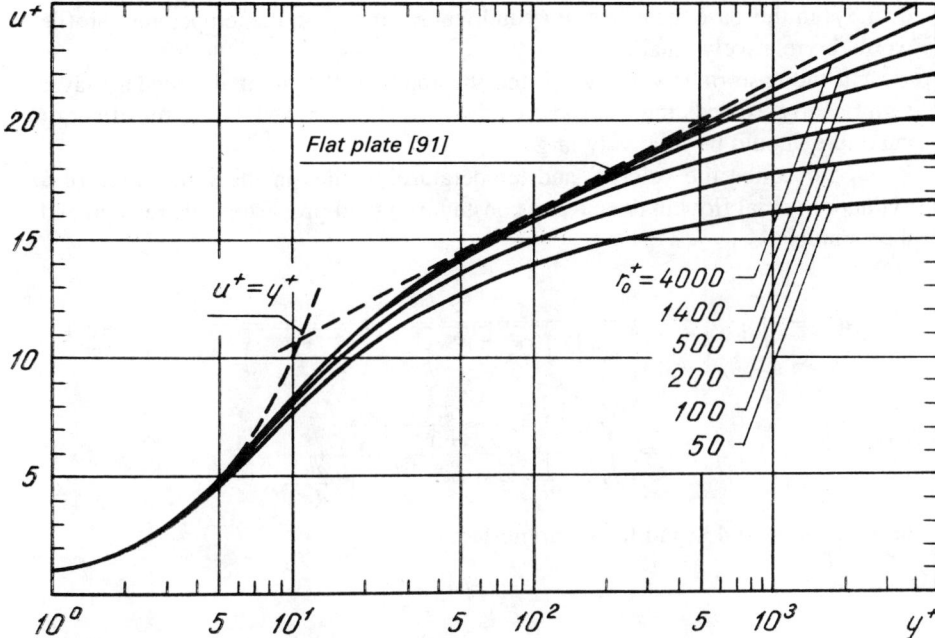

**FIG. 8.1** Law of the wall for a cylinder in axial flow.

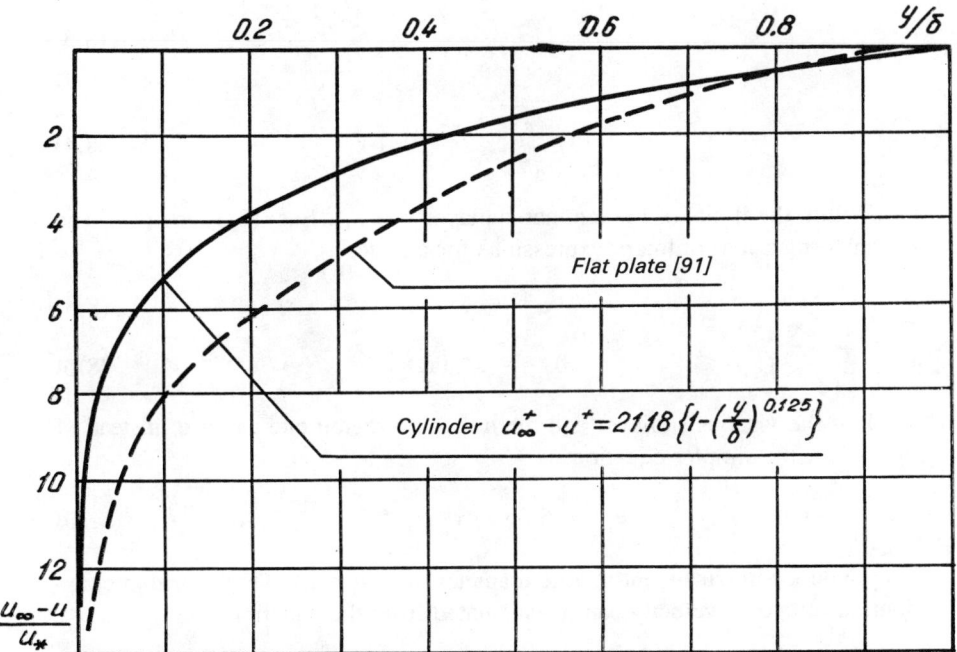

**FIG. 8.2** Excess velocity distribution. The solid curve is for a cylinder with $d = 15.5$ mm.

quently, in this case the effect of cylindrical geometry on the temperature profile would be relatively small.

In the case of low Pr, when the thickness of the thermal boundary layer significantly exceeds the thickness of the velocity boundary layer, the effect of curvature should be relatively large.

To describe the velocity and temperature profiles in the wall region of a cylinder in axial flow in the case of constant physical properties, we recommend the formulas

$$u^+ = \int_0^{y^+} \frac{2\,dy^+}{\frac{r_0^+ + y^+}{r_0^+} + \sqrt{\left(\frac{r_0^+ + y^+}{r_0^+}\right)^2 + 4\left(\frac{r_0^+ + y^+}{r_0^+}\right)\varkappa^2 n^2 y^{+2}}} \quad (3.17)$$

$$\vartheta^+ = \int_0^{y^+} \frac{\frac{r_0^+}{r_0^+ + y^+}\,dy^+}{\frac{1}{\Pr_f} + \varkappa^2 y^{+2}\,\frac{1}{\Pr_t}\,n^2\,\frac{du^+}{dy^+}} \quad (8.1)$$

in which $\kappa = 0.43$, and the damping factor

$$n = 1 - \exp\left(-\frac{y^+}{23}\right) \quad (8.2)$$

At low $y^+$ we have $n^2 y^{+2} \to 0$, and for the laminar boundary layer we have

$$u^+ = \int_0^{y^+} \frac{r_0^+}{r_0^+ + y^+}\,dy^+ \quad (8.3)$$

$$\vartheta^+ = \Pr_f \int_0^{y^+} \frac{r_0^+}{r_0^+ + y^+}\,dy^+ \quad (8.4)$$

Within the limits of the laminar sublayer $y \ll r_0^+$, Eqs. (8.3) and (8.4) can be transformed into ordinary expressions for a plate:

$$u^+ = y^+ \quad (8.5)$$

$$\vartheta^+ = y^+ \Pr_f \quad (8.6)$$

Past the viscous sublayer and the transition region one can use, instead of Eq. (3.17), the simpler equation:

$$u^+ = 5.35 \lg y^+ + 5.2 \quad (3.20)$$

for cylinders with sufficiently large diameter ($d > 4$ mm). The logarithmic distribution of excess velocity can be calculated from the equation

$$u_+^\infty - u^+ = -5.35 \lg \frac{y}{\delta^* u_\infty^+} + B \quad (8.7)$$

where $B$ is a function of the cylinder diameter. For cylinders with $d = 15.5$ and 4 mm, $B = -3.25$ and $-4.35$, respectively.

Most experiments were performed at freestream turbulence intensity $\overline{Tu_\infty} \approx 1.2\%$, and hence the relations obtained correspond to the above value.

## 8.2 EFFECT OF PHYSICAL PROPERTIES VARIATION ON VELOCITY AND TEMPERATURE PROFILES

The influence of physical properties variation is basically by means of changes in density and viscosity. Our experiments show that the variation in these quantities greatly distorts the velocity profiles along the entire boundary layer, especially in the transition region (Fig. 8.3).

The distortion of the velocity profile depends significantly on the freestream turbulence, and also on induced instability of the boundary layer.

If the freestream turbulence intensity is $1.0-1.5\%$, the velocity profile over a cylinder in axial flow can be calculated from the expression

$$u_q^+ = \int_0^{y_{fq}^+} \frac{2 dy_{fq}^+}{\frac{\mu_y}{\mu_f}\left(\frac{r_0^+ + y_{fq}^+}{r_0^+}\right) + \sqrt{\left(\frac{\mu_y}{\mu_f}\right)^2 \left(\frac{r_0^+ + y_{fq}^+}{r_0^+}\right)^2 + 4 \frac{\rho_y}{\rho_f}\left(\frac{r_0^+ + y_{fq}^+}{r_0^+}\right) \varkappa^2 n^2 y_{fq}^{+2}}}$$

(3.34)

Equation (3.34) satisfactorily describes the experimental data up to the outer part of the boundary layer.

## 8.3 DRAG AND HEAT TRANSFER OF A CYLINDER IN AXIAL FLOW

Measurements of velocity and temperature profiles in the boundary layer made it possible to determine the frictional drag of a cylinder at constant as well as at variable physical properties of the heated flow. The local heat-transfer coefficients were determined directly by measuring the wall temperature. Skin friction for an unheated duct was also measured directly by the projecting plank technique.

As expected, both heat transfer and friction at the surface of a cylinder in axial flow (in the entrance region of the annulus) depend on the cylinder radius. More precisely, they depend on the ratio of the thickness of the boundary layer to the cylinder radius $\delta/r_0$, and ratio of wall to flow temperature $\psi$. In addition, heat transfer and friction under these conditions are rather sensitive to the freestream turbulence.

Within the limits of experimental error, we did not notice any influence of

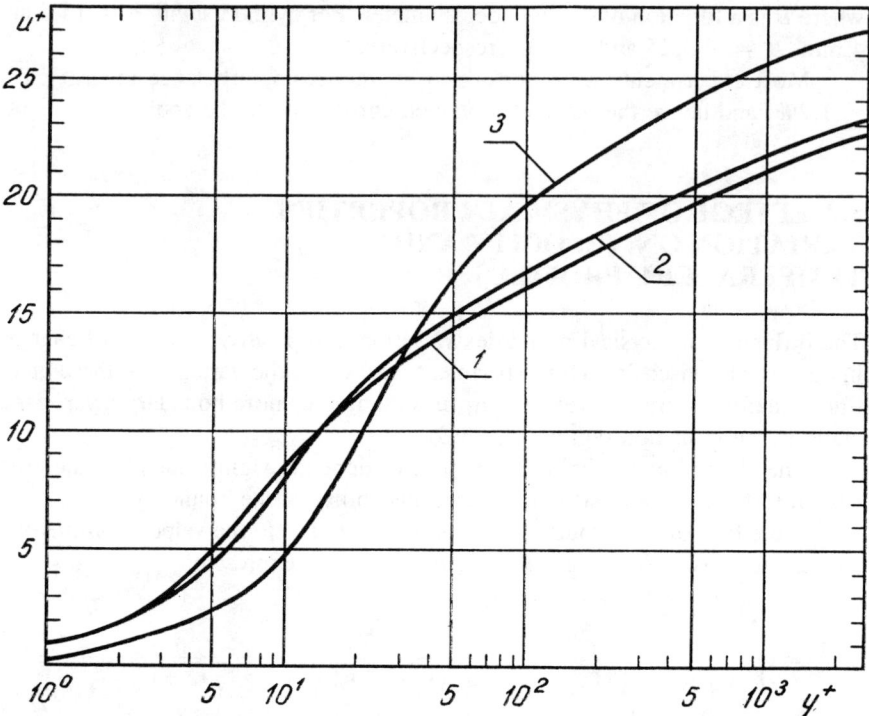

**FIG. 8.3** Distortion of velocity profile by physical properties variation of the coolant. (1)–(3) Data at $\psi = 1.0$, 1.19, and 3.15, respectively.

temperature variation in the boundary layer on the manner in which the surface curvature affects heat transfer and friction.

Figure 8.4 shows how the curvature affects the friction and heat transfer according to estimates of a number of investigators.

On the basis of our experimental data we can make a number of design recommendations for determining friction and heat transfer parameters for a cylinder in axial flow at constant heat flux and large temperature differences.

The skin friction of a cylinder at $Tu_\infty = 0.8-1.4$ can be calculated from the expression

$$c_f = 0.329 \, (\lg Re_{x-x_0})^{-2.45} \psi^{-0.3} \Omega^{0.14} \qquad (4.22)$$

Under the same conditions the heat transfer coefficients are expressed, within ±6%, by the equation

$$St = 0.183 \, (\lg Re_x)^{-2.45} \psi^{-0.25} \Omega^{0.14} \qquad (5.12)$$

The accuracy can be improved somewhat by using

$$Nu_x = 0.0253 \, Re_x^{0.8} \psi^{-0.25} \Omega^{0.14} \qquad (5.9)$$

at $Re_x < 2 \times 10^6$, and

$$Nu_x = 0.0106 \, Re_x^{0.86} \psi^{-0.25} \Omega^{0.14} \tag{5.11}$$

at $Re_x = 2 \times 10^6 - 5 \times 10^7$.

It is seen by comparing Eqs. (4.22) and (5.12) that the surface curvature has the same effect on both friction and heat transfer. However, $\psi$ has a somewhat greater effect on friction than on heat transfer. These expressions can be used for obtaining a direct relationship between heat transfer and friction:

$$\frac{St}{c_f} = 0.556 \left( \frac{\lg Re_{x-x_0}}{\lg Re_x} \right)^{2.45} \Psi^{0.05} \tag{8.8}$$

Equation (8.8) can be used for calculating friction coefficients from known values of the heat transfer coefficient. This is helpful since determination of friction directly, particularly under conditions of large temperature variation in the boundary layer, involves great difficulties. The quantity $x_0$, which represents the downstream distance from the virtual origin, can be determined with sufficient accuracy from Fig. 8.5, taken from the book by Žukauskas and Šlančiauskas [52].

Equations (4.22) and (5.12) can also be used for calculating heat transfer and skin friction in the case of a gas flow over a flat plate, at high temperature differences.

Numerous experiments showed that the freestream turbulence has a significant effect on heat transfer and friction. At small temperature differences it increases both these quantities. Turbulence intensity effects are relatively large at $Re_x > 10^6$, whereas at low $Re_x$ its effect may even go unnoticed (Fig. 8.6).

At high temperature differences the flow turbulence lessens the effect of physical properties variation on heat transfer.

Our studies show that in the entrance region of an annulus the heat transfer from the inner tube (a cylinder in axial flow) is governed by many factors: flow variables, geometry, physical properties variation, and the freestream turbulence

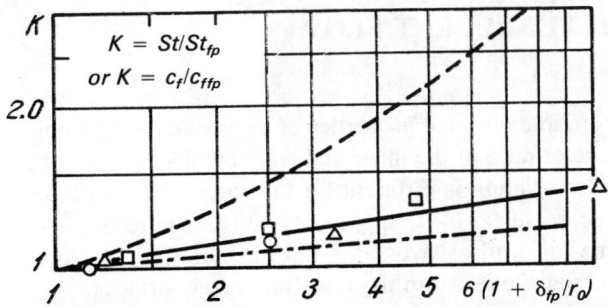

**FIG. 8.4** Effect of transverse curvature on heat transfer and friction. The solid line is plotted from our data; the dashed from ref. [53]; the symbols are from [55] (squares at $Re_x = 10^5$, triangles at $10^6$, and circles at $10^7$); and the dash dotted line represents friction according to ref. [54].

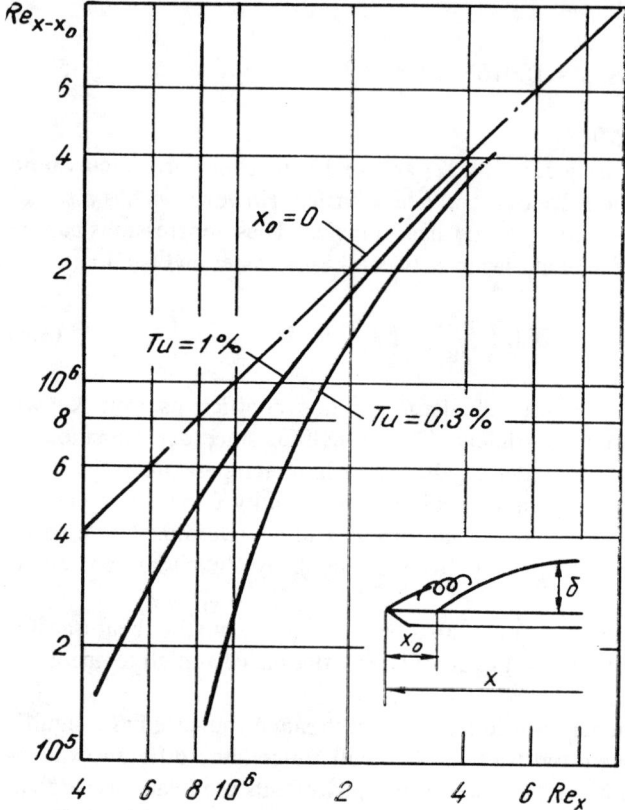

**FIG. 8.5** Graph for determining the location of the virtual origin of a turbulent boundary layer [52].

intensity. Sufficiently accurate data on heat transfer and friction can be obtained only by making allowance for the effect of each of these factors.

## 8.4 HEAT TRANSFER AND HYDRAULIC DRAG IN HYDRODYNAMICALLY FULLY DEVELOPED TURBULENT FLOW IN ANNULI

Heat transfer in annuli at variable physical properties of the coolant is a complex function of geometry (diameter ratio of the inner and outer cylinders $d_1/d_2$). The effect of physical properties variation is different for the inner and outer tubes. However, this difference decreases as $d_1/d_2$ approaches unity, manifesting itself in the same manner in one- as well as two-sided heating, that is, it becomes independent of the ratio of heat fluxes from the inner and outer surfaces.

The length of stabilization of $\psi$ is a function of the ratio $d_1/d_2$ and is different for the inner and outer walls.

## CONCLUSIONS AND PRACTICAL RECOMMENDATIONS 159

Heat transfer in annuli at specified heat fluxes on the surfaces and large temperature differences should be calculated from the expression

$$\frac{Nu_i}{Nu_{i,\,\psi=1}} = 1 - \{1 - \exp[-K_{fi}(a\varphi_i + n\mu\,\Phi_i\,K_{fi})]\}\omega_i \qquad (6.23)$$

where $i = 1; 2$,

$$a = -0.53\, n_\rho - \frac{1}{3} n_\lambda - \frac{1}{4} n_c$$

$$\varphi_i = 1 - \exp(-0.1\, \tilde{x}_i),$$

$$\Phi_1 = 0.37\,[1 - \exp(-1.32 \cdot 10^{-3}\, \tilde{x}_1^{1.66})]$$

$$\Phi_2 = \frac{1.25\,(0.01\,\tilde{x}_2)^2}{1 + (0.01\,\tilde{x}_2)^2}$$

$$\tilde{x}_1 = \frac{x_1}{d}\frac{d_2}{d_1}$$

$$\tilde{x}_2 = \frac{x_2}{d}$$

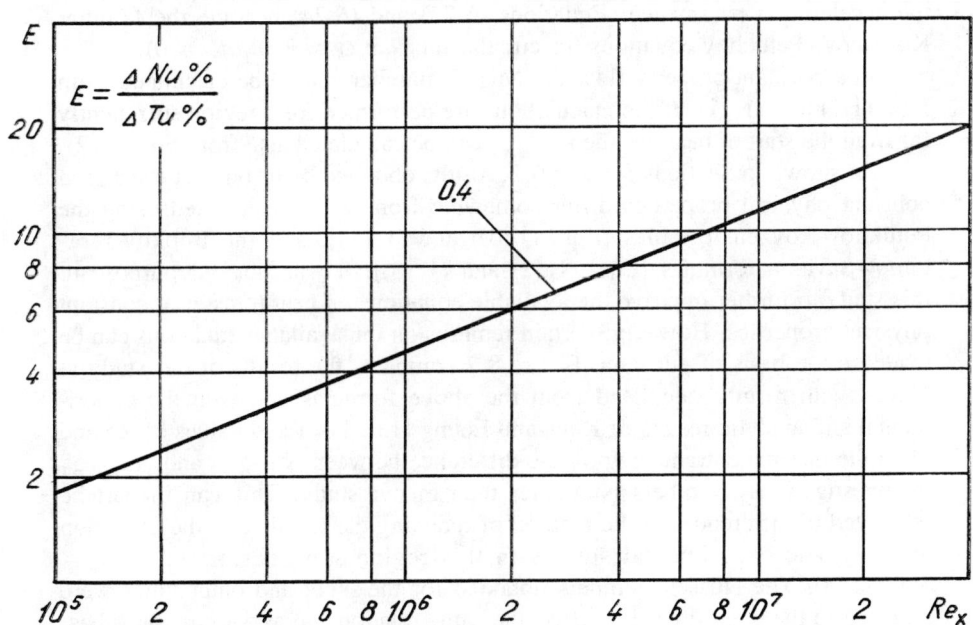

**FIG. 8.6** Effectiveness of the freestream turbulence intensity.

$$K_{fi} = \frac{q_i d}{\lambda T_f \mathrm{Nu}_{i,\,\psi=1}}$$

$$\omega_1 = 0.44 + 0.6 \exp\left(-0.68 \frac{d_2}{d_1}\right)$$

$$\omega_2 = 1 - 0.262 \exp\left[-0.022 \left(\frac{d_2}{d_1}\right)^{2.74}\right]$$

If one must determine a specific wall temperature for given values of $q$, $\overline{\rho w}$, $T_f$, $d_1$, and $d_2$, we can recommend Eq. (6.23) in modified form:

$$\psi_i = 1 + \frac{K_{fi}}{1 - \{1 - \exp[-K_{fi}(a\varphi_i + n_\mu \Phi_i K_{fi})]\}\omega_i} \quad (6.24)$$

These formulas are suitable over a wide range of flow variables for annuli with any ratio $d_1/d_2$, including parallel flat plates ($d_1 = d_2$). The geometric parameter for parallel plates is equal to 0.744, and this means that the effect of $\psi$ on heat transfer is smaller than in a tube ($\omega = 1$).

At low Re (Re $< 8 \times 10^3$–$10^4$) and high heat flux, the flow is laminarized. For this reason the use of Eqs. (6.23) and (6.24) is recommended in the case of $q_{\mathrm{in}}^+ \leq 0.007$ and $K' \leq 4 \times 10^{-7}$ ($q_{\mathrm{in}}^+$ is the heat-flux parameter and $K'$ is the acceleration parameter). Equations (6.23) and (6.24) become the familiar Kurganov–Petukhov equations for circular tubes at $\omega = 1$ ($d_1/d_2 = 0$).

In a constant-property flow the Nusselt number should be calculated from Eqs. (1.7) and (1.11). If the calculations are performed for a region sufficiently far from the start of heating, then $\mathrm{Nu}_{\psi=1}$ can be calculated also from Eq. (1.12).

As shown in Sections 6.1 and 6.2, results obtained by us on heat transfer at constant physical properties differ somewhat from values calculated using the Petukhov-Royzen formulas [Eqs. (1.10) and (1.11)], and the Bobkov-Ibragimov-Savanin formulas [Eqs. (1.12) and (1.13)]. It was not our purpose in this study to further improve the available equations on heat transfer at constant physical properties. However, certain remarks on the available equations can be made on the basis of our data. Figure 8.7 compares the results of our study at $\mathrm{Nu}_{\psi=1}$ with results calculated from the above formulas and with the experimental and analytic results of Kays and Leung [13]. For fixed values of Re and Pr in the thermal entrance region the difference between $\mathrm{Nu}_{1,\,\psi=1}$ and $\mathrm{Nu}_{2,\,\psi=1}$ in investigations by others is smaller than in our study. This can in part be attributed to differences in the method of reducing the test data to the condition $\psi = 1$. The experimental studies in this region were performed at $\psi = 1.15$–$1.20$. The Nusselt numbers obtained for the inner and outer tubes were adjusted to the conditions $\psi = 1$ by the same relationship as for circular tubes. In this case the Nusselt numbers at constant physical properties are found to be

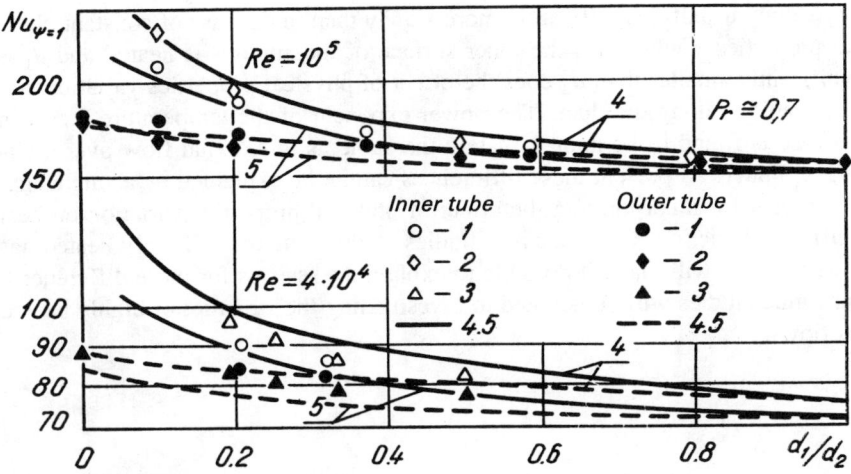

**FIG. 8.7** Stabilized heat transfer versus $d_1/d_2$ for an annulus heated from one side. (1) Our experiments, (2) analytic [13], (3) experimental [13], (4) from Eqs. (1.7) and (1.8) [17], (5) from Eqs. (1.12) and (1.14) [18].

on the high side, with $Nu_{1,\psi=1}$ overestimated to a greater degree than $Nu_{2,\psi=1}$. This is due to the fact that physical properties variation has a greater effect on heat transfer from the outer tube than from the inner tube. This is observed particularly at $d_1/d_2 < 0.4$. It was calculated that at $d_1/d_2 < 0.2$ the value of $Nu_{2,\psi=1}$ is overestimated by 1% or less, and that of $Nu_{1,\psi=1}$ by 3.5–4.5%. Correction for $\psi$ from our results would have made possible a moderate improvement of the available design equations.

The annulus hydraulic drag was investigated by us at Re from $4 \times 10^3$ to $3 \times 10^5$ and $\psi$ from 1 to 2.5. The experimental results show that over the range of flow variables under study the drag of an annulus is independent of $\psi$. It can be calculated from the familiar equation for circular tubes,

$$\xi = (1.82 \lg Re - 1.64)^{-2} \tag{6.33}$$

if the hydraulic diameter $d = d_2 - d_1$ is used as the reference dimension.

## 8.5 CONCLUSION

Experimental studies of heat transfer and hydraulic drag performed on a cylinder in axial flow (external flow over a heated surface) and in annuli showed that physical properties variation of the gaseous coolant affects each of these two cases differently. Whereas heat transfer and friction in external flows have a similar dependence on the temperature factor, in a duct flow the effect of properties variation on these two flow variables is different. In addition, under these conditions, stabilization of heat transfer along a duct, including annuli, occurs very slowly usually, significantly more slowly than in the case of constant physical properties. Only when the inner surface of the annulus is heated and $d_1$ is significantly smaller than $d_2$ does the effect of physical properties variation stabilize rapidly along the duct. The power exponent of the temperature factor in this case is found to be approximately the same as in external flow over a circular cylinder. At present these differences cannot be explained by a single analytic model for incorporating the effect of physical properties variation on heat transfer and friction. Only detailed studies of the structure of highly heated gas flows in ducts will make it possible to explain the reasons for these differences. Our future studies will be devoted to investigating the structure of highly heated gas flows.

# APPENDIXES

## 1. EXPERIMENTAL DATA ON ISOTHERMAL VELOCITY PROFILES IN THE BOUNDARY LAYER OF A CYLINDER IN AXIAL FLOW

**Table 1** Cylinder with $d = 15.5$ mm, $\overline{Tu}_\infty \approx 1.2\%$

| $y$, mm | $u$, m/sec | $y^+$ | $u^+$ | $\dfrac{y \cdot 10^2}{\delta^* u_\infty^+}$ | $u_\infty^+ - u^+$ |
|---|---|---|---|---|---|
| Run 1–1 | | $x = 0.125$ m; $T_f = 308.0$ K; $\delta^* = 0.452$ mm; $\delta^{**} = 0.3756$ mm; $u_* = 2.392$ m/sec; $c_f = 0.00542$; $\text{Re}_{x-x_0} = 345200$; $\text{Re}_{\delta^{**}} = 1502$ | | | |
| 0.22 | 34.8 | 42.0 | 14.54 | 2.49 | 4.67 |
| 0.32 | 35.5 | 60.9 | 14.85 | 3.64 | 4.36 |
| 0.42 | 36.9 | 80.2 | 15.44 | 4.79 | 3.77 |
| 0.5 | 37.3 | 96.4 | 15.58 | 5.77 | 3.63 |
| 0.6 | 38.3 | 116 | 16.02 | 6.92 | 3.19 |
| 0.7 | 39.1 | 135 | 16.36 | 8.07 | 2.85 |
| 0.8 | 39.6 | 154 | 16.57 | 9.23 | 2.64 |
| 0.9 | 40.1 | 174 | 16.79 | 10.4 | 2.42 |
| 1 | 40.7 | 193 | 17.04 | 11.5 | 2.17 |
| 2 | 43.8 | 386 | 18.31 | 23.1 | 0.9 |
| 3 | 45.0 | 579 | 18.8 | 34.6 | 0.41 |
| 4 | 45.4 | 772 | 18.97 | 46.1 | 0.24 |
| 5 | 45.8 | 964 | 19.09 | 57.7 | 0.12 |
| 6 | 45.8 | 1160 | 19.15 | 69.2 | 0.06 |
| 7 | 45.9 | 1350 | 19.18 | 80.7 | 0.03 |
| 8 | 45.9 | 1540 | 19.21 | 92.3 | 0 |
| | | $x = 0.325$ m; $T_f = 808.5$ K; $\delta^* = 0.7716$ mm; $\delta^{**} = 0.6655$ mm; $u_* = 2.309$ m/sec; $c_f = 0.00473$; $\text{Re}_{x-x_0} = 747900$; $\text{Re}_{\delta^{**}} = 2547$ | | | | |
| 0.12 | 29.2 | 21.6 | 12.63 | 0.732 | 7.93 |
| 0.22 | 32.5 | 40.2 | 14.08 | 1.36 | 6.48 |
| 0.32 | 34.6 | 58.7 | 14.99 | 1.99 | 5.57 |
| 0.42 | 35.5 | 77.3 | 15.39 | 2.62 | 5.17 |
| 0.5 | 36.3 | 93.0 | 15.70 | 3.15 | 4.86 |
| 0.6 | 37.0 | 112 | 16.00 | 3.78 | 4.56 |
| 0.7 | 37.6 | 130 | 16.30 | 4.41 | 4.26 |
| 0.8 | 38.2 | 149 | 16.52 | 5.04 | 4.04 |
| 0.9 | 38.7 | 167 | 16.74 | 5.67 | 3.82 |
| 1.0 | 39.3 | 186 | 17.03 | 6.30 | 3.53 |
| 2.0 | 42.6 | 372 | 18.46 | 12.6 | 2.10 |
| 3.0 | 44.3 | 558 | 19.16 | 18.9 | 1.40 |

**Table 1** Cylinder with $d = 15.5$ mm, $\overline{Tu_\infty} \approx 1.2\%$ (*Continued*)

| $y$ mm | $u$ m/sec | $y^+$ | $u^+$ | $\dfrac{y \cdot 10^2}{\delta^* u_\infty^+}$ | $u_\infty^+ - u^+$ |
|---|---|---|---|---|---|
| 4.0 | 45.5 | 744 | 19.72 | 25.2 | 0.84 |
| 5.0 | 46.2 | 930 | 20.04 | 31.5 | 0.53 |
| 6.0 | 46.8 | 1120 | 20.27 | 37.8 | 0.29 |
| 7.0 | 47.2 | 1300 | 20.44 | 44.1 | 0.12 |
| 8.0 | 47.3 | 1490 | 20.50 | 50.4 | 0.06 |
| 9.0 | 47.5 | 1670 | 20.56 | 56.7 | 0 |

$x = 0.525$ m; $T_f = 308.5$ K; $\delta^* = 1.177$ mm; $\delta^{**} = 0.9931$ mm; $u_* = 2.282$ m/sec; $c_f = 0.00446$; $Re_{x-x_0} = 1235000$; $Re_{\delta^{**}} = 3863$

| $y$ mm | $u$ m/sec | $y^+$ | $u^+$ | $\dfrac{y \cdot 10^2}{\delta^* u_\infty^+}$ | $u_\infty^+ - u^+$ |
|---|---|---|---|---|---|
| 0.12 | 25.1 | 27.3 | 11.02 | 0.46 | 10.16 |
| 0.22 | 31.1 | 39.7 | 13.63 | 0.866 | 7.54 |
| 0.32 | 33.5 | 58.0 | 14.68 | 1.27 | 6.49 |
| 0.42 | 34.8 | 76.3 | 15.26 | 1.67 | 5.92 |
| 0.6 | 35.9 | 110 | 15.74 | 2.41 | 5.44 |
| 0.7 | 36.6 | 129 | 16.05 | 2.81 | 5.12 |
| 0.8 | 37.3 | 147 | 16.36 | 3.21 | 4.82 |
| 0.9 | 37.7 | 165 | 16.51 | 3.61 | 4.67 |
| 1.0 | 38.0 | 184 | 16.66 | 4.01 | 4.52 |
| 2.0 | 41.3 | 367 | 18.08 | 8.02 | 3.09 |
| 3.0 | 43.8 | 551 | 19.21 | 12.0 | 1.96 |
| 4.0 | 44.6 | 735 | 19.53 | 16.0 | 1.64 |
| 5.0 | 46.0 | 918 | 20.16 | 20.0 | 1.02 |
| 6.0 | 46.5 | 1100 | 20.40 | 24.1 | 0.77 |
| 7.0 | 47.2 | 1290 | 20.70 | 28.1 | 0.47 |
| 8.0 | 47.6 | 1470 | 20.88 | 32.1 | 0.29 |
| 10.0 | 47.9 | 1840 | 21.00 | 40.0 | 0.18 |
| 12.0 | 48.3 | 2200 | 21.18 | 48.1 | 0 |

Run 1–5

$x = 0.125$ m; $T_f = 309.3$ K; $\delta^* = 0.3284$ mm; $\delta^{**} = 0.2949$ mm; $u_* = 2.232$ m/sec; $c_f = 0.00444$; $Re_{x-x_0} = 1045000$; $Re_{\delta^{**}} = 3601$

| $y$ mm | $u$ m/sec | $y^+$ | $u^+$ | $\dfrac{y \cdot 10^2}{\delta^* u_\infty^+}$ | $u_\infty^+ - u^+$ |
|---|---|---|---|---|---|
| 0.12 | 35.3 | 66.8 | 15.82 | 1.67 | 5.33 |
| 0.22 | 36.6 | 124 | 16.39 | 3.10 | 4.83 |
| 0.32 | 37.8 | 182 | 16.92 | 4.53 | 4.30 |
| 0.42 | 38.9 | 239 | 17.41 | 5.96 | 3.81 |
| 0.5 | 39.9 | 288 | 17.89 | 7.17 | 3.33 |
| 0.6 | 40.6 | 345 | 18.20 | 8.60 | 3.02 |
| 0.7 | 41.7 | 403 | 18.66 | 10.0 | 2.56 |
| 0.8 | 42.4 | 460 | 19.00 | 11.5 | 2.22 |
| 1.0 | 43.5 | 575 | 19.50 | 14.3 | 1.72 |
| 2.0 | 46.1 | 1150 | 20.65 | 28.7 | 0.57 |

**Table 1** Cylinder with $d = 15.5$ mm, $\overline{Tu}_\infty \approx 1.2\%$ (Continued)

| $y$ mm | $u$ m/sec | $y^+$ | $u^+$ | $\dfrac{y \cdot 10^2}{\delta^* u_\infty^+}$ | $u_\infty^+ - u^+$ |
|---|---|---|---|---|---|
| 3.0 | 46.8 | 1730 | 21.00 | 43.0 | 0.22 |
| 4.0 | 47.0 | 2300 | 21.07 | 57.4 | 0.15 |
| 5.0 | 47.2 | 2880 | 21.15 | 71.7 | 0.07 |
| 6.0 | 47.3 | 3450 | 21.17 | 86.1 | 0.05 |
| 7.0 | 47.4 | 4030 | 21.22 | 100.0 | 0 |

$x = 0.325$ m; $T_f = 309.3$ K; $\delta^* = 0.5857$ mm; $\delta^{**} = 0.5124$ mm; $u_* = 2.136$ m/sec; $c_f = 0.09392$; $\text{Re}_{x-x_0} = 2229000$; $\text{Re}_{\delta^{**}} = 6369$

| $y$ mm | $u$ m/sec | $y^+$ | $u^+$ | $\dfrac{y \cdot 10^2}{\delta^* u_\infty^+}$ | $u_\infty^+ - u^+$ |
|---|---|---|---|---|---|
| 0.12 | 31.4 | 63.9 | 14.71 | 0.877 | 7.88 |
| 0.22 | 35.0 | 119 | 16.38 | 1.63 | 6.21 |
| 0.32 | 36.7 | 174 | 17.17 | 2.38 | 5.42 |
| 0.42 | 37.7 | 229 | 17.65 | 3.14 | 4.94 |
| 0.5 | 38.7 | 275 | 18.11 | 3.78 | 4.48 |
| 0.6 | 39.4 | 330 | 18.45 | 4.53 | 4.14 |
| 0.7 | 39.9 | 385 | 18.69 | 5.29 | 3.90 |
| 0.8 | 40.6 | 440 | 18.99 | 6.05 | 3.60 |
| 0.9 | 41.1 | 495 | 19.24 | 6.80 | 3.35 |
| 1.0 | 41.2 | 550 | 19.29 | 7.56 | 3.30 |
| 2.0 | 44.4 | 1100 | 20.80 | 15.1 | 1.79 |
| 3.0 | 45.9 | 1650 | 21.49 | 22.7 | 1.10 |
| 4.0 | 47.0 | 2200 | 21.99 | 30.2 | 0.60 |
| 5.0 | 47.6 | 2750 | 22.27 | 37.8 | 0.32 |
| 6.0 | 48.0 | 3300 | 22.49 | 45.3 | 0.10 |
| 7.0 | 48.2 | 3850 | 22.59 | 52.9 | 0 |

$x = 0.525$ m; $T_f = 309.3$ K; $\delta^* = 0.8382$ mm; $\delta^{**} = 0.733$ mm; $u_* = 2.057$ m/sec; $c_f = 0.00364$; $\text{Re}_{x-x_0} = 3617000$; $\text{Re}_{\delta^{**}} = 9101$

| $y$ mm | $u$ m/sec | $y^+$ | $u^+$ | $\dfrac{y \cdot 10^2}{\delta^* u_\infty^+}$ | $u_\infty^+ - u^+$ |
|---|---|---|---|---|---|
| 0.12 | 28.0 | 61.5 | 13.60 | 0.591 | 9.84 |
| 0.22 | 33.8 | 114 | 16.44 | 0.988 | 6.99 |
| 0.32 | 35.5 | 167 | 17.26 | 1.61 | 6.18 |
| 0.42 | 36.7 | 220 | 17.83 | 2.10 | 5.61 |
| 0.5 | 37.7 | 265 | 18.33 | 2.54 | 5.11 |
| 0.6 | 38.2 | 318 | 18.56 | 3.05 | 4.88 |
| 0.7 | 38.6 | 371 | 18.79 | 3.56 | 4.65 |
| 0.8 | 39.3 | 424 | 19.09 | 4.07 | 4.35 |
| 0.9 | 39.6 | 477 | 19.24 | 4.58 | 4.20 |
| 1.0 | 40.2 | 530 | 19.56 | 5.09 | 3.88 |
| 2.0 | 43.3 | 1060 | 21.05 | 10.2 | 2.39 |

**Table 1** Cylinder with $d = 15.5$ mm, $\overline{Tu}_\infty \approx 1.2\%$ (*Continued*)

| $y$ mm | $u$ m/sec | $y^+$ | $u^+$ | $\dfrac{y \cdot 10^2}{\delta^* u_\infty^+}$ | $u_\infty^+ - u^+$ |
|---|---|---|---|---|---|
| 3.0 | 44.9 | 1570 | 21.83 | 15.3 | 1.61 |
| 4.0 | 45.9 | 2120 | 22.34 | 20.4 | 1.10 |
| 5.0 | 46.6 | 2650 | 22.64 | 25.4 | 0.80 |
| 6.0 | 47.2 | 3180 | 22.97 | 30.5 | 0.47 |
| 7.0 | 47.5 | 3710 | 23.11 | 35.6 | 0.33 |
| 8.0 | 47.8 | 4240 | 23.26 | 40.7 | 0.18 |
| 9.0 | 48.0 | 4770 | 23.32 | 45.8 | 0.12 |
| 10.0 | 48.1 | 5300 | 23.38 | 50.9 | 0.06 |
| 11.0 | 48.2 | 5830 | 23.44 | 56.0 | 0 |

$x = 0.825$ m; $T_f = 309.3$ K; $\delta^* = 1.0506$ mm; $\delta^{**} = 0.9272$ mm; $u_* = 2.019$ m/sec; $c_f = 0.00348$; $\text{Re}_{x-x_0} = 4938000$; $\text{Re}_{\delta^{**}} = 11540$

| 0.12 | 31.4 | 60.3 | 15.57 | 0.460 | 8.40 |
| 0.22 | 33.8 | 112 | 16.76 | 0.857 | 7.21 |
| 0.32 | 35.4 | 164 | 17.54 | 1.25 | 6.43 |
| 0.42 | 36.3 | 216 | 17.98 | 1.65 | 5.99 |
| 0.5 | 37.1 | 260 | 18.36 | 1.98 | 5.61 |
| 0.6 | 37.5 | 312 | 18.60 | 2.38 | 5.37 |
| 0.7 | 38.4 | 364 | 19.05 | 2.78 | 4.92 |
| 0.8 | 38.8 | 456 | 19.20 | 3.18 | 4.77 |
| 0.9 | 39.4 | 467 | 19.53 | 3.57 | 4.44 |
| 1.0 | 39.6 | 519 | 19.61 | 3.97 | 4.36 |
| 2.0 | 42.7 | 1040 | 21.15 | 7.94 | 2.82 |
| 3.0 | 44.1 | 1560 | 21.84 | 11.9 | 2.13 |
| 4.0 | 45.3 | 2080 | 22.47 | 15.9 | 1.50 |
| 5.0 | 46.1 | 2600 | 22.86 | 19.8 | 1.11 |
| 6.0 | 46.7 | 3120 | 23.11 | 23.8 | 0.86 |
| 7.0 | 47.2 | 3640 | 23.37 | 27.8 | 0.60 |
| 8.0 | 47.6 | 4160 | 23.60 | 31.8 | 0.37 |
| 9.0 | 47.8 | 4670 | 23.70 | 35.7 | 0.67 |
| 10.0 | 47.9 | 5190 | 23.72 | 39.7 | 0.25 |
| 11.0 | 48.2 | 5710 | 23.87 | 43.7 | 0.10 |
| 12.0 | 48.4 | 6230 | 23.97 | 47.6 | 0 |

**Table 2** Cylinder with $d = 4$ mm, $\overline{Tu}_\infty \approx 1.2\%$

| $y$ mm | $u$ m/sec | $y^+$ | $u^+$ | $\dfrac{y \cdot 10^2}{\delta^* u_\infty^+}$ | $u_\infty^+ - u^+$ |
|---|---|---|---|---|---|

Run 11–1     $x = 0.125$ m; $T_f = 299.4$ K; $\delta^* = 0.257$ mm; $\delta^{**} = 0.188$ mm;
$u_* = 2.8$ m/sec; $c_f = 0.00716$; $\text{Re}_{x-x_0} = 165600$; $\text{Re}_{\delta^{**}} = 744$

| $y$ mm | $u$ m/sec | $y^+$ | $u^+$ | $\dfrac{y \cdot 10^2}{\delta^* u_\infty^+}$ | $u_\infty^+ - u^+$ |
|---|---|---|---|---|---|
| 0.12 | 35.9 | 27.5 | 12.82 | 2.69 | 3.89 |
| 0.22 | 39.4 | 51.2 | 14.07 | 5.02 | 2.64 |
| 0.32 | 40.9 | 74.9 | 14.62 | 7.34 | 2.09 |
| 0.42 | 42.0 | 98.6 | 15.00 | 9.67 | 1.71 |
| 0.5 | 42.6 | 118 | 15.21 | 11.6 | 1.50 |
| 0.6 | 43.0 | 142 | 15.37 | 13.9 | 1.34 |
| 0.7 | 43.5 | 165 | 15.52 | 16.2 | 1.19 |
| 0.8 | 44.2 | 189 | 15.78 | 18.6 | 0.93 |
| 1.0 | 44.7 | 237 | 15.98 | 23.2 | 0.73 |
| 1.2 | 45.2 | 284 | 16.13 | 27.9 | 0.58 |
| 1.4 | 45.6 | 331 | 16.28 | 32.5 | 0.44 |
| 1.6 | 45.9 | 379 | 16.38 | 37.2 | 0.34 |
| 2.0 | 46.1 | 474 | 16.47 | 46.5 | 0.24 |
| 3.0 | 46.5 | 711 | 16.62 | 69.7 | 0.10 |
| 4.0 | 46.7 | 948 | 16.67 | 93.0 | 0.05 |
| 5.0 | 46.8 | 1185 | 16.71 | 116.2 | 0.00 |

$x = 0.325$ m; $T_f = 301$ K; $\delta^* = 0.544$ mm; $\delta^{**} = 0.455$ mm;
$u_* = 2.601$ m/sec; $c_f = 0.00590$; $\text{Re}_{x-x_0} = 490000$; $\text{Re}_{\delta^{**}} = 1827$

| $y$ mm | $u$ m/sec | $y^+$ | $u^+$ | $\dfrac{y \cdot 10^2}{\delta^* u_\infty^+}$ | $u_\infty^+ - u^+$ |
|---|---|---|---|---|---|
| 0.12 | 32.6 | 25.2 | 12.53 | 1.15 | 5.88 |
| 0.22 | 36.7 | 47.0 | 14.11 | 2.15 | 4.0 |
| 0.32 | 38.3 | 68.8 | 14.71 | 3.15 | 3.0 |
| 0.42 | 39.3 | 90.6 | 15.10 | 4.14 | 3.32 |
| 0.5 | 40.6 | 108 | 15.60 | 4.98 | 2.82 |
| 0.6 | 41.4 | 130 | 15.90 | 5.98 | 2.51 |
| 0.7 | 42.0 | 152 | 16.14 | 6.98 | 2.27 |
| 0.8 | 42.4 | 174 | 16.32 | 7.98 | 2.10 |
| 1.0 | 43.3 | 217 | 16.66 | 9.97 | 1.75 |
| 1.2 | 44.2 | 261 | 17.01 | 11.9 | 1.41 |
| 1.4 | 44.5 | 304 | 17.12 | 13.9 | 1.29 |
| 1.6 | 45.1 | 348 | 17.34 | 15.9 | 1.07 |
| 2.0 | 45.8 | 435 | 17.61 | 19.9 | 0.80 |
| 3.0 | 46.9 | 653 | 18.04 | 29.9 | 0.37 |
| 4.0 | 47.3 | 871 | 18.20 | 39.9 | 0.21 |
| 5.0 | 47.5 | 1089 | 18.25 | 49.8 | 0.16 |
| 6.0 | 47.6 | 1306 | 18.31 | 59.8 | 0.10 |
| 7.0 | 47.9 | 1524 | 18.41 | 69.8 | 0.00 |

**Table 2** Cylinder with $d = 4$ mm, $\overline{Tu_\infty} \approx 1.2\%$ *(Continued)*

| $y$ mm | $u$ m/sec | $y^+$ | $u^+$ | $\dfrac{y \cdot 10^3}{\delta^* u_\infty^+}$ | $u_\infty^+ - u^+$ |
|---|---|---|---|---|---|
| Run 11-6 | | $x = 0.125$ m; $T_f = 313.5$ K; $\delta^* = 0.192$ mm; $\delta^{**} = 0.126$ mm; $u_* = 2.011$ m/sec; $c_f = 0.00425$; $\mathrm{Re}_{x-x_0} = 1551000$; $\mathrm{Re}_{\delta^{**}} = 4203$ | | | |
| 0.12 | 34.8 | 178  | 17.31 | 2.78 | 4.39 |
| 0.22 | 38.0 | 332  | 18.87 | 5.18 | 2.82 |
| 0.32 | 39.3 | 486  | 19.54 | 7.59 | 2.15 |
| 0.42 | 40.1 | 640  | 19.92 | 9.99 | 1.77 |
| 0.5  | 40.4 | 769  | 20.09 | 12.0 | 1.61 |
| 0.6  | 41.1 | 923  | 20.41 | 14.4 | 1.28 |
| 0.7  | 41.3 | 1077 | 20.53 | 16.8 | 1.16 |
| 0.8  | 41.7 | 1230 | 20.71 | 19.2 | 0.98 |
| 0.9  | 41.9 | 1384 | 20.85 | 21.6 | 0.84 |
| 1.0  | 42.1 | 1538 | 20.94 | 24.0 | 0.75 |
| 1.2  | 42.6 | 1846 | 21.16 | 28.8 | 0.53 |
| 1.4  | 42.8 | 2154 | 21.30 | 33.6 | 0.40 |
| 1.8  | 43.3 | 2769 | 21.54 | 43.2 | 0.15 |
| 2.0  | 43.4 | 3077 | 21.56 | 48.0 | 0.13 |
| 3.0  | 43.5 | 4615 | 21.65 | 72.0 | 0.04 |
| 4.0  | 43.6 | 6154 | 21.69 | 96.1 | 0.00 |
| | | $x = 0.325$ m; $T_f = 324$ K; $\delta^* = 0.266$ mm; $\delta^{**} = 0.199$ mm; $u_* = 2.142$ m/sec; $c_f = 0.00415$; $\mathrm{Re}_{x-x_0} = 2887500$; $\mathrm{Re}_{\delta^{**}} = 7126$ | | | |
| 0.12 | 36.8 | 189  | 17.22 | 1.98  | 4.74 |
| 0.22 | 40.2 | 352  | 18.77 | 3.69  | 3.18 |
| 0.32 | 42.3 | 515  | 19.75 | 5.40  | 2.20 |
| 0.42 | 43.6 | 679  | 20.35 | 7.11  | 1.60 |
| 0.5  | 43.8 | 816  | 20.47 | 8.54  | 1.48 |
| 0.6  | 44.3 | 979  | 20.68 | 10.2  | 1.27 |
| 0.7  | 44.6 | 1142 | 20.83 | 11.9  | 1.12 |
| 0.8  | 44.8 | 1305 | 20.93 | 13.6  | 1.02 |
| 0.9  | 44.9 | 1469 | 20.97 | 15.3  | 0.98 |
| 1.0  | 45.1 | 1632 | 21.07 | 17.1  | 0.88 |
| 1.2  | 45.4 | 1958 | 21.20 | 20.5  | 0.76 |
| 1.4  | 45.6 | 2285 | 21.30 | 23.9  | 0.66 |
| 1.6  | 45.8 | 2611 | 21.40 | 27.3  | 0.55 |
| 2.0  | 46.2 | 3264 | 21.56 | 34.1  | 0.39 |
| 4.0  | 46.8 | 6529 | 21.84 | 68.3  | 0.12 |
| 5.0  | 46.9 | 8161 | 21.89 | 85.4  | 0.06 |
| 6.0  | 47.0 | 9794 | 21.95 | 102.5 | 0.00 |

## 2. EXPERIMENTAL DATA ON DISTORTION OF VELOCITY AND TEMPERATURE PROFILES IN THE BOUNDARY LAYER DUE TO PHYSICAL PROPERTIES VARIATIONS OF THE COOLANT

**Table 3** Cylinder with $d = 15.5$ mm, $\overline{Tu}_\infty \approx 1.2\%$

| $y$ mm | $T$ K | $u$ m/sec | $y^+$ | $u^+$ | $\dfrac{y \cdot 10^2}{\delta^* u_\infty^+}$ | $u_\infty^+ - u^+$ | $\vartheta^+$ |
|---|---|---|---|---|---|---|---|

Run 2–1  $\quad x = 0.526$ m; $T_w = 372.7$ K; $\delta^* = 1.101$ mm; $\delta^{**} = 0.8027$ mm; $\mathrm{Re}_{\delta^{**}} = 3097$; $c_f = 0.00413$; $u_* = 2.206$ m/sec; $q_w = 9998$ W/m$^2$; $\vartheta_* = 2.97$ K

| $y$ mm | $T$ K | $u$ m/sec | $y^+$ | $u^+$ | $\dfrac{y \cdot 10^2}{\delta^* u_\infty^+}$ | $u_\infty^+ - u^+$ | $\vartheta^+$ |
|---|---|---|---|---|---|---|---|
| 0.12 | 339.0 | 24.6 | 20.4 | 11.15 | 0.479 | 10.86 | 11.3 |
| 0.32 | 334.0 | 34.7 | 55.4 | 15.75 | 1.30 | 6.26 | 13.0 |
| 0.42 | 332.0 | 36.0 | 72.9 | 16.33 | 1.71 | 5.68 | 13.7 |
| 0.5 | 329.5 | 36.6 | 87.7 | 16.60 | 2.06 | 5.41 | 14.5 |
| 0.6 | 327.5 | 37.8 | 105 | 17.13 | 2.48 | 4.88 | 15.1 |
| 0.7 | 326.4 | 37.9 | 123 | 17.18 | 2.89 | 4.83 | 15.5 |
| 0.9 | 324.7 | 38.5 | 158 | 17.46 | 3.71 | 4.55 | 16.1 |
| 1.0 | 323.8 | 39.3 | 175 | 17.83 | 4.13 | 4.18 | 16.4 |
| 2.0 | 319.3 | 42.3 | 351 | 19.18 | 8.25 | 2.83 | 18.0 |
| 3.0 | 317.1 | 44.4 | 526 | 20.15 | 12.4 | 1.86 | 18.7 |
| 4.0 | 314.7 | 45.6 | 701 | 20.66 | 16.5 | 1.35 | 19.5 |
| 5.0 | 313.7 | 46.8 | 877 | 21.21 | 20.6 | 0.80 | 19.9 |
| 6.0 | 312.8 | 47.2 | 1050 | 21.43 | 24.8 | 0.58 | 20.2 |
| 8.0 | 312.0 | 48.1 | 1400 | 21.83 | 33.0 | 0.18 | 20.4 |
| 10.0 | 311.8 | 48.5 | 1750 | 22.01 | 41.3 | 0 | 20.5 |

Run 2–5  $\quad x = 0.526$ m; $T_w = 368.8$ K; $\delta^* = 0.9803$ mm; $\delta^{**} = 0.7342$ mm; $\mathrm{Re}_{\delta^{**}} = 8992$; $c_f = 0.00337$; $u_* = 1.968$ m/sec; $q_w = 24983$ W/m$^2$; $\vartheta_* = 2.58$ K

| $y$ mm | $T$ K | $u$ m/sec | $y^+$ | $u^+$ | $\dfrac{y \cdot 10^2}{\delta^* u_\infty^+}$ | $u_\infty^+ - u^+$ | $\vartheta^+$ |
|---|---|---|---|---|---|---|---|
| 0.216 | 330.3 | 34.2 | 108 | 17.37 | 0.90 | 6.99 | 14.9 |
| 0.32 | 327.5 | 35.7 | 159 | 18.16 | 1.32 | 6.20 | 16.0 |
| 0.42 | 325.4 | 36.7 | 209 | 18.65 | 1.74 | 5.71 | 16.8 |
| 0.5 | 324.1 | 37.7 | 251 | 19.17 | 2.09 | 5.19 | 17.3 |
| 0.6 | 323.1 | 38.3 | 302 | 19.48 | 2.51 | 4.88 | 17.7 |
| 0.8 | 322.0 | 39.4 | 402 | 20.09 | 3.35 | 4.27 | 18.1 |
| 0.9 | 321.2 | 39.5 | 452 | 20.30 | 3.77 | 4.06 | 18.4 |
| 1.0 | 320.8 | 40.0 | 503 | 21.90 | 4.19 | 2.46 | 18.6 |
| 2.0 | 316.4 | 43.1 | 1000 | 22.64 | 8.38 | 1.72 | 20.3 |
| 3.0 | 314.7 | 44.6 | 1510 | 23.19 | 12.6 | 1.17 | 21.0 |
| 4.0 | 312.8 | 45.6 | 2010 | 23.53 | 16.8 | 0.83 | 21.7 |
| 5.0 | 312.3 | 46.3 | 2510 | 23.77 | 20.9 | 0.59 | 21.9 |
| 6.0 | 312.0 | 46.8 | 3020 | 24.01 | 25.1 | 0.35 | 22.0 |
| 7.0 | 311.6 | 47.2 | 3520 | 24.12 | 29.3 | 0.24 | 22.2 |
| 8.0 | 311.1 | 47.5 | 4020 | 24.19 | 33.5 | 0.17 | 22.4 |
| 9.0 | 310.6 | 47.6 | 4520 | 24.27 | 37.7 | 0.09 | 22.6 |
| 10.0 | 310.6 | 47.8 | 5030 | 24.36 | 41.9 | 0 | 22.6 |

**Table 3** Cylinder with $d = 15.5$ mm, $\overline{Tu}_\infty \approx 1.2\%$ (*Continued*)

| $y$ mm | $T$ K | $u$ m/sec | $y^+$ | $u^+$ | $\dfrac{y \cdot 10^2}{\delta^* u_\infty^+}$ | $u_\infty^+ - u^+$ | $\vartheta^+$ |
|---|---|---|---|---|---|---|---|
| Run 5-1 | | \multicolumn{6}{l}{$x = 0.536$ m; $T_w = 994$ K; $\delta^* = 2.8$ mm; $\delta^{**} = 0.6391$ mm; $Re_{\delta^{**}} = 2489$; $c_f = 0.00318$; $u_* = 1.954$ m/sec; $q_w = 83221$ W/m²; $\vartheta_* = 28$ K} | | | | | |
| 0.22 | 535.0 | 27.8 | 33.5  | 14.22 | 0.307 | 10.87 | 16.4 |
| 0.32 | 526.0 | 32.6 | 49.0  | 16.68 | 0.449 | 8.41  | 16.7 |
| 0.42 | 516.0 | 35.7 | 64.5  | 18.28 | 0.591 | 6.81  | 17.1 |
| 0.5  | 507.0 | 36.6 | 77.6  | 18.74 | 0.712 | 6.35  | 17.4 |
| 0.7  | 486.8 | 38.5 | 109   | 19.73 | 0.996 | 5.36  | 18.1 |
| 0.8  | 476.8 | 39.7 | 124   | 20.30 | 1.14  | 4.79  | 18.5 |
| 1.0  | 455.9 | 40.2 | 155   | 20.57 | 1.42  | 4.52  | 19.2 |
| 2.0  | 411.3 | 43.0 | 311   | 22.00 | 2.85  | 3.09  | 20.8 |
| 4.0  | 362.5 | 46.9 | 621   | 24.00 | 5.69  | 1.09  | 22.5 |
| 5.0  | 348.6 | 47.3 | 777   | 24.23 | 7.12  | 0.86  | 23.0 |
| 7.0  | 332.3 | 48.3 | 1090  | 24.70 | 9.96  | 0.39  | 23.6 |
| 9.0  | 322.7 | 48.8 | 1400  | 24.99 | 12.8  | 0.10  | 24.0 |
| 10.0 | 320.0 | 48.7 | 1550  | 24.95 | 14.2  | 0.14  | 24.0 |
| 11.0 | 318.0 | 48.8 | 1710  | 25.00 | 15.6  | 0.09  | 24.1 |
| 13.0 | 315.1 | 49.0 | 2020  | 25.09 | 18.5  | 0     | 24.2 |
| Run 5-5 | | \multicolumn{6}{l}{$x = 0.536$ m; $T_w = 985.5$ K; $\delta^* = 2.122$ mm; $\delta^{**} = 0.5835$ mm; $Re_{\delta^{**}} = 7309$; $c_f = 0.0026$; $u_* = 1.784$ m/sec; $q_w = 219668$ W/m²; $\vartheta_* = 25.09$ K} | | | | | |
| 0.12 | 544.2 | 21.6 | 52.4  | 12.09 | 0.197 | 15.65 | 17.6 |
| 0.22 | 526.3 | 34.3 | 97.5  | 19.23 | 0.367 | 8.51  | 18.3 |
| 0.32 | 499.5 | 37.2 | 142   | 20.89 | 0.536 | 6.85  | 19.4 |
| 0.42 | 481.0 | 38.2 | 188   | 21.44 | 0.706 | 6.30  | 20.1 |
| 0.6  | 458.0 | 40.4 | 271   | 22.66 | 1.02  | 5.08  | 21.0 |
| 0.8  | 442.9 | 41.3 | 361   | 23.16 | 1.36  | 4.58  | 21.6 |
| 0.9  | 439.2 | 41.9 | 406   | 23.48 | 1.53  | 4.26  | 21.8 |
| 2.0  | 383.8 | 45.2 | 903   | 25.38 | 3.40  | 2.36  | 24.0 |
| 3.0  | 358.2 | 46.5 | 1350  | 26.07 | 5.10  | 1.67  | 25.0 |
| 4.0  | 345.4 | 47.8 | 1810  | 26.78 | 6.80  | 0.96  | 25.5 |
| 5.0  | 335.2 | 48.2 | 2260  | 27.02 | 8.49  | 0.72  | 25.9 |
| 7.0  | 323.6 | 48.8 | 3160  | 27.38 | 11.9  | 0.36  | 26.4 |
| 8.0  | 320.2 | 48.9 | 3610  | 27.40 | 13.6  | 0.34  | 26.5 |
| 10.0 | 317.1 | 49.1 | 4520  | 27.57 | 17.0  | 0.17  | 26.6 |
| 12.0 | 314.2 | 49.3 | 5420  | 27.66 | 20.4  | 0.08  | 26.8 |
| 15.0 | 312.9 | 49.5 | 6770  | 27.74 | 25.5  | 0     | 26.8 |

## APPENDIXES 171

**Table 4** Cylinder with $d = 15.5$ mm and Turbulence Promoters 1 and 2; $P = 0.183$ MPa, $x = 130$ mm

| $y$ mm | $T$ K | $u$ m/sec | $y^+$ | $u^+$ | $\dfrac{y \cdot 10^2}{\delta^* u_\infty^+}$ | $u_\infty^+ - u^+$ | $\vartheta^+$ |
|---|---|---|---|---|---|---|---|

### Run 4–1 — Cylinder with turbulence promoter No 1

$T_w = 361$ K; $\delta^* = 0.644$ mm; $\delta^{**} = 0.481$ mm; $Re_{\delta^{**}} = 2533$;
$c_f = 0.00488$; $u_* = 2.41$ m/sec; $q_w = 13700$ W/m²; $\vartheta_* = 2.76$ K

| $y$ mm | $T$ K | $u$ m/sec | $y^+$ | $u^+$ | $\dfrac{y \cdot 10^2}{\delta^* u_\infty^+}$ | $u_\infty^+ - u^+$ | $\vartheta^+$ |
|---|---|---|---|---|---|---|---|
| 0.32 | 319.8 | 38.1 | 83.3 | 15.77 | 2.40 | 4.47 | 14.9 |
| 0.42 | 318.4 | 39.2 | 109 | 16.21 | 3.17 | 4.03 | 15.4 |
| 0.5 | 317.2 | 39.9 | 132 | 16.49 | 3.81 | 3.75 | 15.8 |
| 0.6 | 316.6 | 40.6 | 158 | 16.78 | 4.57 | 3.46 | 16.0 |
| 0.8 | 314.7 | 41.3 | 211 | 17.08 | 6.09 | 3.16 | 16.7 |
| 1.0 | 313.5 | 42.4 | 263 | 17.53 | 7.61 | 2.71 | 17.2 |
| 1.4 | 311.8 | 43.2 | 369 | 17.86 | 10.66 | 2.39 | 17.8 |
| 1.8 | 310.8 | 44.4 | 474 | 18.38 | 13.70 | 1.87 | 18.1 |
| 3.0 | 309.6 | 45.8 | 790 | 18.92 | 22.83 | 1.33 | 18.6 |
| 5.0 | 309.0 | 48.2 | 1320 | 19.92 | 38.06 | 0.33 | 18.8 |
| 7.0 | 309.0 | 48.9 | 1841 | 20.2 | 53.28 | 0.04 | 18.8 |

$T_w = 603.7$ K; $\delta^* = 0.748$ mm; $\delta^{**} = 0.405$ mm; $Re_{\delta^{**}} = 2106$;
$c_f = 0.00447$; $u_* = 2.29$ m/sec; $q_w = 69700$ W/m²; $\vartheta_* = 14.3$ K

| $y$ mm | $T$ K | $u$ m/sec | $y^+$ | $u^+$ | $\dfrac{y \cdot 10^2}{\delta^* u_\infty^+}$ | $u_\infty^+ - u^+$ | $\vartheta^+$ |
|---|---|---|---|---|---|---|---|
| 0.32 | 354.0 | 36.9 | 77.6 | 16.13 | 2.02 | 5.02 | 16.7 |
| 0.42 | 350.0 | 37.6 | 102 | 16.44 | 2.66 | 4.71 | 17.0 |
| 0.5 | 355.5 | 38.8 | 123 | 16.97 | 3.19 | 4.18 | 17.3 |
| 0.6 | 354.0 | 39.4 | 147 | 17.25 | 3.83 | 3.90 | 17.4 |
| 0.8 | 344.5 | 40.9 | 196 | 17.89 | 5.11 | 3.26 | 18.1 |
| 1.0 | 339.6 | 42.0 | 245 | 18.38 | 6.38 | 2.77 | 18.5 |
| 1.4 | 330.4 | 43.3 | 344 | 18.93 | 8.94 | 2.22 | 19.1 |
| 1.8 | 324.4 | 44.4 | 442 | 19.41 | 11.49 | 1.74 | 19.5 |
| 3.0 | 312.5 | 46.4 | 736 | 20.29 | 19.15 | 0.86 | 20.2 |
| 5.0 | 312.4 | 49.9 | 1227 | 20.96 | 31.92 | 0.2 | 20.4 |
| 8.0 | 311.6 | 48.4 | 1964 | 21.15 | 51.07 | 0 | 20.4 |

### Run 4–2 — Cylinder with turbulence promoter No 2

$T_w = 364$ K; $\delta^* = 0.409$ mm; $\delta^{**} = 0.261$ mm; $Re_{\delta^{**}} = 1441$;
$c_f = 0.00476$; $u_* = 2.36$ m/sec; $q_w = 16100$ W/m²; $\vartheta_* = 3.25$ K

| $y$ mm | $T$ K | $u$ m/sec | $y^+$ | $u^+$ | $\dfrac{y \cdot 10^2}{\delta^* u_\infty^+}$ | $u_\infty^+ - u^+$ | $\vartheta^+$ |
|---|---|---|---|---|---|---|---|
| 0.12 | 313.0 | 31.9 | 31.3 | 13.49 | 1.41 | 7.01 | 15.7 |
| 0.22 | 312.2 | 35.9 | 58.2 | 15.16 | 2.63 | 5.34 | 16.0 |
| 0.32 | 311.4 | 37.8 | 85.2 | 16.0 | 3.85 | 4.5 | 16.2 |
| 0.42 | 310.6 | 39.0 | 112 | 16.5 | 5.07 | 4.0 | 16.5 |
| 0.5 | 309.8 | 40.4 | 135 | 17.10 | 6.10 | 3.4 | 16.7 |
| 0.7 | 308.7 | 42.2 | 189 | 17.83 | 8.54 | 2.67 | 17.1 |

**Table 4** Cylinder with $d = 15.5$ mm and Turbulence Promoters 1 and 2; $P = 0.183$ MPa, $x = 130$ mm (*Continued*)

| $y$ mm | $T$ K | $u$ m/sec | $y^+$ | $u^+$ | $\dfrac{y \cdot 10^2}{\delta^* u_\infty^+}$ | $u_\infty^+ - u^+$ | $\vartheta^+$ |
|---|---|---|---|---|---|---|---|
| 0.9 | 307.3 | 43.5 | 242 | 18.40 | 10.97 | 2.1 | 17.5 |
| 1.2 | 306.0 | 45.0 | 323 | 19.0 | 12.63 | 1.5 | 17.9 |
| 1.6 | 304.8 | 45.9 | 431 | 19.4 | 19.51 | 1.1 | 18.3 |
| 2.0 | 304.0 | 46.9 | 539 | 19.8 | 24.38 | 0.7 | 18.6 |
| 4.0 | 302.6 | 46.2 | 1078 | 20.4 | 48.77 | 0.12 | 19.0 |
| 5.0 | 302.5 | 48.4 | 1347 | 20.46 | 60.96 | 0.04 | 19.0 |
| 6.0 | 302.4 | 48.5 | 1617 | 20.50 | 73.15 | 0 | 19.1 |

$T_w = 624.2$ K; $\delta^* = 0.661$ mm; $\delta^{**} = 0.268$ mm; $\text{Re}_{\delta^{**}} = 1429$; $c_f = 0.00407$; $u_* = 2.2$ m/sec; $q_w = 74100$ W/m²; $\vartheta_* = 15.6$ K

| $y$ mm | $T$ K | $u$ m/sec | $y^+$ | $u^+$ | $\dfrac{y \cdot 10^2}{\delta^* u_\infty^+}$ | $u_\infty^+ - u^+$ | $\vartheta^+$ |
|---|---|---|---|---|---|---|---|
| 0.12 | 374.0 | 31.2 | 27.9 | 14.1 | 0.84 | 8.07 | 16.2 |
| 0.22 | 368.0 | 34.8 | 51.9 | 15.8 | 1.57 | 6.37 | 16.4 |
| 0.32 | 362.5 | 37.2 | 76 | 16.95 | 2.30 | 5.22 | 16.8 |
| 0.42 | 357 | 38.5 | 100 | 17.56 | 3.03 | 4.6 | 17.1 |
| 0.5 | 351.5 | 40.0 | 120 | 18.26 | 3.64 | 3.91 | 17.5 |
| 0.7 | 342.1 | 41.1 | 168 | 18.74 | 5.10 | 3.43 | 18.1 |
| 0.9 | 336.2 | 42.5 | 216 | 19.38 | 6.55 | 2.79 | 18.5 |
| 1.2 | 330.0 | 43.9 | 288 | 20.02 | 8.74 | 2.14 | 18.9 |
| 1.6 | 322.8 | 45.3 | 384 | 20.63 | 11.65 | 1.53 | 19.3 |
| 2.0 | 319.1 | 46.5 | 481 | 21.21 | 14.57 | 0.96 | 19.0 |
| 4.0 | 309.9 | 48.5 | 961 | 22.12 | 29.14 | 0.05 | 20.2 |
| 6.0 | 307.8 | 48.6 | 1442 | 22.18 | 43.7 | −0.01 | 20.3 |
| 7.0 | 307.5 | 48.6 | 1683 | 22.17 | 50.99 | 0 | 20.3 |

## 3. EXPERIMENTAL DATA ON FRICTION ON A CYLINDER IN AXIAL AIR FLOW

**Table 5** Cylinder with $d = 15.5$ mm, $\overline{Tu_\infty} \approx 1.2\%$

| $x$ m | $T_w$ K | $c_f \cdot 10^3$ | $\text{Re}_{x-x_0} \cdot 10^{-5}$ | $\text{Re}_\delta^{**} \cdot 10^{-3}$ | $\text{Re}_{r_0} \cdot 10^{-3}$ | $\Psi$ | $\dfrac{c_f \cdot 10^3}{\Psi^{*-0.3}\Omega^{0.18}}$ |
|---|---|---|---|---|---|---|---|
| $P = 0.135$ MPa | | | | | | | |
| 0.125 | 308.0 | 5.42 | 3.4 | 1.39 | 28.7 | 1.0 | 4.75 |
| 0.325 | 308.5 | 4.73 | 7.8 | 2.55 | 29.7 | 1.0 | 4.10 |
| 0.525 | 308.5 | 4.46 | 12.3 | 3.86 | 30.1 | 1.0 | 3.75 |
| $P = 0.167$ MPa | | | | | | | |
| 0.125 | 308.8 | 5.42 | 3.8 | 1.56 | 35.7 | 1.0 | 4.81 |
| 0.325 | 308.7 | 4.55 | 9.9 | 3.27 | 37.1 | 1.0 | 3.90 |
| 0.525 | 308.8 | 4.23 | 15.5 | 4.57 | 37.6 | 1.0 | 3.55 |
| 0.825 | 308.8 | 4.12 | 18.7 | 5.20 | 37.0 | 1.0 | 3.44 |
| $P = 0.224$ MPa | | | | | | | |
| 0.125 | 309.0 | 5.11 | 5.4 | 2.11 | 48.8 | 1.0 | 4.54 |
| 0.325 | 308.9 | 4.44 | 12.1 | 3.89 | 49.9 | 1.0 | 3.89 |
| 0.525 | 308.9 | 4.07 | 20.5 | 5.78 | 50.4 | 1.0 | 3.42 |
| 0.825 | 308.9 | 3.98 | 23.7 | 6.43 | 50.0 | 1.0 | 3.34 |
| $P = 0.309$ MPa | | | | | | | |
| 0.125 | 308.3 | 4.82 | 8.4 | 2.47 | 67.10 | 1.0 | 4.33 |
| 0.325 | 308.3 | 3.98 | 20.6 | 5.75 | 69.5 | 1.0 | 3.42 |
| 0.525 | 308.3 | 3.78 | 27.7 | 7.28 | 69.3 | 1.0 | 3.18 |
| 0.825 | 308.3 | 3.66 | 36.1 | 8.78 | 69.2 | 1.0 | 3.06 |
| $P = 0.433$ MPa | | | | | | | |
| 0.125 | 309.3 | 4.44 | 10.4 | 3.60 | 94.6 | 1.0 | 3.94 |
| 0.325 | 309.3 | 3.92 | 22.3 | 6.36 | 96.3 | 1.0 | 3.48 |
| 0.525 | 309.3 | 3.64 | 36.1 | 9.10 | 96.2 | 1.0 | 3.09 |
| 0.825 | 309.3 | 3.48 | 49.4 | 11.5 | 96.5 | 1.0 | 2.92 |
| $P = 0.899$ MPa | | | | | | | |
| 0.325 | 310.5 | 3.77 | 34.0 | 5.6 | 186.7 | 1.0 | 3.44 |
| 0.525 | 310.4 | 3.58 | 68.0 | 11.1 | 183.0 | 1.0 | 3.22 |
| 0.825 | 310.7 | 3.37 | 62.0 | 14.6 | 195.0 | 1.0 | 2.96 |

**Table 5** Cylinder with $d = 15.5$ mm, $\overline{Tu}_\infty \approx 1.2\%$ *(Continued)*

| $x$ m | $T_w$ K | $c_f \cdot 10^3$ | $\mathrm{Re}_{x-x_0} \cdot 10^{-5}$ | $\mathrm{Re}_\delta{**} \cdot 10^{-3}$ | $\mathrm{Re}_{r_0} \cdot 10^{-3}$ | $\Psi$ | $\dfrac{c_f \cdot 10^3}{\Psi^{-0.3}\Omega^{0.18}}$ |
|---|---|---|---|---|---|---|---|
| $P = 1.36$ MPa | | | | | | | |
| 0.325 | 309.2 | 3.57 | 40.0 | 9.48 | 276.6 | 1.0 | 3.06 |
| 0.525 | 309.2 | 3.42 | 56.0 | 10.7 | 271.9 | 1.0 | 3.0 |
| 0.825 | 309.4 | 3.24 | 76.0 | 14.6 | 285.6 | 1.0 | 2.84 |
| $P = 0.136$ MPa | | | | | | | |
| 0.126 | 363.5 | 5.02 | 2.82 | 1.13 | 28.7 | 1.18 | 4.8 |
| 0.326 | 370.6 | 4.38 | 8.20 | 2.56 | 29.6 | 1.2 | 4.04 |
| 0.526 | 372.7 | 4.13 | 10.5 | 3.10 | 29.7 | 1.21 | 3.77 |
| 0.826 | 374.2 | 3.86 | 12.0 | 3.29 | 29.4 | 1.21 | 3.54 |
| $P = 0.168$ MPa | | | | | | | |
| 0.126 | 364.4 | 5.02 | 3.09 | 1.27 | 36.1 | 1.18 | 4.87 |
| 0.326 | 371.7 | 4.21 | 9.91 | 2.96 | 37.1 | 1.2 | 3.89 |
| 0.526 | 373.7 | 3.92 | 13.6 | 3.75 | 37.3 | 1.21 | 3.55 |
| 0.826 | 375.2 | 3.81 | 17.8 | 4.62 | 36.8 | 1.22 | 3.38 |
| $P = 0.224$ MPa | | | | | | | |
| 0.126 | 358 | 4.73 | 4.68 | 1.79 | 48.0 | 1.16 | 4.5 |
| 0.326 | 364.5 | 4.11 | 12.5 | 3.71 | 49.4 | 1.18 | 3.77 |
| 0.526 | 366.3 | 3.77 | 21.3 | 5.49 | 49.7 | 1.18 | 3.36 |
| 0.826 | 367.6 | 3.69 | 26.2 | 6.50 | 49.3 | 1.19 | 3.26 |
| $P = 0.309$ MPa | | | | | | | |
| 0.126 | 358 | 4.46 | 6.22 | 2.25 | 66.7 | 1.16 | 4.31 |
| 0.326 | 364.5 | 3.68 | 20.8 | 5.34 | 69.0 | 1.18 | 3.31 |
| 0.526 | 366.3 | 3.5 | 35.3 | 8.65 | 68.7 | 1.19 | 3.16 |
| 0.826 | 367.6 | 3.39 | 32.0 | 7.54 | 68.4 | 1.19 | 3.03 |
| $P = 0.433$ MPa | | | | | | | |
| 0.126 | 360.5 | 4.11 | 8.9 | 2.99 | 93.3 | 1.16 | 3.82 |
| 0.326 | 367.2 | 3.63 | 26.5 | 6.84 | 95.5 | 1.19 | 3.35 |
| 0.526 | 368.8 | 3.37 | 39.1 | 8.99 | 94.6 | 1.19 | 3.08 |
| 0.826 | 370.4 | 3.22 | 53.6 | 11.63 | 95.0 | 1.2 | 2.83 |
| $P = 0.897$ MPa | | | | | | | |
| 0.326 | 377.8 | 3.71 | 18.0 | 5.54 | 187.2 | 1.22 | 3.6 |
| 0.526 | 378.8 | 3.45 | 37.5 | 8.33 | 183.5 | 1.22 | 3.28 |
| 0.826 | 378.2 | 3.34 | 47.0 | 10.22 | 193.7 | 1.22 | 3.15 |

**Table 5** Cylinder with $d = 15.5$ mm, $\overline{Tu}_\infty \approx 1.2\%$ (*Continued*)

| $x$ m | $T_w$ K | $c_f \cdot 10^3$ | $\mathrm{Re}_{x-x_0} \cdot 10^{-5}$ | $\mathrm{Re}_\delta^{**} \cdot 10^{-3}$ | $\mathrm{Re}_{r_0} \cdot 10^{-3}$ | $\Psi$ | $\dfrac{c_f \cdot 10^3}{\Psi^{*-0.3}\Omega^{0.18}}$ |
|---|---|---|---|---|---|---|---|
| $P = 1.36$ MPa | | | | | | | |
| 0.326 | 377.2 | 3.5  | 33.0 | 7.64  | 269.0 | 1.20 | 3.33 |
| 0.526 | 377.6 | 3.23 | 42.8 | 12.09 | 263.1 | 1.20 | 3.04 |
| 0.826 | 377.9 | 3.1  | 67.0 | 14.85 | 277.6 | 1.12 | 2.81 |
| $P = 0.135$ MPa | | | | | | | |
| 0.128 | 447.8 | 4.84 | 2.54 | 1.0 | 29.1 | 1.45 | 4.94 |
| 0.328 | 465.9 | 4.22 | 7.08 | 2.2 | 29.7 | 1.51 | 4.18 |
| 0.528 | 471.7 | 3.98 | 11.9 | 3.3 | 30.0 | 1.53 | 3.86 |
| 0.828 | 475.5 | 3.72 | 13.5 | 3.4 | 29.5 | 1.54 | 3.61 |
| $P = 0.167$ MPa | | | | | | | |
| 0.128 | 448.5 | 4.84 | 5.64 | 1.95 | 35.8 | 1.45 | 4.79 |
| 0.328 | 466.2 | 4.06 | 10.7 | 3.06 | 36.9 | 1.51 | 4.02 |
| 0.528 | 471.9 | 3.78 | 14.7 | 3.86 | 37.4 | 1.53 | 3.67 |
| 0.828 | 474.8 | 3.68 | 21.0 | 5.07 | 37.4 | 1.54 | 3.48 |
| $P = 0.225$ MPa | | | | | | | |
| 0.128 | 449.4 | 4.92 | 4.02 | 1.6  | 49.5 | 1.45 | 4.96 |
| 0.328 | 468.1 | 3.96 | 15.2 | 4.22 | 51.0 | 1.51 | 3.81 |
| 0.528 | 473.8 | 3.63 | 22.3 | 5.55 | 50.5 | 1.53 | 3.46 |
| 0.828 | 476.6 | 3.55 | 26.4 | 6.31 | 50.4 | 1.54 | 3.33 |
| $P = 0.309$ MPa | | | | | | | |
| 0.328 | 464.1 | 3.55 | 17.4 | 4.43 | 67.7 | 1.50 | 3.54 |
| 0.528 | 469.2 | 3.37 | 39.6 | 8.83 | 69.2 | 1.52 | 3.2  |
| 0.828 | 472.1 | 3.37 | 33.5 | 7.46 | 69.3 | 1.53 | 3.18 |
| $P = 0.434$ MPa | | | | | | | |
| 0.128 | 445.6 | 3.96 | 9.26 | 2.88 | 92.5 | 1.44 | 3.96 |
| 0.328 | 462.4 | 3.5  | 28.8 | 7.09 | 95.1 | 1.49 | 3.34 |
| 0.528 | 467.0 | 3.25 | 45.0 | 9.72 | 96.1 | 1.51 | 3.08 |
| 0.828 | 469.9 | 3.11 | 48.3 | 10.2 | 96.4 | 1.52 | 2.94 |
| $P = 0.897$ MPa | | | | | | | |
| 0.328 | 476.0 | 3.5  | 21.9 | 5.17 | 182.7 | 1.54 | 3.58 |
| 0.828 | 480.7 | 3.19 | 49.0 | 8.4  | 189.2 | 1.55 | 3.16 |

**Table 5** Cylinder with $d = 15.5$ mm, $\overline{Tu}_\infty \approx 1.2\%$ (*Continued*)

| $x$ m | $T_w$ K | $c_f \cdot 10^3$ | $Re_{x-x_0} \cdot 10^{-5}$ | $Re_\delta^{**} \cdot 10^{-3}$ | $Re_{r_0} \cdot 10^{-3}$ | $\Psi$ | $\dfrac{c_f \cdot 10^3}{\Psi^{-0.3}\Omega^{0.13}}$ |
|---|---|---|---|---|---|---|---|
| $P = 1.35$ MPa | | | | | | | |
| 0.328 | 474.5 | 3.25 | 34.1 | 7.48 | 264.3 | 1.52 | 3.31 |
| 0.528 | 478.2 | 3.0  | 46.1 | 12.4 | 258.9 | 1.53 | 2.99 |
| 0.828 | 480.3 | 2.95 | 69.0 | 11.7 | 271.6 | 1.53 | 2.91 |
| $P = 0.134$ MPa | | | | | | | |
| 0.131 | 602.4 | 4.34 | 2.6  | 0.93 | 28.3 | 1.95 | 4.91 |
| 0.331 | 648.2 | 3.73 | 10.7 | 2.81 | 29.4 | 2.1  | 4.19 |
| 0.531 | 663.1 | 3.57 | 11.8 | 2.94 | 29.9 | 2.15 | 3.93 |
| 0.831 | 675.0 | 3.34 | 16.6 | 3.78 | 29.7 | 2.19 | 3.65 |
| $P = 0.167$ MPa | | | | | | | |
| 0.131 | 608.7 | 4.34 | 3.1   | 1.12 | 35.7 | 1.97 | 4.84 |
| 0.331 | 655.2 | 3.64 | 11.4  | 2.98 | 36.8 | 2.12 | 3.9  |
| 0.531 | 669.6 | 3.38 | 16.71 | 3.83 | 37.6 | 2.17 | 3.57 |
| 0.831 | 681.7 | 3.3  | 19.0  | 4.28 | 37.3 | 2.21 | 3.48 |
| $P = 0.224$ MPa | | | | | | | |
| 0.131 | 599.0 | 4.41 | 4.8  | 1.66 | 48.6 | 1.94 | 4.83 |
| 0.331 | 641.2 | 3.55 | 15.9 | 3.95 | 49.8 | 2.08 | 3.78 |
| 0.531 | 661.2 | 3.26 | 19.6 | 4.33 | 50.4 | 2.14 | 3.47 |
| 0.831 | 672.7 | 3.18 | 23.1 | 4.99 | 49.4 | 2.18 | 3.36 |
| $P = 0.308$ MPa | | | | | | | |
| 0.131 | 599.7 | 3.86 | 5.8  | 1.86 | 66.4 | 1.84 | 4.29 |
| 0.331 | 644.6 | 3.18 | 21.9 | 4.88 | 68.6 | 2.09 | 3.39 |
| 0.531 | 659.7 | 3.02 | 26.0 | 5.41 | 69.1 | 2.14 | 3.21 |
| 0.831 | 671.0 | 2.93 | 31.5 | 6.46 | 67.7 | 2.82 | 3.04 |
| $P = 0.434$ MPa | | | | | | | |
| 0.131 | 598.2 | 3.55 | 8.6  | 2.54 | 93.9 | 1.94 | 3.88 |
| 0.331 | 641.6 | 3.14 | 28.7 | 6.28 | 96.0 | 2.08 | 3.42 |
| 0.531 | 655.5 | 2.91 | 37.7 | 7.72 | 94.7 | 2.13 | 3.15 |
| 0.831 | 666.4 | 2.78 | 47.3 | 5.37 | 96.0 | 2.16 | 2.89 |
| $P = 0.898$ MPa | | | | | | | |
| 0.331 | 659.1 | 3.15 | 16.9 | 5.24 | 180.5 | 2.12 | 3.59 |
| 0.531 | 661.1 | 2.88 | 35.3 | 8.34 | 188.5 | 2.12 | 3.16 |

**Table 5** Cylinder with $d = 15.5$ mm, $\overline{Tu}_\infty \approx 1.2\%$ (*Continued*)

| $\dfrac{x}{m}$ | $\dfrac{T_w}{K}$ | $c_f \cdot 10^3$ | $\text{Re}_{x-x_0} \cdot 10^{-5}$ | $\text{Re}_\delta^{**} \cdot 10^{-3}$ | $\text{Re}_{r_0} \cdot 10^{-3}$ | $\Psi^*$ | $\dfrac{c_f \cdot 10^3}{\Psi^{*-0.3}\Omega^{0.18}}$ |
|---|---|---|---|---|---|---|---|
| *P* = 1.36 MPa | | | | | | | |
| 0.331 | 658,3 | 3.06 | 43.5 | 6.09 | 265.1 | 2.11 | 3.26 |
| 0.531 | 667,6 | 2.69 | 72.1 | 12.0 | 261.0 | 2.13 | 2.96 |
| *P* = 0.136 MPa | | | | | | | |
| 0.136 | 929 | 3,85 | 3.65 | 1,08 | 28,9 | 3.0 | 4.68 |
| 0.336 | 982 | 3.39 | 8.95 | 2.20 | 29.6 | 3.17 | 4.1 |
| 0.536 | 994 | 3.18 | 11.3 | 2.49 | 29.9 | 3.21 | 3,76 |
| 0.836 | 1007 | 2.98 | 15,1 | 3.11 | 30.1 | 3.26 | 3.46 |
| *P* = 0.168 MPa | | | | | | | |
| 0.136 | 921 | 3.87 | 2.93 | 0.96 | 36.0 | 2.98 | 4,76 |
| 0.336 | 979 | 3,25 | 11.1 | 2,55 | 37,1 | 3.17 | 3.89 |
| 0.536 | 993 | 3.02 | 14.1 | 2,95 | 37.4 | 3,21 | 3.56 |
| 0.836 | 1009 | 2,94 | 19.7 | 3.81 | 37.4 | 3,26 | 3.4 |
| *P* = 0.226 MPa | | | | | | | |
| 0.136 | 891 | 3,94 | 4,4 | 1,39 | 48,6 | 2.88 | 4.8 |
| 0.336 | 962 | 3,17 | 14.5 | 3.26 | 50.0 | 3.11 | 3.74 |
| 0.536 | 980 | 2,91 | 19.3 | 3.80 | 50.4 | 3.17 | 3.39 |
| 0.836 | 998 | 2,84 | 19.0 | 3.83 | 49,8 | 3,22 | 3.31 |
| *P* = 0.311 MPa | | | | | | | |
| 0,136 | 892 | 3.44 | 5.7 | 1.25 | 66,4 | 2,88 | 4.24 |
| 0.336 | 966 | 2.84 | 20.1 | 4.06 | 68.5 | 3,12 | 3.42 |
| 0.536 | 990 | 2,7 | 24.8 | 4.60 | 68.8 | 3,2 | 3.18 |
| 0,836 | 1006 | 2.61 | 32.7 | 5.91 | 68.6 | 3,25 | 3.03 |
| *P* = 0.435 MPa | | | | | | | |
| 0.136 | 892 | 3.17 | 6.9 | 1,97 | 92.5 | 2,88 | 3,94 |
| 0.336 | 966 | 2.8 | 28.0 | 5.4 | 95.2 | 3,12 | 3.37 |
| 0.536 | 986 | 2.6 | 42.5 | 7.3 | 97.1 | 3.19 | 3,01 |
| 0.836 | 1002 | 2.48 | 38,7 | 6.74 | 95,0 | 3.24 | 2.89 |
| *P* = 0.897 MPa | | | | | | | |
| 0.336 | 970 | 2.8 | 22.5 | 4.93 | 174.7 | 3.05 | 3.62 |
| 0.536 | 986 | 2.32 | 76.0 | 11.4 | 171,4 | 3.08 | 2.89 |
| *P* = 1.36 MPa | | | | | | | |
| 0.336 | 975 | 2.75 | 27.0 | 5.55 | 256.2 | 3.04 | 3.45 |
| 0.536 | 991 | 2.53 | 51.0 | 9.27 | 251.8 | 3,09 | 3.12 |

**178** HEAT TRANSFER OF GAS COOLANTS

**Table 6** Cylinder with $d = 15.5$ mm, $\overline{Tu}_\infty \approx 1.2\%$, $\psi = 1$

| $P$ MPa | $x$ m | $T_f$ K | $c_f \cdot 10^3$ | $\text{Re}_{x-x_0} \cdot 10^{-5}$ | $\text{Re}_\delta^{**} \cdot 10^{-3}$ | $\text{Re}_{r_0} \cdot 10^{-3}$ |
|---|---|---|---|---|---|---|
| 0.135 | 0.125 | 299.0 | 7.16 | 1.66 | 0.74 | 30.7 |
|       | 0.325 | 300.8 | 5.90 | 4.9  | 1.83 |      |
| 0.167 | 0.125 | 301.3 | 6.66 | 2.11 | 0.88 | 37.7 |
|       | 0.325 | 301.0 | 5.49 | 6.09 | 2.11 |      |
| 0.224 | 0.125 | 301.6 | 6.03 | 2.96 | 1.12 | 60.8 |
|       | 0.325 | 302.0 | 4.95 | 9.85 | 3.05 |      |
| 0.309 | 0.125 | 303.6 | 5.29 | 5.75 | 1.91 | 98.2 |
|       | 0.325 | 303.8 | 4.62 | 14.5 | 4.16 |      |
| 0.433 | 0.125 | 307.8 | 4.95 | 8.98 | 2.72 | 156.0 |
|       | 0.325 | 307.9 | 4.34 | 22.7 | 5.88 |      |
| 0.899 | 0.125 | 313.0 | 4.25 | 15.5 | 4.20 | 259.0 |
|       | 0.325 | 313.8 | 4.15 | 28.9 | 7.13 |      |

**Table 7** Cylinder with $d = 15.5$ mm, $\overline{Tu}_\infty \approx 5\%$, $\psi = 1$

| $P$ MPa | $x$ m | $T_f$ K | $c_f \cdot 10^3$ | $\text{Re}_{x-x_0} \cdot 10^{-5}$ | $\text{Re}_\delta^{**} \cdot 10^{-3}$ | $\text{Re}_{r_0} \cdot 10^{-3}$ |
|---|---|---|---|---|---|---|
| 0.135 | 0.325 | 309.2 | 7.03 | 1.96 | 1.03 | 27.8 |
|       | 0.825 | 308.7 | 6.70 | 2.06 | 2.48 | 29.1 |
| 0.224 | 0.325 | 309.2 | 6.26 | 4.49 | 1.80 | 46.9 |
|       | 0.825 | 309.4 | 5.65 | 5.09 | 3.23 | 47.0 |
| 0.433 | 0.325 | 309.6 | 5.48 | 9.87 | 3.87 | 89.4 |
|       | 0.825 | 309.2 | 4.84 | 11.2 | 6.03 | 90.7 |

## 4. EXPERIMENTAL DATA ON LOCAL HEAT TRANSFER FROM A CYLINDER WITH $d = 15.5$ mm AND BLUNT LEADING EDGE

**Table 8**

| $L \cdot 10^3$ m | $u_\infty$ m/sec | $\text{Re}_x \cdot 10^{-5}$ | $T_w$ K | $a$ W/m²·K | $\text{Nu}_x$ | $\text{Nu}_x \Psi^{0.25}$ | $\dfrac{\text{Nu}_x \Psi^{0.25}}{\Omega^{0.14}}$ |
|---|---|---|---|---|---|---|---|
| Run 1 | | $P_0 = 0.14$ MPa, $T_f = 308$ K | | | | | |
| 20.0  | 49.0 | 0.74  | 341.8 | 284 | 188  | 193  | 188  |
| 40.5  | 49.1 | 1.59  | 348.8 | 247 | 350  | 361  | 353  |
| 68.0  | 49.3 | 2.73  | 353.0 | 222 | 539  | 558  | 540  |
| 100.5 | 49.6 | 4.10  | 356.9 | 205 | 745  | 772  | 740  |
| 150.5 | 49.9 | 6.19  | 359.3 | 195 | 1047 | 1088 | 1028 |
| 251.0 | 50.6 | 10.57 | 362.3 | 184 | 1687 | 1755 | 1623 |
| 350.0 | 51.2 | 14.93 | 364.3 | 181 | 2316 | 2413 | 2192 |
| 499.5 | 52.2 | 21.76 | 365.1 | 178 | 3262 | 3402 | 3027 |
| 650.5 | 53.0 | 28.81 | 366.8 | 176 | 4199 | 4384 | 3835 |
| 749.5 | 53.6 | 33.53 | 367.8 | 173 | 4763 | 4977 | 4313 |
| 859.0 | 54.0 | 38.70 | 368.0 | 168 | 5310 | 5549 | 4763 |
| Run 2 | | $P_0 = 0.181$ MPa, $T_f = 308.3$ K | | | | | |
| 20.0  | 49.3 | 0.96  | 342.0 | 362 | 240  | 246  | 243  |
| 40.5  | 49.8 | 2.08  | 348.9 | 304 | 431  | 444  | 430  |
| 100.5 | 50.7 | 5.41  | 357.7 | 253 | 919  | 953  | 915  |
| 251.0 | 52.3 | 14.02 | 363.6 | 229 | 2103 | 2189 | 2030 |
| 499.5 | 53.1 | 28.61 | 366.5 | 219 | 4006 | 4182 | 3737 |
| 859.0 | 54.8 | 50.76 | 371.2 | 215 | 6785 | 7097 | 6128 |
| Run 3 | | $P_0 = 0.264$ MPa, $T_f = 308.4$ K | | | | | |
| 20.0  | 48.1 | 1.37  | 342.9 | 482 | 320  | 328  | 324  |
| 40.5  | 48.4 | 2.94  | 349.1 | 410 | 580  | 597  | 585  |
| 68.0  | 48.4 | 5.04  | 352.9 | 375 | 911  | 941  | 914  |
| 100.5 | 48.6 | 7.57  | 357.0 | 358 | 1298 | 1346 | 1295 |
| 150.0 | 48.6 | 11.35 | 360.4 | 334 | 1818 | 1887 | 1794 |
| 251.0 | 49.5 | 19.46 | 363.2 | 314 | 2875 | 2991 | 2785 |
| 350.0 | 50.1 | 27.55 | 365.2 | 309 | 3962 | 4130 | 3782 |
| 499.5 | 50.9 | 39.99 | 365.9 | 305 | 5586 | 5829 | 5232 |
| 650.5 | 51.7 | 52.86 | 367.9 | 295 | 7043 | 7360 | 6513 |
| 749.5 | 52.1 | 61.48 | 367.2 | 298 | 8218 | 8580 | 7520 |
| 859.0 | 52.6 | 71.06 | 369.2 | 290 | 9130 | 9550 | 8297 |
| Run 4 | | $P_0 = 0.389$ MPa, $T_f = 308$ K | | | | | |
| 20.0  | 48.7 | 2.04  | 343.3 | 643 | 426  | 438  | 433  |
| 40.5  | 48.9 | 4.39  | 349.0 | 559 | 793  | 817  | 802  |
| 100.5 | 49.3 | 11.35 | 357.1 | 494 | 1792 | 1857 | 1791 |

**180** HEAT TRANSFER OF GAS COOLANTS

**Table 8** (*Continued*)

| $L \cdot 10^3$ m | $u_\infty$ m/sec | $Re_x \cdot 10^{-5}$ | $T_w$ K | $a$ W/m²·K | $Nu_x$ | $Nu_x \Psi^{0.25}$ | $\dfrac{Nu_x \Psi^{0.25}}{\Omega^{0.14}}$ |
|---|---|---|---|---|---|---|---|
| 251.0 | 50.2 | 29.14 | 362.5 | 439 | 4029 | 4192 | 3921 |
| 499.5 | 51.5 | 59.70 | 364.8 | 428 | 7842 | 8179 | 7415 |
| 859.0 | 53.1 | 105.93 | 368.5 | 412 | 13025 | 13612 | 11890 |
| Run 5 | | | $P_0 = 0.557$ MPa, $T_f = 308$ K | | | | |
| 20.0 | 49.3 | 2.96 | 344.3 | 841 | 558 | 572 | 566 |
| 40.5 | 49.5 | 6.37 | 349.6 | 755 | 1071 | 1103 | 1084 |
| 68.0 | 49.7 | 10.95 | 353.6 | 717 | 1744 | 1805 | 1744 |
| 100.5 | 50.0 | 16.44 | 357.6 | 659 | 2392 | 2478 | 2394 |
| 150.0 | 50.4 | 24.85 | 360.9 | 618 | 3369 | 3504 | 3350 |
| 251.0 | 50.9 | 42.26 | 363.3 | 592 | 5430 | 5653 | 5313 |
| 350.0 | 51.5 | 59.66 | 365.0 | 579 | 7423 | 7735 | 7195 |
| 499.5 | 52.3 | 86.64 | 364.7 | 579 | 10621 | 11078 | 10062 |
| 650.5 | 52.9 | 114.21 | 366.4 | 568 | 13572 | 14169 | 12685 |
| 749.5 | 53.2 | 132.66 | 367.4 | 555 | 15294 | 15982 | 14181 |
| 859.0 | 53.5 | 152.93 | 368.4 | 559 | 17664 | 18476 | 16250 |
| Run 6 | | | $P_0 = 0.766$ MPa, $T_f = 308$ K | | | | |
| 20.0 | 49.2 | 4.07 | 344.2 | 1127 | 747 | 767 | 759 |
| 40.5 | 49.3 | 8.72 | 349.9 | 987 | 1400 | 1443 | 1418 |
| 100.5 | 49.7 | 22.48 | 357.9 | 849 | 3075 | 3190 | 3088 |
| 251.0 | 50.8 | 58.06 | 362.5 | 778 | 7139 | 7492 | 7004 |
| 499.5 | 52.0 | 118.74 | 364.5 | 751 | 13764 | 14356 | 13090 |
| 859.0 | 53.0 | 207.8 | 368.7 | 710 | 22424 | 23430 | 20710 |
| Run 7 | | | $P_0 = 1.015$ MPa, $T_f = 308$ K | | | | |
| 20.0 | 48.0 | 5.24 | 345.3 | 1408 | 933 | 958 | 948 |
| 40.5 | 48.2 | 11.25 | 350.4 | 1236 | 1752 | 1807 | 1776 |
| 68.0 | 48.4 | 19.35 | 353.7 | 1161 | 2821 | 2920 | 2851 |
| 100.5 | 48.6 | 29.04 | 358.5 | 1081 | 3920 | 4069 | 3946 |
| 150.0 | 49.0 | 43.92 | 361.0 | 1029 | 5607 | 5831 | 5601 |
| 251.0 | 49.6 | 74.84 | 362.6 | 998 | 9153 | 9524 | 8997 |
| 350.0 | 50.1 | 105.87 | 363.5 | 978 | 12546 | 13048 | 12160 |
| 499.5 | 50.8 | 153.43 | 364.3 | 964 | 17683 | 18436 | 16880 |
| 650.5 | 51.5 | 202.2 | 366.4 | 933 | 22278 | 23258 | 20970 |
| 749.5 | 51.8 | 234.5 | 368.0 | 908 | 24997 | 26122 | 23380 |
| 859.0 | 52.1 | 279.2 | 369.5 | 903 | 28480 | 29790 | 26450 |
| Run 8 | | | $P_0 = 1.341$ MPa, $T_f = 308$ K | | | | |
| 20.0 | 48.2 | 6.94 | 344.6 | 1794 | 1189 | 1221 | 1210 |
| 40.5 | 48.3 | 14.88 | 349.9 | 1558 | 2209 | 2278 | 2244 |
| 100.5 | 48.8 | 38.48 | 357.4 | 1361 | 4937 | 5118 | 4974 |

## Table 8 (Continued)

| $L \cdot 10^3$ m | $u_\infty$ m/sec | $Re_x \cdot 10^{-5}$ | $T_w$ K | $a$ W/m²·K | $Nu_x$ | $Nu_x \Psi'^{0.25}$ | $\dfrac{Nu_x \Psi'^{0.25}}{\Omega^{0.14}}$ |
|---|---|---|---|---|---|---|---|
| 251.0 | 49.6 | 98.88 | 362.0 | 1259 | 11554 | 12016 | 11379 |
| 499.5 | 50.8 | 201.8 | 363.6 | 1238 | 22690 | 23640 | 21728 |
| 859.0 | 51.8 | 370.9 | 368.1 | 1145 | 36156 | 37780 | 33702 |
| Run 9 | | | $P_0 = 1.753$ MPa, $T_f = 308$ K | | | | |
| 20.0 | 46.4 | 8.69 | 344.9 | 2168 | 1437 | 1476 | 1465 |
| 40.5 | 46.6 | 18.67 | 349.7 | 1919 | 2721 | 2803 | 2761 |
| 68.0 | 46.8 | 32.16 | 352.7 | 1817 | 4417 | 4567 | 4473 |
| 100.5 | 47.0 | 48.19 | 357.9 | 1630 | 3913 | 6132 | 5964 |
| 150.0 | 47.5 | 73.19 | 359.1 | 1625 | 8858 | 9195 | 8866 |
| 251.0 | 47.9 | 124.28 | 360.9 | 1567 | 14380 | 14941 | 14170 |
| 350.0 | 48.4 | 175.49 | 362.1 | 1530 | 19620 | 20405 | 19110 |
| 499.5 | 48.9 | 253.5 | 362.7 | 1513 | 27740 | 28877 | 26610 |
| 650.5 | 49.5 | 333.9 | 364.4 | 1470 | 35110 | 36585 | 33250 |
| 749.5 | 49.7 | 386.9 | 365.8 | 1441 | 39660 | 41365 | 37260 |
| 859.0 | 50.1 | 448.7 | 366.9 | 1440 | 45390 | 47387 | 42000 |
| Run 10 | | | $P_0 = 2.332$ MPa, $T_f = 308$ K | | | | |
| 20.0 | 45.4 | 11.30 | 344.4 | 2726 | 1807 | 1855 | 1840 |
| 40.5 | 45.5 | 24.19 | 349.3 | 2425 | 3438 | 3545 | 3496 |
| 100.5 | 45.9 | 62.46 | 356.6 | 2104 | 7633 | 7912 | 7704 |
| 251.0 | 46.7 | 160.67 | 360.4 | 1972 | 18086 | 18806 | 17893 |
| 499.5 | 47.3 | 325.4 | 362.0 | 1895 | 34737 | 36150 | 33441 |
| 859.0 | 48.7 | 575.8 | 366.8 | 1744 | 55046 | 57500 | 51616 |

## Table 9

| $L \cdot 10^3$ m | $u_\infty$ m/sec | $Re_x \cdot 10^{-5}$ | $T_w$, K | $a$ W/m²·K | $Nu_x$ | $Nu_x \Psi'^{0.25}$ | $\dfrac{Nu_x \Psi'^{0.25}}{\Omega^{0.14}}$ |
|---|---|---|---|---|---|---|---|
| Run 11 | | | $P_0 = 0.132$ MPa, $T_f = 308$ K | | | | |
| 20.0 | 48.6 | 0.70 | 394.0 | 263 | 175 | 186 | 181 |
| 40.6 | 48.9 | 1.50 | 411.7 | 229 | 326 | 350 | 342 |
| 68.2 | 49.3 | 2.60 | 424.5 | 208 | 508 | 550 | 532 |
| 100.7 | 49.6 | 3.90 | 437.0 | 188 | 684 | 746 | 715 |
| 150.4 | 50.1 | 5.92 | 446.8 | 176 | 960 | 1054 | 996 |
| 251.7 | 51.1 | 10.13 | 456.5 | 168 | 1549 | 1710 | 1580 |
| 350.9 | 51.8 | 14.40 | 462.5 | 164 | 2107 | 2332 | 2118 |
| 500.8 | 52.7 | 20.86 | 466.5 | 160 | 2938 | 3259 | 2899 |
| 652.3 | 53.3 | 27.55 | 471.0 | 158 | 3778 | 4201 | 3675 |
| 751.7 | 54.0 | 32.03 | 473.8 | 158 | 4347 | 4838 | 4192 |
| 861.5 | 54.5 | 37.03 | 477.0 | 156 | 4946 | 5515 | 4734 |

## Table 9 (Continued)

| $L \cdot 10^3$ m | $u_\infty$ m/sec | $Re_x \cdot 10^{-5}$ | $T_w$ K | $a$ W/m²·K | $Nu_x$ | $Nu_x \Psi'^{0.25}$ | $\dfrac{Nu_x \Psi'^{0.25}}{\Omega^{0.14}}$ |
|---|---|---|---|---|---|---|---|
| **Run 12** | | | $P_0 = 0.182$ MPa, $T_f = 308$ K | | | | |
| 20.0  | 49.0 | 0.96  | 391.6 | 351 | 233  | 250  | 247  |
| 40.6  | 49.4 | 2.07  | 410.9 | 294 | 418  | 450  | 440  |
| 100.7 | 50.1 | 5.41  | 434.4 | 245 | 893  | 974  | 935  |
| 251.7 | 51.6 | 14.08 | 453.6 | 220 | 2022 | 2226 | 2067 |
| 500.8 | 52.9 | 29.83 | 463.9 | 206 | 3795 | 4205 | 3764 |
| 861.5 | 54.5 | 51.12 | 475.7 | 194 | 6154 | 6852 | 5912 |
| **Run 13** | | | $P_0 = 0.265$ MPa, $T_f = 308.4$ K | | | | |
| 20.0  | 48.2 | 1.38  | 394.7 | 471 | 313  | 333  | 329  |
| 40.6  | 48.4 | 2.96  | 412.2 | 398 | 565  | 608  | 595  |
| 68.2  | 48.7 | 5.11  | 424.3 | 358 | 872  | 944  | 916  |
| 100.7 | 49.0 | 7.67  | 435.2 | 336 | 1221 | 1332 | 1286 |
| 150.4 | 49.4 | 11.64 | 445.2 | 314 | 1716 | 1879 | 1786 |
| 251.8 | 50.2 | 19.86 | 454.5 | 299 | 2745 | 3025 | 2819 |
| 350.9 | 50.8 | 28.21 | 459.8 | 287 | 3700 | 4088 | 3747 |
| 500.8 | 51.7 | 40.92 | 463.0 | 281 | 5162 | 5709 | 5129 |
| 652.3 | 52.4 | 54.04 | 467.2 | 274 | 6551 | 7265 | 6429 |
| 751.7 | 52.9 | 62.78 | 470.2 | 269 | 7417 | 8240 | 7222 |
| 861.5 | 53.3 | 72.50 | 473.5 | 266 | 8418 | 9369 | 8133 |
| **Run 14** | | | $P_0 = 0.391$ MPa, $T_f = 308$ K | | | | |
| 20.0  | 48.8 | 2.04   | 396.6 | 638 | 423   | 451   | 446   |
| 40.6  | 49.9 | 4.38   | 413.1 | 548 | 778   | 837   | 821   |
| 100.7 | 49.4 | 11.32  | 436.4 | 464 | 1689  | 1842  | 1777  |
| 251.7 | 50.5 | 29.24  | 454.5 | 412 | 3793  | 4180  | 3910  |
| 500.8 | 51.8 | 60.02  | 463.6 | 393 | 7211  | 7983  | 7211  |
| 861.5 | 53.4 | 106.75 | 473.4 | 376 | 11893 | 13237 | 11570 |
| **Run 15** | | | $P_0 = 0.555$ MPa, $T_f = 308$ K | | | | |
| 20.0  | 49.4 | 2.96   | 387.2 | 936 | 622   | 659   | 652   |
| 40.6  | 49.6 | 6.37   | 414.5 | 709 | 1008  | 1087  | 1068  |
| 68.2  | 49.9 | 10.98  | 425.8 | 662 | 1614  | 1751  | 1706  |
| 100.7 | 50.2 | 16.47  | 437.0 | 618 | 2248  | 2454  | 2373  |
| 150.4 | 50.6 | 24.95  | 445.3 | 583 | 3188  | 3494  | 3343  |
| 251.7 | 51.3 | 42.57  | 454.7 | 548 | 5044  | 5558  | 5218  |
| 350.9 | 52.0 | 60.39  | 459.2 | 538 | 6937  | 7665  | 7097  |
| 500.8 | 52.8 | 87.48  | 462.1 | 527 | 9700  | 10728 | 9740  |
| 652.3 | 53.5 | 115.42 | 465.8 | 516 | 12354 | 13701 | 12260 |
| 751.7 | 53.9 | 133.93 | 470.3 | 502 | 13867 | 15406 | 13660 |
| 861.5 | 54.3 | 154.52 | 473.4 | 499 | 15792 | 17576 | 15450 |

## Table 9 (Continued)

| $L \cdot 10^3$ m | $u_\infty$ m/sec | $Re_x \cdot 10^{-5}$ | $T_w$ K | $a$ W/m²·K | $Nu_x$ | $Nu_x\Psi^{0.25}$ | $\dfrac{Nu_x\Psi^{0.25}}{\Omega^{0.14}}$ |
|---|---|---|---|---|---|---|---|
| Run 16 | | | $P_0 = 0.770$ MPa, $T_f = 308$ K | | | | |
| 20.0  | 48.2 | 3.98   | 401.0 | 1071 | 718   | 768   | 763   |
| 40.6  | 48.4 | 8.54   | 416.6 | 933  | 1327  | 1433  | 1408  |
| 100.7 | 49.3 | 22.06  | 438.4 | 808  | 2939  | 3213  | 3113  |
| 251.7 | 49.9 | 56.98  | 454.2 | 729  | 6709  | 7393  | 6961  |
| 500.8 | 51.1 | 116.9  | 461.1 | 705  | 12952 | 14312 | 13046 |
| 861.5 | 52.4 | 206.1  | 472.6 | 656  | 20782 | 23130 | 20450 |
| Run 17 | | | $P_0 = 0.014$ MPa, $T_f = 308$ K | | | | |
| 20.0  | 48.1 | 5.25   | 401.8 | 1345 | 897   | 960   | 951   |
| 40.6  | 48.2 | 11.27  | 416.9 | 1181 | 1678  | 1812  | 1783  |
| 68.2  | 48.4 | 19.40  | 426.3 | 1098 | 2675  | 2899  | 2833  |
| 100.7 | 48.6 | 29.09  | 438.9 | 1016 | 3693  | 4036  | 3918  |
| 150.4 | 48.8 | 43.90  | 446.9 | 967  | 5290  | 5808  | 5584  |
| 251.7 | 49.5 | 74.97  | 454.9 | 917  | 8437  | 9298  | 8788  |
| 351.9 | 50.0 | 106.11 | 459.2 | 897  | 11563 | 12777 | 11900 |
| 500.8 | 50.7 | 153.48 | 461.9 | 880  | 16191 | 17907 | 16390 |
| 652.3 | 51.4 | 202.4  | 467.2 | 851  | 20408 | 22653 | 20440 |
| 751.7 | 51.7 | 234.4  | 472.4 | 833  | 22985 | 25582 | 22880 |
| 861.5 | 52.1 | 270.6  | 474.9 | 827  | 26181 | 29170 | 25890 |
| Run 18 | | | $P_0 = 1.348$ MPa, $T_f = 308$ K | | | | |
| 20.0  | 47.9 | 6.93   | 401.4 | 1677 | 1114  | 1192  | 1182  |
| 40.6  | 48.1 | 14.9   | 416.5 | 1472 | 2094  | 2261  | 2227  |
| 100.7 | 48.5 | 38.68  | 437.7 | 1276 | 4647  | 5074  | 4931  |
| 251.7 | 49.3 | 99.56  | 453.6 | 1162 | 10701 | 11791 | 11160 |
| 500.8 | 50.5 | 203.8  | 461.0 | 1105 | 20336 | 22470 | 20630 |
| 861.5 | 51.7 | 359.7  | 473.8 | 1070 | 33892 | 37760 | 33650 |
| Run 19 | | | $P_0 = 1.759$ MPa, $T_f = 308.4$ K | | | | |
| 20.0  | 46.8 | 8.82   | 402.4 | 2053 | 1362  | 1457  | 1445  |
| 40.6  | 46.9 | 18.96  | 416.4 | 1829 | 2596  | 2801  | 2760  |
| 68.2  | 47.2 | 32.70  | 425.4 | 1708 | 4147  | 4495  | 4403  |
| 100.7 | 47.4 | 48.99  | 438.2 | 1561 | 5673  | 6198  | 6029  |
| 150.4 | 47.8 | 74.22  | 445.2 | 1496 | 8170  | 8954  | 8634  |
| 251.7 | 48.3 | 126.25 | 452.8 | 1433 | 13174 | 14505 | 13762 |
| 350.9 | 48.8 | 178.83 | 455.4 | 1406 | 18115 | 19963 | 18709 |
| 500.8 | 49.4 | 258.0  | 458.7 | 1375 | 25260 | 27887 | 25726 |
| 652.3 | 49.9 | 339.6  | 464.2 | 1326 | 31760 | 35190 | 31991 |
| 861.5 | 50.5 | 453.6  | 473.2 | 1306 | 41350 | 46020 | 41199 |

## Table 9 (*Continued*)

| $L \cdot 10^3$ m | $u_\infty$ m/sec | $Re_x \cdot 10^{-5}$ | $T_w$ K | $a$ W/m²·K | $Nu_x$ | $Nu_x \Psi^{0.25}$ | $\dfrac{Nu_x \Psi^{0.25}}{\Omega^{0.14}}$ |
|---|---|---|---|---|---|---|---|
| 20.0  | 45.0 | 11.11  | 402.4 | 2565 | 1702  | 1821  | 1806  |
| 40.6  | 45.2 | 23.86  | 416.7 | 2253 | 3199  | 3449  | 3401  |
| 100.7 | 45.6 | 61.73  | 437.3 | 1950 | 7085  | 7730  | 7527  |
| 251.7 | 46.5 | 158.94 | 451.2 | 1805 | 16589 | 18248 | 17338 |
| 500.8 | 47.5 | 324.7  | 458.4 | 1718 | 31522 | 34800 | 32192 |
| 861.5 | 48.6 | 572.1  | 472.2 | 1641 | 51880 | 57690 | 57186 |

## Table 10

| $L \cdot 10^3$ m | $u_\infty$, m/sec | $Re_x \cdot 10^{-5}$ | $T_w$, K | $a$ W/m²·K | $Nu_x$ | $Nu_x \Psi^{0.25}$ | $\dfrac{Nu_x \Psi^{0.25}}{\Omega^{0.14}}$ |
|---|---|---|---|---|---|---|---|
| **Run 21** | | $P_0 = 0.133$ MPa, $T_f = 313$ K | | | | | |
| 20.0  | 48.3 | 0.67  | 495.6 | 247 | 161  | 181  | 175  |
| 40.6  | 48.4 | 1.45  | 535.8 | 211 | 295  | 338  | 330  |
| 68.2  | 48.6 | 2.49  | 566.4 | 189 | 454  | 527  | 510  |
| 100.9 | 49.0 | 3.75  | 590.9 | 175 | 630  | 739  | 708  |
| 150.9 | 49.4 | 5.68  | 615.5 | 160 | 865  | 1024 | 968  |
| 252.0 | 49.9 | 9.64  | 641.0 | 151 | 1371 | 1641 | 1488 |
| 351.9 | 51.3 | 13.87 | 654.2 | 149 | 1890 | 2272 | 2059 |
| 502.0 | 51.6 | 19.94 | 666.5 | 143 | 2601 | 3142 | 2793 |
| 654.5 | 52.5 | 26.44 | 674.0 | 140 | 3312 | 4007 | 3499 |
| 754.0 | 53.0 | 30.73 | 681.0 | 137 | 3737 | 4533 | 3921 |
| 864.0 | 53.9 | 35.77 | 689.2 | 134 | 4180 | 5092 | 4363 |
| **Run 22** | | $P_0 = 0.182$ MPa, $T_f = 313$ K | | | | | |
| 20.0  | 47.5 | 0.91  | 490.7 | 329 | 215  | 241  | 238  |
| 40.6  | 47.6 | 1.95  | 532.5 | 274 | 385  | 440  | 430  |
| 100.9 | 48.1 | 5.04  | 588.0 | 234 | 839  | 983  | 943  |
| 252.3 | 49.1 | 13.05 | 637.7 | 202 | 1830 | 2187 | 2026 |
| 502.4 | 50.8 | 26.86 | 663.5 | 188 | 3411 | 4094 | 3655 |
| 864.5 | 52.7 | 47.86 | 692.8 | 175 | 5475 | 6680 | 5758 |
| **Run 23** | | $P_0 = 0.266$ MPa, $T_f = 314$ K | | | | | |
| 20.0  | 48.2 | 1.35  | 492.0 | 440 | 287  | 321  | 317  |
| 40.6  | 48.3 | 2.88  | 531.9 | 375 | 524  | 598  | 586  |
| 68.2  | 48.6 | 4.96  | 560.3 | 340 | 814  | 940  | 913  |
| 100.9 | 48.8 | 7.43  | 587.9 | 315 | 1128 | 1320 | 1269 |
| 150.7 | 49.2 | 11.27 | 611.7 | 295 | 1585 | 1872 | 1779 |
| 252.0 | 49.9 | 19.18 | 638.1 | 273 | 2472 | 2952 | 2703 |
| 351.9 | 50.5 | 27.18 | 652.0 | 264 | 3348 | 4018 | 3679 |
| 502.0 | 51.4 | 39.49 | 662.1 | 257 | 4660 | 5615 | 5040 |
| 654.5 | 52.2 | 52.22 | 674.1 | 251 | 5921 | 7158 | 6329 |
| 754.0 | 52.6 | 60.72 | 681.9 | 247 | 6703 | 8131 | 7126 |
| 864.0 | 53.2 | 70.26 | 690.4 | 241 | 7498 | 9129 | 7924 |

**Table 10** (*Continued*)

| $L \cdot 10^3$ m | $u_\infty$ m/sec | $\mathrm{Re}_x \cdot 10^{-5}$ | $T_w$ K | $a$ W/m²·K | $\mathrm{Nu}_x$ | $\mathrm{Nu}_x \Psi'^{0.25}$ | $\dfrac{\mathrm{Nu}_x \Psi'^{0.25}}{\Omega^{0.14}}$ |
|---|---|---|---|---|---|---|---|
| Run 24 | | $P_0 = 0.390$ MPa, $T_f = 314.6$ K | | | | | |
| 20.0 | 49.2 | 2.01 | 494.3 | 611 | 398 | 446 | 441 |
| 40.6 | 49.3 | 4.31 | 531.4 | 523 | 731 | 834 | 818 |
| 100.9 | 50.7 | 11.36 | 588.7 | 427 | 1531 | 1793 | 1730 |
| 252.3 | 50.9 | 28.76 | 638.2 | 381 | 3447 | 4120 | 3854 |
| 502.4 | 52.3 | 59.10 | 659.0 | 356 | 6458 | 7774 | 7022 |
| 864.5 | 54.0 | 105.1 | 685.2 | 339 | 10596 | 12874 | 11250 |
| Run 25 | | $P_0 = 0.558$ MPa, $T_f = 313.6$ K | | | | | |
| 20.0 | 49.0 | 2.86 | 506.7 | 765 | 499 | 562 | 556 |
| 40.6 | 49.1 | 6.15 | 542.5 | 671 | 939 | 1077 | 1058 |
| 68.2 | 49.4 | 10.6 | 569.0 | 608 | 1459 | 1694 | 1651 |
| 100.9 | 49.6 | 15.91 | 599.6 | 569 | 2038 | 2395 | 2314 |
| 150.7 | 50.0 | 24.09 | 622.7 | 538 | 2901 | 3443 | 3291 |
| 252.3 | 50.7 | 41.02 | 648.1 | 499 | 4523 | 5424 | 5093 |
| 351.9 | 51.3 | 58.08 | 659.5 | 486 | 6157 | 7413 | 6864 |
| 502.0 | 52.1 | 84.28 | 669.0 | 474 | 8580 | 10369 | 9417 |
| 654.5 | 52.6 | 111.18 | 680.3 | 458 | 10837 | 13156 | 11767 |
| 754.0 | 53.1 | 129.24 | 689.9 | 446 | 12163 | 14815 | 13134 |
| 864.0 | 53.6 | 149.28 | 699.9 | 440 | 13751 | 16804 | 14779 |
| Run 26 | | $P_0 = 0.764$ MPa, $T_f = 313.5$ K | | | | | |
| 20.0 | 48.7 | 3.89 | 505.0 | 1005 | 656 | 739 | 732 |
| 40.6 | 49.1 | 8.39 | 538.1 | 889 | 1244 | 1424 | 1400 |
| 100.9 | 49.3 | 21.62 | 593.6 | 760 | 2726 | 3201 | 3101 |
| 252.3 | 50.4 | 55.80 | 640.6 | 658 | 5964 | 7130 | 6713 |
| 502.5 | 51.8 | 114.53 | 661.6 | 625 | 11335 | 13659 | 12450 |
| 864.5 | 53.1 | 202.3 | 690.7 | 593 | 18508 | 22550 | 19920 |
| Run 27 | | $P_0 = 1.015$ MPa, $T_f = 313.7$ K | | | | | |
| 20.0 | 48.3 | 5.11 | 505.7 | 1264 | 824 | 928 | 921 |
| 40.6 | 48.4 | 10.99 | 550.6 | 1056 | 1477 | 1701 | 1674 |
| 68.2 | 48.7 | 18.93 | 564.4 | 1032 | 2475 | 2868 | 2800 |
| 100.9 | 49.0 | 28.44 | 593.3 | 941 | 3371 | 3958 | 3838 |
| 150.7 | 49.1 | 42.92 | 616.4 | 881 | 4745 | 5629 | 5407 |

## Table 10 (Continued)

| $L \cdot 10^3$ m | $u_\infty$ m/sec | $Re_x \cdot 10^{-5}$ | $T_w$ K | $a$ W/m²·K | $Nu_x$ | $Nu_x \Psi'^{0.25}$ | $\dfrac{Nu_x \Psi'^{0.25}}{\Omega^{0.14}}$ |
|---|---|---|---|---|---|---|---|
| 252.0 | 49.9 | 73.23  | 640.2 | 825 | 7475  | 8934  | 8428  |
| 351.9 | 50.5 | 103.53 | 651.2 | 807 | 10233 | 12280 | 11440 |
| 502.0 | 51.3 | 150.29 | 661.0 | 784 | 14205 | 17103 | 15640 |
| 654.5 | 51.8 | 198.31 | 671.5 | 760 | 17973 | 21729 | 19590 |
| 754.0 | 52.2 | 230.0  | 681.8 | 739 | 20125 | 24432 | 21850 |
| 864.0 | 52.6 | 265.1  | 692.3 | 728 | 22710 | 27695 | 24570 |

Run 28    $P_0 = 1.347$ MPa, $T_f = 313.6$ K

| $L \cdot 10^3$ m | $u_\infty$ m/sec | $Re_x \cdot 10^{-5}$ | $T_w$ K | $a$ W/m²·K | $Nu_x$ | $Nu_x \Psi'^{0.25}$ | $\dfrac{Nu_x \Psi'^{0.25}}{\Omega^{0.14}}$ |
|---|---|---|---|---|---|---|---|
| 20.0  | 47.4 | 6.65   | 507.8 | 1559 | 1017  | 1147  | 1137  |
| 40.6  | 47.6 | 14.30  | 540.6 | 1373 | 1921  | 2202  | 2169  |
| 100.9 | 48.0 | 36.98  | 593.6 | 1175 | 4212  | 4945  | 4800  |
| 252.3 | 49.9 | 95.30  | 638.8 | 1036 | 9388  | 11219 | 10620 |
| 502.4 | 50.2 | 195.17 | 657.9 | 987  | 17889 | 21530 | 19770 |
| 864.0 | 51.5 | 344.5  | 689.2 | 933  | 29123 | 35470 | 31580 |

Run 29    $P_0 = 1.759$ MPa, $T_f = 313$ K

| $L \cdot 10^3$ m | $u_\infty$ m/sec | $Re_x \cdot 10^{-5}$ | $T_w$ K | $a$ W/m²·K | $Nu_x$ | $Nu_x \Psi'^{0.25}$ | $\dfrac{Nu_x \Psi'^{0.25}}{\Omega^{0.14}}$ |
|---|---|---|---|---|---|---|---|
| 20.0  | 46.8 | 8.56   | 505.7 | 1913 | 1248  | 1406  | 1396  |
| 40.6  | 46.9 | 18.40  | 538.9 | 1691 | 2367  | 2710  | 2669  |
| 68.2  | 47.2 | 31.74  | 560.7 | 1573 | 3764  | 4355  | 4265  |
| 100.9 | 47.3 | 47.59  | 593.7 | 1442 | 5171  | 6071  | 5905  |
| 150.7 | 47.6 | 71.94  | 612.9 | 1349 | 7273  | 8604  | 8297  |
| 252.0 | 48.2 | 122.48 | 635.5 | 1277 | 11575 | 13815 | 13110 |
| 351.9 | 48.8 | 173.79 | 646.4 | 1257 | 15947 | 19105 | 17830 |
| 502.0 | 49.4 | 260.9  | 654.3 | 1228 | 22262 | 26759 | 24660 |
| 654.5 | 49.9 | 330.5  | 665.3 | 1189 | 28120 | 33941 | 30790 |
| 754.0 | 50.2 | 383.5  | 677.9 | 1147 | 31263 | 37922 | 34100 |
| 864.0 | 50.5 | 441.9  | 687.0 | 1119 | 34948 | 42514 | 37680 |

Run 30    $P_0 = 2.318$ MPa, $T_f = 313.7$ K

| $L \cdot 10^3$ m | $u_\infty$ m/sec | $Re_x \cdot 10^{-5}$ | $T_w$ K | $a$ W/m²·K | $Nu_x$ | $Nu_x \Psi'^{0.25}$ | $\dfrac{Nu_x \Psi'^{0.25}}{\Omega^{0.14}}$ |
|---|---|---|---|---|---|---|---|
| 20.0  | 45.4 | 10.87  | 506.1 | 2388 | 1557  | 1754  | 1740  |
| 40.6  | 45.6 | 23.38  | 537.5 | 2093 | 2926  | 3347  | 3300  |
| 100.9 | 46.0 | 60.55  | 590.9 | 1786 | 6401  | 7508  | 7311  |
| 252.0 | 46.9 | 155.9  | 631.8 | 1586 | 14365 | 16979 | 16140 |
| 502.0 | 48.0 | 318.7  | 652.7 | 1516 | 27454 | 32940 | 30444 |
| 864.0 | 49.1 | 561.9  | 686.8 | 1451 | 45307 | 55040 | 49363 |

## Table 11

| $L \cdot 10^3$ m | $u_\infty$ m/sec | $Re_x \cdot 10^{-5}$ | $T_w$ K | $a$, W/m²·K | $Nu_x$ | $Nu_x \Psi^{0.25}$ | $\dfrac{Nu_x \Psi^{0.25}}{\Omega^{0.14}}$ |
|---|---|---|---|---|---|---|---|
| **Run 31** | | $P_0 = 0.131$ MPa, $T_f = 313\text{--}315.5$ K | | | | | |
| 40.7 | 48.1 | 1.43 | 774.0 | 177 | 248 | 312 | 305 |
| 68.5 | 48.4 | 2.47 | 824.4 | 167 | 403 | 514 | 497 |
| 101.3 | 48.8 | 3.72 | 868.2 | 164 | 591 | 763 | 731 |
| 151.3 | 49.5 | 5.68 | 906.1 | 154 | 833 | 1086 | 1026 |
| 253.5 | 50.7 | 9.75 | 940.1 | 140 | 1276 | 1679 | 1552 |
| 353.8 | 51.6 | 13.87 | 954.9 | 136 | 1734 | 2291 | 2079 |
| 505.2 | 52.7 | 20.18 | 967.9 | 132 | 2404 | 3188 | 2831 |
| 658.2 | 53.7 | 26.67 | 977.2 | 128 | 3036 | 4029 | 3512 |
| 758.5 | 54.1 | 31.21 | 983.0 | 126 | 3430 | 4558 | 3943 |
| 869.4 | 54.5 | 35.67 | 988.3 | 123 | 3865 | 5144 | 4408 |
| **Run 32** | | $P_0 = 0.189$ MPa, $T_f = 313.4\text{--}315$ K | | | | | |
| 40.7 | 47.9 | 2.04 | 753.1 | 254 | 357 | 445 | 436 |
| 101.3 | 48.5 | 5.29 | 860.3 | 227 | 799 | 1027 | 992 |
| 253.5 | 49.4 | 13.70 | 939.1 | 184 | 1674 | 2203 | 2043 |
| 505.2 | 50.4 | 27.98 | 967.6 | 176 | 3198 | 4237 | 3786 |
| 869.4 | 53.8 | 51.92 | 993.9 | 170 | 5334 | 7116 | 6142 |
| **Run 33** | | $P_0 = 0.270$ MPa, $T_f = 313\text{--}314$ K | | | | | |
| 20.1 | 48.2 | 1.37 | 653.3 | 381 | 250 | 301 | 297 |
| 40.7 | 48.4 | 2.95 | 742.7 | 326 | 458 | 569 | 558 |
| 68.5 | 48.6 | 5.09 | 805.6 | 311 | 751 | 907 | 880 |
| 101.3 | 49.0 | 7.66 | 854.3 | 297 | 1070 | 1375 | 1322 |
| 151.3 | 49.4 | 11.62 | 896.3 | 273 | 1479 | 1923 | 1828 |
| 253.5 | 50.3 | 19.89 | 934.6 | 259 | 2364 | 3109 | 2895 |
| 353.8 | 50.9 | 28.12 | 954.3 | 250 | 3192 | 4217 | 3862 |
| 505.2 | 51.9 | 40.96 | 968.9 | 231 | 4212 | 5585 | 5017 |
| 658.2 | 52.8 | 54.29 | 980.6 | 226 | 5378 | 7153 | 6330 |
| 758.5 | 53.4 | 63.14 | 991.7 | 238 | 6524 | 8703 | 7627 |
| 869.4 | 54.1 | 73.27 | 998.2 | 236 | 7672 | 10250 | 8990 |
| **Run 34** | | $P_0 = 0.390$ MPa, $T_f = 313\text{--}314$ K | | | | | |
| 20.1 | 48.2 | 1.97 | 652.7 | 507 | 333 | 400 | 391 |
| 40.7 | 48.5 | 4.24 | 738.2 | 443 | 621 | 771 | 757 |
| 101.3 | 48.9 | 11.05 | 851.3 | 399 | 1435 | 1840 | 1776 |
| 253.5 | 49.9 | 28.38 | 938.8 | 345 | 3142 | 4136 | 3869 |
| 505.2 | 50.8 | 57.84 | 967.3 | 328 | 5987 | 7939 | 7171 |
| 869.4 | 53.9 | 105.9 | 1003.1 | 303 | 9515 | 12720 | 11100 |

## Table 11  (Continued)

| $L \cdot 10^3$ m | $u_\infty$ m/sec | $Re_x \cdot 10^{-5}$ | $T_w$ K | $a$ W/m²·K | $Nu_x$ | $Nu_x \Psi^{0.25}$ | $\dfrac{Nu_x \Psi^{0.25}}{\Omega^{0.14}}$ |
|---|---|---|---|---|---|---|---|
| **Run 35** | | | $P_0 = 0.548$ MPa, $T_f = 313$–$315.5$ K | | | | |
| 20.1  | 48.0 | 2.78   | 655.0  | 657  | 431   | 519   | 508   |
| 40.7  | 48.2 | 5.96   | 738.8  | 608  | 854   | 1059  | 1041  |
| 68.5  | 48.4 | 10.29  | 801.0  | 557  | 1343  | 1697  | 1653  |
| 101.3 | 48.7 | 15.45  | 851.3  | 517  | 1862  | 2393  | 2309  |
| 151.3 | 48.9 | 23.3   | 893.5  | 486  | 2635  | 3423  | 3272  |
| 253.5 | 49.8 | 40.0   | 935.5  | 457  | 4176  | 5493  | 5157  |
| 353.8 | 50.4 | 56.63  | 954.4  | 449  | 5730  | 7569  | 7005  |
| 505.2 | 51.6 | 82.86  | 968.2  | 435  | 7955  | 10556 | 9580  |
| 658.2 | 52.2 | 108.95 | 985.7  | 421  | 10026 | 13345 | 11930 |
| 758.5 | 52.7 | 126.76 | 996.2  | 413  | 11331 | 15138 | 13420 |
| 869.4 | 53.2 | 146.78 | 1008.3 | 404  | 12708 | 17029 | 15017 |
| **Run 36** | | | $P_0 = 0.749$ MPa, $T_f = 313$–$314$ K | | | | |
| 20.1  | 48.3 | 3.79   | 654.3  | 835  | 548   | 659   | 652   |
| 40.7  | 48.5 | 8.14   | 733.4  | 769  | 1080  | 1336  | 1314  |
| 101.3 | 49.1 | 21.14  | 845.8  | 651  | 2344  | 3003  | 2907  |
| 253.5 | 50.1 | 54.62  | 929.1  | 576  | 5250  | 6893  | 6490  |
| 505.2 | 50.9 | 110.99 | 961.0  | 549  | 10010 | 13253 | 12081 |
| 869.4 | 53.7 | 201.8  | 1001.9 | 507  | 15906 | 21260 | 18780 |
| **Run 37** | | | $P_0 = 1.002$ MPa, $T_f = 313$–$313.5$ K | | | | |
| 20.1  | 48.0 | 5.05   | 657.1  | 1038 | 682   | 822   | 814   |
| 40.7  | 48.1 | 10.84  | 735.8  | 939  | 1319  | 1634  | 1606  |
| 68.5  | 48.3 | 18.70  | 796.2  | 853  | 2060  | 2600  | 2544  |
| 101.3 | 48.6 | 28.05  | 845.6  | 803  | 2895  | 3711  | 3599  |
| 151.3 | 48.9 | 42.47  | 889.0  | 747  | 4046  | 5248  | 5041  |
| 253.5 | 49.6 | 72.50  | 932.6  | 693  | 6324  | 8316  | 7801  |
| 353.8 | 50.2 | 102.61 | 951.1  | 701  | 8946  | 11809 | 11000 |
| 505.2 | 51.0 | 149.17 | 967.9  | 693  | 12661 | 16801 | 15370 |
| 658.2 | 51.6 | 196.72 | 981.7  | 670  | 15958 | 21240 | 19156 |
| 758.5 | 52.2 | 228.7  | 996.5  | 651  | 17858 | 23858 | 21330 |
| 869.4 | 52.7 | 264.7  | 1009.5 | 635  | 19976 | 26808 | 23786 |

## Table 11  (Continued)

| $L \cdot 10^3$ m | $u_\infty$ m/sec | $Re_x \cdot 10^{-5}$ | $T_w$ K | $a$ W/m²·K | $Nu_x$ | $Nu_x \Psi'^{0,25}$ | $\dfrac{N_{1x} \Psi'^{0,25}}{\Omega^{0,14}}$ |
|---|---|---|---|---|---|---|---|
| Run 38 | | $P_0 = 1.327$ MPa, $T_f = 313$–$314$ K | | | | | |
| 20.1  | 47.3 | 6.57   | 655.6  | 1300 | 854   | 1027  | 1018  |
| 40.7  | 47.5 | 14.08  | 731.3  | 1165 | 1635  | 2021  | 1991  |
| 101.3 | 48.2 | 36.61  | 844.8  | 1012 | 3642  | 4665  | 4529  |
| 253.5 | 49.0 | 94.31  | 928.0  | 890  | 8115  | 10655 | 10080 |
| 505.2 | 49.8 | 191.64 | 961.3  | 870  | 15860 | 21010 | 19290 |
| 869.4 | 52.3 | 345.8  | 1007.8 | 793  | 24910 | 33330 | 29680 |
| Run 39 | | $P_0 = 1.715$ MPa, $T_f = 313$–$314$ K | | | | | |
| 40.7  | 46.1 | 17.75  | 728.9  | 1429 | 2007  | 2481  | 2445  |
| 68.5  | 46.3 | 30.62  | 787.2  | 1308 | 3158  | 3976  | 3894  |
| 101.3 | 46.6 | 45.94  | 838.6  | 1263 | 4552  | 5824  | 5665  |
| 151.3 | 47.1 | 69.90  | 879.6  | 1199 | 6499  | 8410  | 8109  |
| 253.5 | 47.5 | 118.52 | 921.3  | 1123 | 10253 | 13431 | 12780 |
| 353.8 | 48.0 | 167.70 | 941.5  | 1107 | 14135 | 18609 | 17420 |
| 505.2 | 48.8 | 243.5  | 958.0  | 1066 | 19473 | 25743 | 23670 |
| 658.2 | 49.5 | 321.4  | 973.3  | 1018 | 24224 | 32145 | 29190 |
| 758.5 | 50.0 | 372.8  | 990.7  | 1021 | 27955 | 37264 | 33570 |
| 869.4 | 50.5 | 431.9  | 1005.4 | 1058 | 33213 | 44439 | 39730 |
| Run 40 | | $P_0 = 2.318$ MPa, $T_f = 314$ K | | | | | |
| 20.1  | 44.7 | 10.72  | 655.9  | 1814 | 1189  | 1429  | 1417  |
| 40.7  | 44.8 | 22.98  | 728.6  | 1807 | 2532  | 3127  | 3084  |
| 101.3 | 45.4 | 59.77  | 840.2  | 1607 | 5777  | 7392  | 7198  |
| 253.5 | 46.2 | 153.97 | 917.7  | 1443 | 13144 | 17192 | 16340 |
| 505.2 | 46.9 | 312.3  | 954.3  | 1367 | 24910 | 32910 | 30420 |
| 869.4 | 49.2 | 563.3  | 1005.0 | 1311 | 41160 | 54900 | 49240 |

## Table 12

| $L \cdot 10^3$ m | $u_\infty$ m/sec | $Re_x \cdot 10^{-5}$ | $T_w$ K | $a$ W/m² · K | $Nu_x$ | $Nu_x \Psi^{0.25}$ | $\dfrac{Nu_x \Psi^{0.25}}{\Omega^{0.14}}$ |
|---|---|---|---|---|---|---|---|
| **Run 41** | | | $P_0 = 0.132$ MPa, $T_f = 314-322$ K | | | | |
| 41.0  | 50.9 | 1.52  | 1146.2 | 171 | 242  | 335  | 328  |
| 102.1 | 51.8 | 3.94  | 1223.3 | 143 | 516  | 723  | 692  |
| 254.9 | 53.7 | 10.14 | 1249.9 | 133 | 1207 | 1805 | 1671 |
| 508.2 | 55.7 | 20.68 | 1265.2 | 128 | 2310 | 3259 | 2899 |
| 874.9 | 58.0 | 37.02 | 1275.2 | 125 | 3866 | 5460 | 4683 |
| **Run 42** | | | $P_0 = 0.198$ MPa, $T_f = 313.5-318$ K | | | | |
| 41.0  | 49.9 | 2.24  | 1122.6 | 248 | 351  | 482  | 472  |
| 102.1 | 50.8 | 5.83  | 1214.2 | 207 | 747  | 1046 | 1004 |
| 254.9 | 52.4 | 15.02 | 1249.7 | 191 | 1735 | 2450 | 2274 |
| 508.2 | 54.4 | 30.71 | 1270.8 | 183 | 3305 | 4670 | 4180 |
| 874.9 | 56.4 | 55.27 | 1291.5 | 174 | 5436 | 7710 | 6663 |
| **Run 43** | | | $P_0 = 0.288$ MPa, $T_f = 314-317$ K | | | | |
| 41.0  | 48.4 | 3.16  | 1076.6 | 329 | 466  | 635  | 623  |
| 102.1 | 49.0 | 8.27  | 1197   | 279 | 1010 | 1411 | 1359 |
| 254.0 | 50.4 | 21.09 | 1244.7 | 249 | 2273 | 3203 | 2985 |
| 508.2 | 52.3 | 43.37 | 1267.3 | 245 | 4458 | 6310 | 5674 |
| 874.9 | 54.3 | 77.71 | 1294.9 | 232 | 7274 | 10350 | 9000 |
| **Run 44** | | | $P_0 = 0.407$ MPa, $T_f = 314-317$ K | | | | |
| 41.0  | 47.7 | 4.39   | 1042.6 | 446 | 631  | 852   | 836   |
| 102.1 | 48.0 | 11.31  | 1173.6 | 372 | 1344 | 1817  | 1754  |
| 254.9 | 49.6 | 29.26  | 1240.6 | 331 | 3018 | 4250  | 3976  |
| 508.2 | 51.5 | 60.52  | 1264.5 | 333 | 6050 | 8570  | 7742  |
| 874.9 | 53.4 | 107.8  | 1307.3 | 312 | 9803 | 14020 | 12265 |
| **Run 45** | | | $P_0 = 0.565$ MPa, $T_f = 313-315.6$ K | | | | |
| 41.0  | 48.1 | 6.18   | 1024.7 | 521 | 819   | 1104  | 1084  |
| 102.1 | 48.4 | 15.86  | 1161.1 | 460 | 1666  | 2314  | 2235  |
| 254.9 | 49.6 | 40.09  | 1199.3 | 439 | 4017  | 5620  | 5277  |
| 508.2 | 51.4 | 84.39  | 1261.8 | 396 | 7233  | 10250 | 9309  |
| 874.9 | 53.2 | 150.26 | 1360.2 | 380 | 11965 | 17190 | 15118 |

# 5. EXPERIMENTAL DATA ON LOCAL HEAT TRANSFER FROM A CYLINDER ($d = 4.02$ mm) WITH SIGNIFICANT SURFACE CURVATURE

**Table 13**

| $L \cdot 10^3$ m | $u_\infty$ m/sec | $Re_x \cdot 10^{-5}$ | $T_w$ K | $a$ W/m²·K | $Nu_x$ | $Nu_x \Psi^{0.25}$ | $\dfrac{Nu_x \Psi^{0.25}}{\Omega^{0.14}}$ |
|---|---|---|---|---|---|---|---|
| Run 1 | | $P_0 = 0.141$ MPa, $T_f = 304.5$ K | | | | | |
| 32.0 | 48.0 | 1.24 | 380.3 | 158 | 177 | 187 | |
| 65.0 | 49.7 | 2.70 | 373.3 | 181 | 424 | 446 | |
| 95.0 | 48.7 | 3.91 | 367.3 | 192 | 665 | 697 | |
| 130.0 | 49.0 | 5.42 | 366.3 | 197 | 940 | 984 | |
| 180.0 | 49.5 | 7.60 | 366.4 | 200 | 1328 | 1390 | |
| 250.0 | 50.0 | 10.70 | 367.2 | 197 | 1815 | 1902 | |
| 350.0 | 50.6 | 15.20 | 367.8 | 193 | 2506 | 2626 | |
| 500.0 | 51.3 | 22.06 | 368.0 | 196 | 3642 | 3817 | |
| 680.0 | 52.2 | 30.52 | 368.7 | 196 | 4948 | 5190 | 3925 |
| 860.0 | 53.0 | 39.22 | 369.0 | 195 | 6230 | 6536 | 4833 |
| Run 2 | | $P_0 = 0.305$ MPa, $T_f = 303.3$ K | | | | | |
| 14.0 | 49.1 | 1.11 | 361.1 | 425 | 190 | 198 | |
| 22.0 | 49.1 | 1.85 | 363.1 | 389 | 291 | 304 | |
| 32.0 | 49.2 | 2.77 | 363.3 | 392 | 440 | 460 | |
| 65.0 | 49.5 | 5.86 | 363.5 | 386 | 909 | 951 | |
| 95.0 | 49.8 | 8.70 | 364.1 | 382 | 1329 | 1390 | |
| 130.0 | 50.1 | 12.04 | 365.6 | 381 | 1823 | 1909 | |
| 180.0 | 50.4 | 16.87 | 367.2 | 370 | 2464 | 2585 | |
| 250.0 | 50.9 | 23.72 | 367.7 | 369 | 3420 | 3588 | 3028 |
| 350.0 | 51.5 | 33.67 | 369.7 | 360 | 4682 | 4921 | 4027 |
| 500.0 | 52.2 | 48.84 | 369.5 | 361 | 6717 | 7060 | 5603 |
| 680.0 | 52.9 | 67.35 | 370.1 | 356 | 9027 | 9487 | 7314 |
| 860.0 | 53.5 | 86.28 | 370.0 | 362 | 11614 | 12206 | 9198 |
| Run 3 | | $P_0 = 0.60$ MPa, $T_f = 306.7$ K | | | | | |
| 9.0 | 47.6 | 1.21 | 349.3 | 986 | 255 | 263 | |
| 14.0 | 47.7 | 2.07 | 352.7 | 880 | 390 | 404 | |
| 22.0 | 47.8 | 3.46 | 355.4 | 804 | 595 | 617 | |
| 32.0 | 47.9 | 5.20 | 357.4 | 780 | 866 | 899 | |
| 65.0 | 48.3 | 11.01 | 359.2 | 766 | 1785 | 1856 | |
| 95.0 | 48.5 | 16.34 | 363.8 | 713 | 2451 | 2556 | |

**192** HEAT TRANSFER OF GAS COOLANTS

## Table 13  (*Continued*)

| $L \cdot 10^3$ m | $u_\infty$ m/sec | $Re_x \cdot 10^{-5}$ | $T_w$ K | $a$ W/m²·K | $Nu_x$ | $Nu_x \Psi'^{0.25}$ | $\dfrac{Nu_x \Psi'^{0.25}}{\Omega^{0.14}}$ |
|---|---|---|---|---|---|---|---|
| 130.0 | 48.9 | 22.66 | 366.3 | 695 | 3292 | 3440 | 3063 |
| 180.0 | 49.4 | 31.80 | 368.2 | 678 | 4465 | 4670 | 4075 |
| 250.0 | 49.9 | 44.82 | 368.8 | 652 | 5978 | 6259 | 5331 |
| 350.0 | 50.6 | 63.72 | 369.6 | 658 | 8468 | 8866 | 7345 |
| 500.0 | 51.3 | 92.46 | 369.6 | 654 | 12049 | 12615 | 10140 |
| 680.0 | 51.9 | 127.31 | 369.8 | 663 | 16630 | 17428 | 13636 |
| 860.0 | 52.4 | 162.56 | 369.7 | 649 | 20590 | 21560 | 16508 |

Run 4        $P_0 = 1.143$ MPa, $T_f = 306.2$ K

| $L \cdot 10^3$ m | $u_\infty$ m/sec | $Re_x \cdot 10^{-5}$ | $T_w$ K | $a$ W/m²·K | $Nu_x$ | $Nu_x \Psi'^{0.25}$ | $\dfrac{Nu_x \Psi'^{0.25}}{\Omega^{0.14}}$ |
|---|---|---|---|---|---|---|---|
| 9.0 | 46.6 | 2.25 | 347.4 | 1678 | 435 | 4489 | |
| 14.0 | 46.7 | 3.86 | 350.4 | 1486 | 660 | 6824 | |
| 22.0 | 46.7 | 6.44 | 353.4 | 1477 | 1094 | 1133 | |
| 32.0 | 46.8 | 9.68 | 355.8 | 1331 | 1479 | 1535 | |
| 65.0 | 47.0 | 20.42 | 357.9 | 1290 | 3011 | 3128 | 2915 |
| 95.0 | 47.2 | 30.28 | 362.4 | 1199 | 4129 | 4307 | 3940 |
| 130.0 | 47.5 | 41.86 | 364.5 | 1160 | 5498 | 5740 | 5157 |
| 180.0 | 47.8 | 58.58 | 366.3 | 1142 | 7528 | 7874 | 6937 |
| 250.0 | 48.1 | 82.24 | 366.9 | 1131 | 10387 | 10865 | 9350 |
| 350.0 | 48.6 | 116.51 | 368.1 | 1109 | 14292 | 14964 | 12543 |
| 500.0 | 49.1 | 168.59 | 368.0 | 1101 | 20312 | 21270 | 17334 |
| 680.0 | 49.7 | 232.2 | 368.2 | 1104 | 27734 | 29030 | 23021 |
| 860.0 | 50.2 | 296.8 | 368.9 | 1111 | 35311 | 36970 | 28726 |

Run 5        $P_0 = 2.32$ MPa, $T_f = 307.7$ K

| $L \cdot 10^3$ m | $u_\infty$ m/sec | $Re_x \cdot 10^{-5}$ | $T_w$ K | $a$ W/m²·K | $Nu_x$ | $Nu_x \Psi'^{0.25}$ | $\dfrac{Nu_x \Psi'^{0.25}}{\Omega^{0.14}}$ |
|---|---|---|---|---|---|---|---|
| 9.0 | 44.2 | 4.15 | 351.3 | 2749 | 709 | 733 | |
| 14.0 | 44.2 | 7.12 | 353.9 | 2579 | 1141 | 1181 | |
| 22.0 | 44.3 | 11.88 | 356.8 | 2420 | 1785 | 1849 | |
| 32.0 | 44.3 | 17.85 | 358.9 | 2282 | 2524 | 2622 | |
| 65.0 | 44.5 | 37.66 | 362.7 | 2127 | 4940 | 5147 | 4828 |
| 95.0 | 44.7 | 55.84 | 364.4 | 2112 | 7243 | 7554 | 6969 |
| 130.0 | 44.9 | 77.18 | 366.6 | 2046 | 9655 | 10080 | 9139 |
| 180.0 | 45.2 | 107.94 | 368.1 | 2002 | 13140 | 13744 | 12228 |
| 230.0 | 45.5 | 151.47 | 368.3 | 2002 | 18306 | 19148 | 16680 |
| 350.0 | 45.9 | 214.4 | 368.6 | 1992 | 25560 | 26740 | 22699 |
| 500.0 | 46.4 | 310.1 | 368.4 | 2011 | 36920 | 38620 | 31891 |
| 680.0 | 46.9 | 426.8 | 368.5 | 2026 | 50660 | 52990 | 42596 |
| 860.0 | 47.4 | 545.3 | 368.9 | 1908 | 60367 | 68140 | 49756 |

## Table 14

| $L \cdot 10^3$ m | $u_\infty$ m/sec | $Re_x \cdot 10^{-5}$ | $T_w^2$ K | $a$ W/m²·K | $Nu_x$ | $Nu_x \Psi^{'0.25}$ | $\dfrac{Nu_x \Psi^{'0.25}}{\Omega^{0.14}}$ |
|---|---|---|---|---|---|---|---|
| Run 6 | | $P_0 = 0.322$ MPa, $T_f = 304.7$ K | | | | | |
| 14.0 | 49.2 | 1.16 | 436.0 | 461 | 206 | 225 | |
| 22.0 | 49.2 | 1.94 | 440.0 | 447 | 333 | 365 | |
| 32.1 | 49.3 | 2.92 | 438.5 | 447 | 500 | 548 | |
| 65.1 | 49.5 | 6.16 | 442.6 | 433 | 1019 | 1118 | |
| 95.2 | 49.8 | 9.14 | 448.3 | 421 | 1460 | 1606 | |
| 130.3 | 50.0 | 12.64 | 454.9 | 403 | 1923 | 2127 | |
| 180.5 | 50.3 | 17.69 | 459.8 | 388 | 2579 | 2860 | |
| 250.7 | 50.8 | 24.88 | 462.7 | 388 | 3592 | 3987 | 3356 |
| 351.0 | 51.4 | 35.30 | 466.9 | 381 | 4956 | 5516 | 4521 |
| 501.5 | 52.1 | 51.23 | 466.4 | 380 | 7065 | 7856 | 6240 |
| 682.1 | 52.8 | 70.77 | 467.7 | 378 | 9556 | 10636 | 8213 |
| 862.7 | 53.5 | 90.71 | 469.2 | 376 | 12056 | 13430 | 10236 |
| Run 7 | | $P_0 = 0.624$ MPa, $T_f = 306$ K | | | | | |
| 9.0 | 48.1 | 1.27 | 410.3 | 977 | 253 | 272 | |
| 14.0 | 48.2 | 2.18 | 418.2 | 867 | 386 | 417 | |
| 22.0 | 48.2 | 3.64 | 425.8 | 846 | 628 | 682 | |
| 32.1 | 48.2 | 5.48 | 431.7 | 812 | 905 | 985 | |
| 65.1 | 48.5 | 11.56 | 443.5 | 736 | 1721 | 1886 | |
| 95.2 | 48.7 | 17.14 | 449.8 | 711 | 2457 | 2705 | |
| 130.3 | 48.9 | 23.71 | 456.6 | 683 | 3247 | 3588 | 3197 |
| 180.5 | 49.2 | 33.18 | 461.4 | 670 | 4430 | 4913 | 4291 |
| 250.7 | 49.7 | 46.63 | 463.0 | 666 | 6134 | 6803 | 5799 |
| 351.0 | 50.2 | 66.09 | 466.9 | 651 | 8421 | 9339 | 7750 |
| 501.5 | 50.8 | 95.82 | 466.8 | 652 | 12076 | 13416 | 10810 |
| 682.1 | 51.5 | 132.19 | 469.2 | 649 | 16345 | 18192 | 14257 |
| 862.7 | 52.1 | 169.37 | 470.5 | 643 | 20510 | 22850 | 17536 |
| Run 8 | | $P_0 = 1.183$ MPa, $T_f = 307$ K | | | | | |
| 9.0 | 47.1 | 2.34 | 409.9 | 1521 | 394 | 423 | |
| 14.0 | 47.1 | 4.02 | 417.4 | 1493 | 663 | 716 | |
| 22.0 | 47.2 | 6.70 | 425.4 | 1368 | 1012 | 1098 | |
| 32.1 | 47.3 | 10.07 | 431.7 | 1327 | 1473 | 1603 | |
| 65.1 | 47.5 | 21.26 | 442.7 | 1228 | 2863 | 3135 | 2924 |
| 95.2 | 47.7 | 31.51 | 448.8 | 1193 | 4106 | 4512 | 4132 |
| 130.3 | 48.9 | 43.59 | 454.6 | 1148 | 5440 | 6000 | 5396 |
| 180.5 | 48.2 | 60.98 | 459.3 | 1119 | 7377 | 8159 | 7195 |
| 250.7 | 48.6 | 85.64 | 461.2 | 1116 | 10244 | 11340 | 9776 |
| 351.0 | 49.0 | 121.34 | 465.7 | 1089 | 14038 | 15568 | 13082 |
| 501.5 | 49.6 | 175.75 | 465.0 | 1095 | 20199 | 22400 | 18285 |
| 682.1 | 50.2 | 242.2 | 466.7 | 1083 | 27199 | 30190 | 23998 |
| 862.7 | 50.8 | 309.8 | 468.1 | 1078 | 34260 | 38060 | 29618 |

## Table 14 (Continued)

| $L \cdot 10^3$ m | $u_\infty$ m/sec | $Re_x \cdot 10^{-5}$ | $T_w$ K | $a$ W/m²·K | $Nu_x$ | $Nu_x\Psi'^{0.25}$ | $\dfrac{Nu_x\Psi'^{0.25}}{\Omega^{0.14}}$ |
|---|---|---|---|---|---|---|---|
| Run 9 | | $P_0 = 2.279$ MPa, $T_f = 307.7$ K | | | | | |
| 9.0 | 45.1 | 4.28 | 413.5 | 2656 | 686 | 739 | |
| 14.0 | 45.2 | 7.35 | 420.0 | 2513 | 1114 | 1204 | |
| 22.0 | 45.2 | 12.27 | 428.3 | 2251 | 1664 | 1807 | |
| 32.1 | 45.3 | 18.43 | 433.3 | 2226 | 2467 | 2687 | |
| 65.1 | 45.5 | 38.92 | 443.8 | 2055 | 4785 | 5240 | 4920 |
| 95.2 | 45.7 | 57.69 | 448.9 | 1995 | 6860 | 7539 | 6955 |
| 130.3 | 45.9 | 79.75 | 454.9 | 1957 | 9264 | 10209 | 9264 |
| 180.5 | 46.2 | 111.57 | 460.2 | 1877 | 12354 | 13664 | 12156 |
| 250.7 | 46.5 | 156.56 | 461.9 | 1866 | 17119 | 18951 | 16522 |
| 351.0 | 46.9 | 221.6 | 466.6 | 1820 | 23421 | 25970 | 22083 |
| 501.5 | 47.4 | 320.5 | 465.4 | 1839 | 33874 | 37570 | 31075 |
| 682.1 | 47.9 | 441.3 | 467.5 | 1827 | 45825 | 50870 | 41024 |
| 862.7 | 48.4 | 564.1 | 469.6 | 1786 | 56694 | 62990 | 49715 |

## Table 15

| $L \cdot 10^3$ m | $u_\infty$ m/sec | $Re_x \cdot 10^{-5}$ | $T_w$, K | $a$ W/m²·K | $Nu_x$ | $Nu_x\Psi'^{0.25}$ | $\dfrac{Nu_x\Psi'^{0.25}}{\Omega^{0.14}}$ |
|---|---|---|---|---|---|---|---|
| Run 10 | | $P_0 = 0.304$ MPa, $T_f = 307$ K | | | | | |
| 14.0 | 49.3 | 1.11 | 588.5 | 440 | 196 | 231 | |
| 22.1 | 49.3 | 1.85 | 584.9 | 400 | 297 | 349 | |
| 32.1 | 49.7 | 2.80 | 580.7 | 432 | 481 | 564 | |
| 65.3 | 50.0 | 5.92 | 603.3 | 398 | 932 | 1103 | |
| 95.5 | 50.3 | 8.78 | 618.6 | 379 | 1308 | 1559 | |
| 130.7 | 50.6 | 12.18 | 632.6 | 367 | 1746 | 2092 | |
| 181.0 | 51.0 | 17.08 | 644.8 | 360 | 2381 | 2867 | |
| 251.4 | 51.6 | 24.04 | 651.9 | 352 | 3246 | 3918 | 3298 |
| 352.1 | 52.3 | 34.21 | 665.2 | 343 | 4444 | 5391 | 4419 |
| 503.1 | 53.0 | 49.62 | 663.9 | 342 | 6341 | 7685 | 6099 |
| 684.3 | 53.8 | 68.59 | 670.6 | 339 | 8542 | 10379 | 8008 |
| 865.6 | 54.0 | 87.10 | 674.6 | 335 | 10706 | 13051 | 9842 |
| Run 11 | | $P_0 = 0.599$ MPa, $T_f = 307.5$ K | | | | | |
| 9.0 | 47.8 | 1.21 | 522.6 | 813 | 210 | 240 | |
| 14.0 | 47.9 | 2.07 | 537.0 | 854 | 379 | 435 | |
| 22.1 | 47.9 | 3.46 | 553.9 | 782 | 579 | 671 | |
| 32.1 | 48.0 | 5.20 | 568.2 | 756 | 840 | 979 | |
| 65.3 | 48.3 | 10.98 | 598.0 | 672 | 1570 | 1854 | |
| 95.4 | 48.5 | 16.29 | 613.3 | 659 | 2272 | 2699 | |
| 130.1 | 48.7 | 22.52 | 627.9 | 635 | 3013 | 3604 | 3209 |

## Table 15  (*Continued*)

| $L \cdot 10^3$ m | $u_\infty$ m/sec | $Re_x \cdot 10^{-5}$ | $T_w$ K | $a$ W/m²·K | $Nu_x$ | $Nu_x \Psi^{0.25}$ | $\dfrac{Nu_x \Psi^{0.25}}{\Omega^{0.14}}$ |
|---|---|---|---|---|---|---|---|
| 180.9 | 49.0 | 31.54 | 641.9 | 619 | 4088 | 4914 | 4292 |
| 251.4 | 49.4 | 44.32 | 648.3 | 608 | 5587 | 6732 | 5734 |
| 352.0 | 49.9 | 62.84 | 662.0 | 585 | 7557 | 9152 | 7589 |
| 503.0 | 50.6 | 91.08 | 660.3 | 590 | 10904 | 13194 | 10615 |
| 684.3 | 51.3 | 125.76 | 665.5 | 584 | 14702 | 17819 | 13932 |
| 865.5 | 52.0 | 161.33 | 668.7 | 578 | 18407 | 22350 | 17126 |

Run 12    $P_0 = 1.142$ MPa, $T_f = 307$ K

| $L \cdot 10^3$ m | $u_\infty$ m/sec | $Re_x \cdot 10^{-5}$ | $T_w$ K | $a$ W/m²·K | $Nu_x$ | $Nu_x \Psi^{0.25}$ | $\dfrac{Nu_x \Psi^{0.25}}{\Omega^{0.14}}$ |
|---|---|---|---|---|---|---|---|
| 9.0 | 46.7 | 2.25 | 517.8 | 1474 | 382 | 435 | |
| 14.0 | 46.8 | 3.86 | 534.7 | 1383 | 615 | 707 | |
| 22.1 | 46.9 | 6.45 | 553.6 | 1295 | 960 | 1115 | |
| 32.1 | 47.0 | 9.71 | 567.9 | 1244 | 1385 | 1615 | |
| 65.3 | 47.3 | 20.52 | 596.6 | 1117 | 2613 | 3086 | 2876 |
| 95.5 | 47.5 | 30.44 | 611.1 | 1096 | 3786 | 4494 | 4115 |
| 130.7 | 48.0 | 42.40 | 626.6 | 1041 | 4950 | 5915 | 5319 |
| 181.0 | 48.4 | 59.45 | 640.3 | 1014 | 6705 | 8059 | 7107 |
| 251.4 | 48.9 | 83.63 | 646.4 | 1015 | 9352 | 11260 | 9698 |
| 352.1 | 49.5 | 118.74 | 659.4 | 971 | 12552 | 15188 | 12752 |
| 503.1 | 50.1 | 172.12 | 657.9 | 980 | 18143 | 21950 | 17918 |
| 684.3 | 50.7 | 237.4 | 664.0 | 964 | 24294 | 29440 | 23383 |
| 865.5 | 50.9 | 301.4 | 668.3 | 950 | 30297 | 36780 | 28600 |

Run 13    $P_0 = 2.324$ MPa, $T_f = 307.7$ K

| $L \cdot 10^3$ m | $u_\infty$ m/sec | $Re_x \cdot 10^{-5}$ | $T_w$ K | $a$ W/m²·K | $Nu_x$ | $Nu_x \Psi^{0.25}$ | $\dfrac{Nu_x \Psi^{0.25}}{\Omega^{0.14}}$ |
|---|---|---|---|---|---|---|---|
| 9.0 | 44.4 | 4.29 | 524.0 | 2550 | 659 | 753 | |
| 14.0 | 44.4 | 7.38 | 540.2 | 2407 | 1068 | 1229 | |
| 22.1 | 44.5 | 12.33 | 558.8 | 2144 | 1587 | 1842 | |
| 32.1 | 44.7 | 18.59 | 570.5 | 2107 | 2341 | 2730 | |
| 65.3 | 44.9 | 39.23 | 598.9 | 1934 | 4513 | 5330 | 5005 |
| 95.4 | 45.1 | 58.18 | 614.0 | 1857 | 6398 | 7601 | 7025 |
| 130.7 | 45.4 | 80.53 | 630.6 | 1786 | 8475 | 10136 | 9198 |
| 181.0 | 45.7 | 112.81 | 643.7 | 1739 | 11473 | 13802 | 12279 |
| 251.4 | 46.1 | 158.58 | 650.9 | 1716 | 15780 | 19031 | 16591 |
| 352.1 | 46.6 | 225.1 | 662.9 | 1665 | 21495 | 26030 | 22134 |
| 503.1 | 47.2 | 326.2 | 661.3 | 1672 | 30890 | 37380 | 30944 |
| 624.3 | 47.7 | 449.5 | 667.1 | 1659 | 41737 | 50630 | 41196 |
| 865.6 | 47.9 | 570.3 | 674.2 | 1626 | 51781 | 63020 | 49739 |

## Table 16

| $L \cdot 10^3$ m | $u_\infty$ m/sec | $Re_x \cdot 10^{-5}$ | $T_w$ K | $a$ W/m²·K | $Nu_x$ | $Nu_x \Psi^{0.25}$ | $\dfrac{Nu_x \Psi^{0.25}}{\Omega^{0.14}}$ |
|---|---|---|---|---|---|---|---|
| **Run 14** | | $P_0 = 0.516$ MPa, $T_f = 306$ K | | | | | |
| 32.2  | 48.4 | 4.56   | 720.4 | 652  | 729   | 903   |       |
| 65.5  | 48.7 | 9.65   | 783.6 | 568  | 1336  | 1690  |       |
| 95.7  | 48.9 | 14.32  | 809.9 | 546  | 1898  | 2422  |       |
| 131.4 | 49.1 | 19.86  | 824.6 | 543  | 2603  | 3334  |       |
| 181.9 | 49.3 | 27.70  | 840.6 | 528  | 3520  | 4534  | 3946  |
| 252.7 | 49.9 | 39.07  | 864.7 | 510  | 4734  | 6135  | 5372  |
| 353.8 | 50.4 | 55.37  | 875.6 | 503  | 6554  | 8520  | 7053  |
| 505.4 | 51.0 | 80.28  | 876.1 | 496  | 9252  | 12037 | 9666  |
| 687.4 | 51.7 | 110.68 | 884.6 | 488  | 12390 | 16157 | 12822 |
| 869.4 | 52.2 | 141.48 | 887.9 | 488  | 15701 | 20490 | 15653 |
| **Run 15** | | $P_0 = 1.223$ MPa, $T_f = 306.7$ K | | | | | |
| 32.2  | 46.8 | 10.39  | 729.8 | 1206 | 1346  | 1672  |       |
| 65.5  | 47.1 | 21.97  | 780.9 | 1095 | 2571  | 3247  | 3023  |
| 95.8  | 47.3 | 32.60  | 808.2 | 1075 | 3727  | 4752  | 4356  |
| 131.1 | 47.5 | 45.12  | 823.3 | 1041 | 4969  | 6360  | 5725  |
| 181.6 | 47.7 | 63.02  | 838.1 | 1017 | 6754  | 8686  | 7666  |
| 252.3 | 48.2 | 88.73  | 859.2 | 992  | 9180  | 11870 | 10197 |
| 353.4 | 48.7 | 125.72 | 872.0 | 975  | 12666 | 16440 | 13815 |
| 505.1 | 49.2 | 182.11 | 870.8 | 969  | 18030 | 23403 | 19120 |
| 687.1 | 49.8 | 251.0  | 877.7 | 960  | 24315 | 31634 | 25170 |
| 869.0 | 50.4 | 321.2  | 880.6 | 970  | 31106 | 40500 | 31540 |
| **Run 16** | | $P_0 = 2.239$ MPa, $T_f = 308$ K | | | | | |
| 32.2  | 44.9 | 18.01  | 797.6 | 1905 | 2119  | 2687  |       |
| 65.5  | 45.1 | 38.04  | 850.6 | 1709 | 3997  | 5149  | 4830  |
| 95.8  | 45.3 | 56.46  | 878.8 | 1657 | 5725  | 7437  | 6866  |
| 131.2 | 45.5 | 78.07  | 903.9 | 1608 | 7650  | 10014 | 9078  |
| 181.8 | 45.8 | 109.25 | 932.8 | 1545 | 10227 | 13489 | 12001 |
| 252.6 | 46.1 | 153.37 | 947.8 | 1522 | 14042 | 18592 | 16195 |
| 353.9 | 46.5 | 217.2  | 971.6 | 1488 | 19278 | 25680 | 21837 |
| 505.8 | 47.1 | 314.9  | 971.2 | 1476 | 27371 | 36460 | 30132 |
| 688.2 | 47.7 | 435.0  | 976.8 | 1476 | 37293 | 49750 | 40088 |
| 870.6 | 48.4 | 558.0  | 985.2 | 1447 | 46290 | 61890 | 48848 |

## 6. EXPERIMENTAL DATA ON HEAT TRANSFER IN AN INTERNALLY HEATED ANNULUS

**Table 17** Annulus with $d_1/d_2 = 0.108$

| $x_1/d$ | $Nu_{12}$ | $Re \cdot 10^{-3}$ | $\Psi_1$ | $q_1 \cdot 10^{-3}$ W/m² | $q_2 \cdot 10^{-3}$ W/m² | $T_{w1}$ K |
|---|---|---|---|---|---|---|
| Run 5403 | | | $G = 0.1022$ kg/sec, $P_{in} = 0.561$ MPa, $T_{fin} = 279.3$ K | | | |
| 1.20 | 552.6 | 238.5 | 1.111 | 16.98 | 0.019 | 310.3 |
| 2.19 | 516.7 | 238.4 | 1.118 | 16.97 | 0.021 | 312.5 |
| 3.79 | 489.3 | 238.4 | 1.125 | 16.96 | 0.022 | 314.3 |
| 6.38 | 465.4 | 238.3 | 1.131 | 16.95 | 0.023 | 316.2 |
| 11.16 | 457.9 | 238.2 | 1.133 | 16.95 | 0.024 | 317.0 |
| 19.14 | 438.5 | 238.0 | 1.138 | 16.94 | 0.025 | 318.9 |
| 27.90 | 431.8 | 237.7 | 1.140 | 16.94 | 0.025 | 319.8 |
| 38.27 | 429.4 | 237.4 | 1.140 | 16.94 | 0.025 | 320.3 |
| 47.04 | 423.1 | 237.2 | 1.142 | 16.93 | 0.026 | 321.2 |
| 55.82 | 418.2 | 237 | 1.143 | 16.93 | 0.025 | 322.0 |
| Run 5407 | | | $G = 0.1028$ kg/sec, $P_{in} = 0.562$ MPa, $T_{fin} = 279.6$ K | | | |
| 1.20 | 548.9 | 239.4 | 1.267 | 40.79 | 0.089 | 354.4 |
| 2.19 | 507.1 | 239.4 | 1.288 | 40.75 | 0.063 | 360.5 |
| 3.79 | 477.8 | 239.3 | 1.305 | 40.74 | 0.044 | 365.5 |
| 6.38 | 448.1 | 239.1 | 1.325 | 40.72 | 0.057 | 371.3 |
| 11.17 | 434.7 | 238.8 | 1.334 | 40.74 | 0.060 | 374.6 |
| 19.15 | 418.3 | 238.2 | 1.346 | 40.78 | 0.063 | 378.8 |
| 27.94 | 409.1 | 237.7 | 1.352 | 40.84 | 0.066 | 381.8 |
| 38.31 | 407.1 | 237.0 | 1.352 | 40.93 | 0.066 | 383.1 |
| 47.09 | 398.3 | 236.4 | 1.358 | 40.99 | 0.068 | 386.1 |
| 55.88 | 394.0 | 235.8 | 1.361 | 41.06 | 0.069 | 388.0 |
| Run 5412 | | | $G = 0.1052$ kg/sec, $P_{in} = 0.572$ MPa, $T_{fin} = 280.9$ K | | | |
| 1.20 | 539.6 | 244.2 | 1.548 | 83.18 | 0.127 | 435.2 |
| 2.20 | 499.2 | 244.0 | 1.590 | 83.03 | 0.143 | 447.4 |
| 3.80 | 465.4 | 243.8 | 1.631 | 82.89 | 0.159 | 459.3 |
| 6.39 | 437.7 | 243.5 | 1.668 | 82.77 | 0.176 | 470.5 |
| 11.19 | 420.4 | 242.8 | 1.690 | 82.73 | 0.190 | 478.5 |
| 19.19 | 397.9 | 241.8 | 1.721 | 82.67 | 0.210 | 489.9 |
| 27.99 | 390.4 | 240.6 | 1.728 | 82.76 | 0.219 | 494.7 |
| 38.39 | 386.6 | 239.3 | 1.727 | 82.90 | 0.224 | 497.8 |
| 47.20 | 380.4 | 238.2 | 1.731 | 82.99 | 0.231 | 502.1 |
| 56.00 | 377.0 | 237.0 | 1.731 | 83.11 | 0.236 | 504.9 |

**Table 17** Annulus with $d_1/d_2 = 0.108$ (*Continued*)

| $x_1/d$ | $Nu_{12}$ | $Re \cdot 10^{-3}$ | $\Psi_1$ | $q_1 \cdot 10^{-3}$ W/m² | $q_2 \cdot 10^{-3}$ W/m² | $T_{w1}$ K |
|---|---|---|---|---|---|---|
| Run 5417 | | $G = 0.1061$ kg/sec, $P_{in} = 0.580$ MPa, $T_{fin} = 281.9$ K | | | | |
| 1.20 | 521.2 | 245.3 | 1.937 | 138.6 | 0.369 | 546.9 |
| 2.20 | 476.2 | 245.1 | 2.019 | 138.0 | 0.432 | 570.7 |
| 3.81 | 438.8 | 244.7 | 2.098 | 137.5 | 0.501 | 594.0 |
| 6.41 | 410.0 | 244.2 | 2.164 | 137.0 | 0.570 | 614.7 |
| 11.23 | 392.5 | 243.0 | 2.204 | 137.0 | 0.625 | 629.3 |
| 19.25 | 373.2 | 241.3 | 2.247 | 137.2 | 0.702 | 647.3 |
| 28.08 | 365.1 | 239.4 | 2.256 | 137.8 | 0.744 | 656.7 |
| 38.52 | 363.7 | 237.2 | 2.244 | 138.8 | 0.760 | 660.6 |
| 47.35 | 357.8 | 235.4 | 2.248 | 139.5 | 0.796 | 668.2 |
| 56.18 | 356.4 | 233.5 | 2.238 | 140.4 | 0.812 | 671.7 |
| Run 5419 | | $G = 0.1040$ kg/sec, $P_{in} = 0.567$ MPa, $T_{fin} = 281.9$ K | | | | |
| 1.20 | 526.1 | 240.4 | 2.074 | 160.5 | 0.531 | 586.0 |
| 2.20 | 464.2 | 240.1 | 2.204 | 159.3 | 0.652 | 623.6 |
| 3.81 | 424.8 | 239.7 | 2.302 | 158.2 | 0.758 | 652.7 |
| 6.42 | 393.4 | 239.1 | 2.388 | 157.2 | 0.878 | 679.4 |
| 11.24 | 374.4 | 237.8 | 2.437 | 156.7 | 0.970 | 697.7 |
| 19.28 | 354.2 | 235.7 | 2.485 | 156.3 | 1.097 | 719.1 |
| 28.12 | 346.3 | 233.6 | 2.489 | 156.6 | 1.156 | 728.6 |
| 38.57 | 343.7 | 231.1 | 2.471 | 157.5 | 1.183 | 733.2 |
| 47.42 | 338.0 | 228.9 | 2.468 | 157.9 | 1.234 | 740.8 |
| 56.26 | 335.6 | 226.8 | 2.454 | 158.6 | 1.260 | 744.9 |
| Run 5422 | | $G = 0.1047$ kg/sec, $P_{in} = 0.573$ MPa, $T_{fin} = 281.9$ K | | | | |
| 1.21 | 500.3 | 241.8 | 2.507 | 214.6 | 1.212 | 708.9 |
| 2.21 | 452.4 | 241.3 | 2.641 | 212.2 | 1.446 | 748.3 |
| 3.82 | 411.5 | 240.8 | 2.771 | 209.4 | 1.710 | 787.4 |
| 6.44 | 379.0 | 239.9 | 2.882 | 206.7 | 1.987 | 822.6 |
| 11.27 | 364.2 | 238.2 | 2.917 | 205.7 | 2.128 | 839.9 |
| 19.34 | 344.1 | 235.4 | 2.961 | 204.1 | 2.394 | 864.8 |
| 28.21 | 338.9 | 232.5 | 2.938 | 204.6 | 2.463 | 871.6 |
| 38.69 | 339.3 | 229.2 | 2.886 | 206.0 | 2.458 | 871.6 |
| 47.56 | 334.5 | 226.3 | 2.863 | 206.5 | 2.534 | 878.2 |
| 56.43 | 333.3 | 223.5 | 2.827 | 207.5 | 2.553 | 880.2 |

**Table 18** Annulus with $d_1/d_2 = 0.205$

| $x_1/d$ | $Nu_{12}$ | $Re \cdot 10^{-3}$ | $\Psi_1$ | $q_1 \cdot 10^{-3}$ W/m² | $q_2 \cdot 10^{-3}$ W/m² | $T_{w1}$ K |
|---|---|---|---|---|---|---|
| Run 3404 | | $G = 0.1007$ kg/sec, $P_{in} = 0.528$ MPa, $T_{fin} = 284$ K | | | | |
| 1.34  | 435.8 | 213.2 | 1.117 | 16.33 | 0.039 | 317.2 |
| 2.46  | 407.0 | 213.2 | 1.124 | 16.35 | 0.042 | 319.7 |
| 4.25  | 392.7 | 213.1 | 1.130 | 16.39 | 0.046 | 321.2 |
| 7.20  | 366.9 | 213.0 | 1.139 | 16.45 | 0.050 | 324.1 |
| 12.53 | 357.8 | 212.8 | 1.143 | 16.57 | 0.053 | 325.7 |
| 21.47 | 340.6 | 212.4 | 1.152 | 16.78 | 0.059 | 328.8 |
| 31.27 | 336.6 | 212.0 | 1.156 | 17.00 | 0.063 | 330.5 |
| 43.09 | 327.7 | 211.5 | 1.161 | 17.28 | 0.066 | 333.1 |
| 43.26 | 326.0 | 211.5 | 1.162 | 17.28 | 0.074 | 333.4 |
| 62.64 | 337.1 | 210.7 | 1.159 | 17.75 | 0.065 | 334.3 |
| Run 3408 | | $G = 0.1016$ kg/sec, $P_{in} = 0.536$ MPa, $T_{fin} = 283.5$ K | | | | |
| 1.34  | 454.1 | 215.3 | 1.250 | 36.50 | 0.075 | 354.8 |
| 2.46  | 423.5 | 215.2 | 1.268 | 36.49 | 0.081 | 360.0 |
| 4.25  | 405.2 | 215.1 | 1.280 | 36.51 | 0.087 | 363.7 |
| 7.21  | 375.9 | 214.8 | 1.301 | 36.54 | 0.097 | 370.3 |
| 12.55 | 361.2 | 214.3 | 1.313 | 36.64 | 0.104 | 374.6 |
| 21.50 | 340.1 | 213.5 | 1.331 | 36.81 | 0.115 | 381.5 |
| 31.31 | 331.9 | 212.6 | 1.338 | 37.03 | 0.122 | 385.4 |
| 43.13 | 318.9 | 211.6 | 1.350 | 37.28 | 0.130 | 391.3 |
| 43.31 | 318.5 | 211.6 | 1.350 | 37.27 | 0.139 | 391.4 |
| 62.71 | 318.2 | 219.9 | 1.348 | 37.74 | 0.133 | 394.8 |
| Run 3412 | | $G = 0.0984$ kg/sec, $P_{in} = 0.534$ MPa, $T_{fin} = 238.8$ K | | | | |
| 1.34  | 429.9 | 208.2 | 1.406 | 59.19 | 0.142 | 399.5 |
| 2.46  | 399.8 | 208.1 | 1.436 | 56.17 | 0.158 | 408.4 |
| 4.26  | 380.5 | 207.8 | 1.457 | 56.22 | 0.171 | 415.0 |
| 7.22  | 349.6 | 207.4 | 1.495 | 56.28 | 0.196 | 427.0 |
| 12.55 | 334.6 | 206.7 | 1.516 | 56.54 | 0.213 | 434.7 |
| 21.52 | 312.2 | 205.4 | 1.549 | 56.95 | 0.245 | 447.6 |
| 31.35 | 305.0 | 204.1 | 1.559 | 57.50 | 0.262 | 454.1 |
| 43.19 | 291.1 | 202.6 | 1.581 | 58.11 | 0.290 | 465.1 |
| 43.37 | 290.9 | 201.8 | 1.582 | 58.12 | 0.303 | 465.2 |
| 62.80 | 292.7 | 199.9 | 1.573 | 59.33 | 0.300 | 470.1 |
| Run 3416 | | $G = 0.0978$ kg/sec, $P_{in} = 0.510$ MPa, $T_{fin} = 283.4$ K | | | | |
| 1.35 | 428.5 | 207.1 | 1.591 | 81.34 | 0.266 | 451.7 |
| 2.47 | 394.5 | 206.9 | 1.639 | 81.24 | 0.303 | 466.1 |
| 4.26 | 371.6 | 206.5 | 1.675 | 81.24 | 0.335 | 477.5 |
| 7.23 | 340.3 | 205.9 | 1.733 | 81.22 | 0.395 | 495.5 |

**Table 18** Annulus with $d_1/d_2 = 0.205$ (*Continued*)

| $x_1/d$ | $Nu_{12}$ | $Re \cdot 10^{-3}$ | $\Psi_1$ | $q_1 \cdot 10^{-3}$ W/m² | $q_2 \cdot 10^{-3}$ W/m² | $T_{w1}$ K |
|---|---|---|---|---|---|---|
| 12.57 | 322.0 | 204.8 | 1.768 | 81.50 | 0.443 | 508.8 |
| 21.56 | 299.1 | 203.0 | 1.815 | 81.99 | 0.521 | 528.0 |
| 31.40 | 289.3 | 201.1 | 1.831 | 82.72 | 0.570 | 539.1 |
| 43.27 | 276.3 | 198.9 | 1.856 | 83.55 | 0.639 | 554.1 |
| 43.45 | 276.6 | 198.9 | 1.856 | 83.55 | 0.631 | 554.0 |
| 62.92 | 274.1 | 195.1 | 1.844 | 85.27 | 0.682 | 563.4 |

Run 3420  $G = 0.100$ kg/sec, $P_{in} = 0.534$ MPa, $T_{fin} = 283.7$ K

| $x_1/d$ | $Nu_{12}$ | $Re \cdot 10^{-3}$ | $\Psi_1$ | $q_1 \cdot 10^{-3}$ W/m² | $q_2 \cdot 10^{-3}$ W/m² | $T_{w1}$ K |
|---|---|---|---|---|---|---|
| 1.35 | 429.9 | 211.2 | 1.860 | 119.3 | 0.576 | 529.0 |
| 2.47 | 391.3 | 210.8 | 1.938 | 118.9 | 0.673 | 552.4 |
| 4.27 | 365.2 | 210.3 | 1.998 | 118.7 | 0.762 | 571.2 |
| 7.24 | 330.4 | 209.4 | 2.088 | 118.3 | 0.932 | 600.1 |
| 12.60 | 310.0 | 207.9 | 2.142 | 118.5 | 1.070 | 621.1 |
| 21.62 | 282.0 | 205.3 | 2.217 | 118.8 | 1.320 | 653.1 |
| 31.49 | 273.0 | 202.5 | 2.233 | 119.8 | 1.458 | 669.1 |
| 43.40 | 261.2 | 199.2 | 2.254 | 120.9 | 1.638 | 688.7 |
| 43.58 | 261.3 | 199.2 | 2.253 | 120.9 | 1.634 | 688.6 |
| 63.11 | 258.0 | 193.9 | 2.220 | 123.7 | 1.765 | 701.6 |

Run 3424  $G = 0.1006$ kg/sec, $P_{in} = 0.538$ MPa, $T_{fin} = 282.9$ K

| $x_1/d$ | $Nu_{12}$ | $Re \cdot 10^{-3}$ | $\Psi_1$ | $q_1 \cdot 10^{-3}$ W/m² | $q_2 \cdot 10^{-3}$ W/m² | $T_{w1}$ K |
|---|---|---|---|---|---|---|
| 1.35 | 423.8 | 212.7 | 2.230 | 167.9 | 1.352 | 633.4 |
| 2.48 | 381.1 | 212.2 | 2.350 | 166.6 | 1.612 | 669.3 |
| 4.28 | 350.2 | 211.5 | 2.447 | 165.4 | 1.882 | 700.0 |
| 7.26 | 313.6 | 210.2 | 2.576 | 163.6 | 2.351 | 742.2 |
| 12.64 | 290.3 | 208.0 | 2.652 | 162.6 | 2.756 | 774.0 |
| 21.69 | 261.6 | 204.4 | 2.744 | 161.1 | 3.455 | 818.5 |
| 31.61 | 251.6 | 200.5 | 2.741 | 161.7 | 3.785 | 837.3 |
| 43.56 | 239.1 | 196.0 | 2.744 | 162.1 | 4.244 | 861.7 |
| 43.74 | 239.2 | 196.0 | 2.744 | 162.1 | 4.244 | 861.6 |
| 63.35 | 238.7 | 189.1 | 2.644 | 166.0 | 4.388 | 869.1 |

**Table 19** Annulus with $d_1/d_2 = 0.373$

| $x_1/d$ | $Nu_{12}$ | $Re \cdot 10^{-3}$ | $\Psi_1$ | $q_1 \cdot 10^{-3}$ W/m² | $q_2 \cdot 10^{-3}$ W/m² | $T_{w1}$ K |
|---|---|---|---|---|---|---|

Run 1503  $G = 0.0494$ kg/sec, $P_{in} = 0.265$ MPa, $T_{fin} = 292.6$ K

| $x_1/d$ | $Nu_{12}$ | $Re \cdot 10^{-3}$ | $\Psi_1$ | $q_1 \cdot 10^{-3}$ W/m² | $q_2 \cdot 10^{-3}$ W/m² | $T_{w1}$ K |
|---|---|---|---|---|---|---|
| 1.68 | 223.4 | 89.86 | 1.058 | 5.534 | 0.024 | 309.7 |
| 3.08 | 203.4 | 89.84 | 1.063 | 5.528 | 0.026 | 311.5 |
| 5.35 | 188.7 | 89.81 | 1.068 | 5.525 | 0.029 | 313.0 |

**Table 19** Annulus with $d_1/d_2 = 0.373$ (*Continued*)

| $x_1/d$ | $Nu_{12}$ | $Re \cdot 10^{-3}$ | $\Psi_1$ | $q_1 \cdot 10^{-3}$ W/m² | $q_2 \cdot 10^{-3}$ W/m² | $T_{w1}$ K |
|---|---|---|---|---|---|---|
| 9.07 | 177.9 | 89.75 | 1.072 | 5.524 | 0.032 | 314.4 |
| 15.87 | 172.1 | 89.63 | 1.075 | 5.525 | 0.034 | 315.6 |
| 27.15 | 159.1 | 89.46 | 1.080 | 5.525 | 0.037 | 318.1 |
| 39.62 | 156.9 | 89.26 | 1.081 | 5.532 | 9.038 | 319.2 |
| 54.36 | 151.9 | 89.03 | 1.083 | 5.540 | 0.039 | 320.9 |
| 66.81 | 150.6 | 88.82 | 1.084 | 5.547 | 0.040 | 321.9 |
| 79.29 | 154.0 | 88.63 | 1.082 | 5.559 | 0.039 | 322.1 |

Run 1510    $G = 0.0497$ kg/sec, $P_{in} = 0.265$ MPa, $T_{fin} = 294.2$ K

| $x_1/d$ | $Nu_{12}$ | $Re \cdot 10^{-3}$ | $\Psi_1$ | $q_1 \cdot 10^{-3}$ W/m² | $q_2 \cdot 10^{-3}$ W/m² | $T_{w1}$ K |
|---|---|---|---|---|---|---|
| 1.69 | 223.0 | 89.89 | 1.178 | 17.20 | 0.086 | 347.1 |
| 3.09 | 201.4 | 89.82 | 1.197 | 17.18 | 0.098 | 352.8 |
| 5.35 | 185.6 | 89.71 | 1.213 | 17.16 | 0.109 | 358.1 |
| 9.08 | 174.0 | 89.53 | 1.226 | 17.16 | 0.120 | 362.8 |
| 15.89 | 165.0 | 89.20 | 1.236 | 17.17 | 0.131 | 367.6 |
| 27.19 | 151.7 | 88.64 | 1.253 | 17.18 | 0.150 | 375.6 |
| 39.67 | 148.8 | 88.03 | 1.255 | 17.23 | 0.158 | 379.3 |
| 54.44 | 143.6 | 87.31 | 1.260 | 17.29 | 0.167 | 384.6 |
| 66.91 | 141.3 | 86.71 | 1.261 | 17.34 | 0.173 | 388.1 |
| 79.41 | 142.0 | 86.13 | 1.257 | 17.40 | 0.174 | 390.0 |

Run 1518    $G = 0.0497$ kg/sec, $P_{in} = 0.262$ MPa, $T_{fin} = 295.3$ K

| $x_1/d$ | $Nu_{12}$ | $Re \cdot 10^{-3}$ | $\Psi_1$ | $q_1 \cdot 10^{-3}$ W/m² | $q_2 \cdot 10^{-3}$ W/m² | $T_{w1}$ K |
|---|---|---|---|---|---|---|
| 1.69 | 224.1 | 89.63 | 1.390 | 38.30 | 0.269 | 411.9 |
| 3.09 | 199.0 | 89.47 | 1.436 | 38.18 | 0.320 | 426.4 |
| 5.36 | 180.9 | 89.23 | 1.476 | 38.09 | 0.369 | 439.6 |
| 9.10 | 167.1 | 88.82 | 1.509 | 38.04 | 0.420 | 452.0 |
| 15.93 | 155.0 | 88.08 | 1.538 | 38.01 | 0.484 | 465.3 |
| 27.26 | 140.5 | 86.85 | 1.574 | 37.98 | 0.584 | 484.5 |
| 39.78 | 136.2 | 85.55 | 1.574 | 38.12 | 0.632 | 493.6 |
| 54.60 | 130.4 | 84.08 | 1.578 | 38.25 | 0.696 | 505.7 |
| 67.11 | 127.5 | 82.90 | 1.574 | 38.40 | 9.736 | 513.4 |
| 79.65 | 127.2 | 81.78 | 1.560 | 38.61 | 0.758 | 518.0 |

Run 1525    $G = 0.0503$ kg/sec, $P_{in} = 0.263$ MPa, $T_{fin} = 293.2$ K

| $x_1/d$ | $Nu_{12}$ | $Re \cdot 10^{-3}$ | $\Psi_1$ | $q_1 \cdot 10^{-3}$ W/m² | $q_2 \cdot 10^{-3}$ W/m² | $T_{w1}$ K |
|---|---|---|---|---|---|---|
| 1.69 | 223.3 | 90.99 | 1.751 | 72.85 | 0.881 | 516.2 |
| 3.10 | 197.2 | 90.70 | 1.838 | 72.34 | 1.079 | 543.9 |
| 5.38 | 176.2 | 90.23 | 1.920 | 71.80 | 1.304 | 571.8 |
| 9.14 | 159.2 | 89.44 | 1.990 | 71.27 | 1.566 | 599.1 |
| 16.00 | 143.5 | 88.01 | 2.050 | 70.71 | 1.916 | 629.2 |
| 27.40 | 126.6 | 85.75 | 2.108 | 69.90 | 2.456 | 668.0 |
| 39.99 | 121.0 | 83.46 | 2.088 | 69.95 | 2.716 | 684.9 |
| 54.89 | 115.2 | 80.97 | 2.063 | 69.98 | 3.028 | 704.1 |
| 67.47 | 112.0 | 79.04 | 2.032 | 70.17 | 3.237 | 716.4 |
| 80.06 | 111.8 | 77.21 | 1.984 | 70.71 | 3.320 | 721.9 |

**Table 19** Annulus with $d_1/d_2 = 0.373$ (*Continued*)

| $x_1/d$ | $Nu_{12}$ | $Re \cdot 10^{-3}$ | $\Psi_1$ | $q_1 \cdot 10^{-3}$ W/m² | $q_2 \cdot 10^{-3}$ W/m² | $T_{w1}$ K |
|---|---|---|---|---|---|---|
| Run 1581 | | $G = 0.0581$ kg/sec, $P_{in} = 0.483$ MPa, $T_{fin} = 293.6$ K | | | | |
| 1.70 | 236.9 | 104.5 | 2.013 | 105.5 | 1.691 | 595.3 |
| 3.11 | 213.9 | 104.1 | 2.102 | 104.6 | 2.029 | 624.8 |
| 5.40 | 190.3 | 103.5 | 2.207 | 103.4 | 2.498 | 660.9 |
| 9.16 | 170.3 | 102.3 | 2.297 | 102.1 | 3.050 | 697.4 |
| 16.05 | 152.7 | 100.3 | 2.363 | 100.7 | 3.754 | 735.1 |
| 27.48 | 132.7 | 97.14 | 2.425 | 98.46 | 4.898 | 784.7 |
| 40.12 | 126.1 | 94.01 | 2.380 | 98.16 | 5.387 | 804.1 |
| 55.06 | 120.4 | 90.68 | 2.321 | 98.01 | 5.899 | 823.0 |
| 67.68 | 117.4 | 88.05 | 2.263 | 98.16 | 6.228 | 834.9 |
| 80.31 | 117.3 | 85.58 | 2.190 | 98.98 | 6.327 | 839.0 |
| Run 1585 | | $G = 0.0590$ kg/sec, $P_{in} = 0.448$ MPa, $T_{fin} = 293.1$ K | | | | |
| 1.70 | 240.2 | 106.0 | 2.452 | 153.8 | 4.496 | 726.3 |
| 3.12 | 207.7 | 105.3 | 2.619 | 150.4 | 5.789 | 781.2 |
| 5.42 | 174.4 | 104.3 | 2.817 | 144.9 | 7.584 | 849.9 |
| 9.21 | 155.2 | 102.6 | 2.911 | 140.7 | 8.907 | 895.6 |
| 16.14 | 135.7 | 99.70 | 2.966 | 135.3 | 10.95 | 945.6 |
| 27.64 | 117.2 | 95.45 | 2.956 | 129.0 | 13.47 | 996.8 |
| 40.35 | 110.8 | 91.30 | 2.832 | 127.4 | 14.34 | 1013.7 |
| 55.37 | 108.1 | 86.93 | 2.668 | 127.6 | 14.65 | 1020.2 |
| 68.03 | 106.0 | 85.50 | 2.547 | 127.6 | 14.95 | 1026.3 |
| 80.70 | 106.6 | 80.52 | 2.420 | 129.0 | 14.79 | 1025.0 |

**Table 20** Annulus with $d_1/d_2 = 0.585$

| $x_1/d$ | $Nu_{12}$ | $Re \cdot 10^{-3}$ | $\Psi_1$ | $q_1 \cdot 10^{-3}$ W/m² | $q_2 \cdot 10^{-3}$ W/m² | $T_{w1}$, K |
|---|---|---|---|---|---|---|
| Run 2504 | | $G = 0.1512$ kg/sec, $P_{in} = 0.480$ MPa, $T_{fin} = 299.1$ K | | | | |
| 2.40 | 428.1 | 233.3 | 1.107 | 31.24 | 0.009 | 331.7 |
| 4.59 | 407.7 | 233.1 | 1.113 | 31.25 | 0.073 | 333.5 |
| 8.19 | 401.4 | 232.8 | 1.114 | 31.29 | 0.075 | 334.5 |
| 13.85 | 374.2 | 232.4 | 1.122 | 31.34 | 0.084 | 337.7 |
| 24.01 | 344.8 | 231.6 | 1.132 | 31.43 | 0.095 | 342.1 |
| 41.04 | 326.3 | 230.3 | 1.138 | 31.61 | 0.107 | 346.4 |
| 60.08 | 313.6 | 228.7 | 1.143 | 31.80 | 0.115 | 350.5 |
| 82.25 | 311.4 | 227.2 | 1.143 | 32.05 | 0.117 | 353.4 |
| 101.3 | 312.4 | 225.8 | 1.141 | 32.26 | 0.117 | 355.6 |
| 119.9 | 311.0 | 224.3 | 1.140 | 32.46 | 0.120 | 358.4 |

**Table 20** Annulus with $d_1/d_2 = 0.585$ (*Continued*)

| $x_1/d$ | $Nu_{12}$ | $Re \cdot 10^{-3}$ | $\Psi'_1$ | $q_1 \cdot 10^{-3}$ W/m² | $q_2 \cdot 10^{-3}$ W/m² | $T_{w1}$ K |
|---|---|---|---|---|---|---|
| Run 2508 | | $G = 0.1448$ kg/sec, $P_{in} = 0.480$ MPa, $T_{fin} = 307.5$ K | | | | |
| 2.40 | 401.4 | 218.0 | 1.222 | 64.00 | 0.194 | 376.9 |
| 4.59 | 383.1 | 217.6 | 1.232 | 64.02 | 0.207 | 380.7 |
| 8.20 | 378.4 | 217.1 | 1.234 | 64.07 | 0.212 | 382.4 |
| 13.87 | 348.0 | 216.2 | 1.252 | 64.10 | 0.243 | 389.9 |
| 24.06 | 312.8 | 214.8 | 1.276 | 64.18 | 0.291 | 400.8 |
| 41.12 | 293.2 | 212.3 | 1.287 | 64.39 | 0.331 | 410.3 |
| 60.22 | 277.8 | 209.7 | 1.296 | 64.63 | 0.370 | 419.5 |
| 82.44 | 272.1 | 206.8 | 1.294 | 64.95 | 0.392 | 426.4 |
| 101.6 | 271.3 | 204.4 | 1.288 | 65.24 | 0.402 | 430.9 |
| 120.3 | 263.3 | 202.1 | 1.289 | 65.50 | 0.431 | 438.0 |
| Run 2514 | | $G = 0.1452$ kg/sec, $P_{in} = 0.478$ MPa, $T_{fin} = 311.7$ K | | | | |
| 2.41 | 391.0 | 215.6 | 1.458 | 133.0 | 0.640 | 457.3 |
| 4.61 | 369.0 | 214.9 | 1.482 | 133.0 | 0.695 | 466.6 |
| 8.23 | 361.9 | 213.8 | 1.486 | 133.2 | 0.721 | 470.9 |
| 13.93 | 329.9 | 212.2 | 1.524 | 133.3 | 0.859 | 487.6 |
| 24.16 | 290.3 | 209.3 | 1.578 | 133.5 | 1.098 | 513.5 |
| 41.33 | 266.2 | 204.8 | 1.601 | 134.2 | 1.325 | 535.9 |
| 60.53 | 248.0 | 200.0 | 1.612 | 135.0 | 1.560 | 556.9 |
| 82.78 | 240.2 | 194.8 | 1.597 | 136.2 | 1.720 | 571.9 |
| 102.1 | 238.6 | 190.5 | 1.574 | 137.2 | 1.806 | 580.6 |
| 120.9 | 229.9 | 186.5 | 1.569 | 138.1 | 2.002 | 595.5 |
| Run 2516 | | $G = 0.1421$ kg/sec, $P_{in} = 0.481$ MPa, $T_{fin} = 312.2$ K | | | | |
| 2.42 | 378.5 | 210.4 | 1.587 | 166.0 | 1.011 | 499.1 |
| 4.62 | 355.8 | 209.5 | 1.618 | 166.0 | 1.106 | 511.7 |
| 8.25 | 347.8 | 208.2 | 1.624 | 166.3 | 1.154 | 517.7 |
| 13.96 | 314.7 | 206.2 | 1.675 | 166.5 | 1.410 | 540.3 |
| 24.23 | 274.6 | 202.7 | 1.743 | 166.8 | 1.859 | 574.9 |
| 41.44 | 249.7 | 197.2 | 1.770 | 167.9 | 2.303 | 604.9 |
| 60.72 | 231.2 | 191.5 | 1.779 | 169.2 | 2.772 | 632.6 |
| 83.14 | 222.9 | 185.2 | 1.753 | 171.1 | 3.108 | 652.0 |
| 102.4 | 221.4 | 180.1 | 1.716 | 172.9 | 3.276 | 662.3 |
| 121.3 | 214.4 | 175.3 | 1.699 | 174.4 | 3.630 | 679.4 |
| Run 2519 | | $G = 0.1401$ kg/sec, $P_{in} = 0.480$ MPa, $T_{fin} = 313.2$ K | | | | |
| 2.42 | 370.6 | 206.4 | 1.791 | 221.8 | 1.888 | 566.6 |
| 4.63 | 344.6 | 205.3 | 1.839 | 221.6 | 2.094 | 585.7 |
| 8.28 | 334.1 | 203.6 | 1.849 | 221.8 | 2.213 | 595.2 |
| 14.01 | 298.5 | 201.0 | 1.921 | 221.6 | 2.813 | 628.3 |

**Table 20**  Annulus with $d_1/d_2 = 0.585$ (*Continued*)

| $x_1/d$ | $Nu_{12}$ | $Re \cdot 10^{-3}$ | $\Psi_1$ | $q_1 \cdot 10^{-3}$ W/m² | $q_2 \cdot 10^{-3}$ W/m² | $T_{w1}$ K |
|---|---|---|---|---|---|---|
| 24.33 | 257.1 | 196.5 | 2.009 | 221.1 | 3.847 | 676.8 |
| 41.64 | 230.2 | 189.6 | 2.033 | 221.6 | 4.903 | 718.6 |
| 61.01 | 212.0 | 182.4 | 2.024 | 222.4 | 5.946 | 753.6 |
| 83.54 | 204.6 | 174.5 | 1.963 | 224.5 | 6.593 | 774.5 |
| 102.9 | 205.3 | 168.3 | 1.891 | 226.8 | 6.798 | 782.1 |
| 121.8 | 199.0 | 163.2 | 1.855 | 228.3 | 7.400 | 801.4 |

Run 2521  $G = 0.1370$ kg/sec, $P_{in} = 0.482$ MPa, $T_{fin} = 313.5$ K

| $x_1/d$ | $Nu_{12}$ | $Re \cdot 10^{-3}$ | $\Psi_1$ | $q_1 \cdot 10^{-3}$ W/m² | $q_2 \cdot 10^{-3}$ W/m² | $T_{w1}$ K |
|---|---|---|---|---|---|---|
| 2.43 | 359.9 | 201.2 | 1.959 | 262.8 | 2.912 | 621.4 |
| 4.65 | 333.6 | 199.9 | 2.016 | 262.3 | 3.242 | 644.9 |
| 8.30 | 321.4 | 197.9 | 2.030 | 262.5 | 3.472 | 657.7 |
| 14.05 | 284.5 | 194.9 | 2.117 | 261.5 | 4.526 | 699.4 |
| 24.41 | 243.3 | 189.7 | 2.214 | 259.8 | 6.284 | 757.9 |
| 41.79 | 217.1 | 181.8 | 2.222 | 259.3 | 8.006 | 805.6 |
| 61.23 | 199.3 | 173.6 | 2.189 | 259.4 | 9.655 | 844.0 |
| 83.83 | 193.6 | 164.8 | 2.092 | 261.5 | 10.47 | 862.8 |
| 103.2 | 194.7 | 158.6 | 1.995 | 264.2 | 10.67 | 868.5 |
| 122.2 | 189.9 | 152.9 | 1.935 | 265.6 | 11.57 | 886.7 |

# 7. EXPERIMENTAL DATA ON HEAT TRANSFER IN AN EXTERNALLY HEATED ANNULUS

**Table 21** Annulus with $d_1/d_2 = 0.108$

| $x_2/d$ | $Nu_{21}$ | $Re \cdot 10^{-3}$ | $\Psi_2$ | $q_1 \cdot 10^{-3}$ W/m² | $q_2 \cdot 10^{-3}$, W/m² | $T_{w2}$ K |
|---|---|---|---|---|---|---|
| Run 5203 | | | $G = 0.1028$ kg/sec, $P_{in} = 0.558$ MPa, $T_{fin} = 279.0$ K | | | |
| 1.20 | 517.1 | 239.7 | 1.073 | 0.072 | 10.49 | 299.7 |
| 2.19 | 472.5 | 239.5 | 1.080 | 0.080 | 10.49 | 301.8 |
| 3.78 | 441.1 | 239.2 | 1.085 | 0.088 | 10.50 | 303.8 |
| 6.37 | 412.1 | 238.9 | 1.091 | 0.098 | 10.52 | 306.0 |
| 11.15 | 391.6 | 238.1 | 1.096 | 0.107 | 10.55 | 308.4 |
| 19.12 | 363.8 | 236.9 | 1.102 | 0.121 | 10.59 | 312.2 |
| 27.88 | 356.0 | 235.5 | 1.103 | 0.130 | 10.63 | 314.8 |
| 38.24 | 350.9 | 234.0 | 1.104 | 0.137 | 10.69 | 317.5 |
| 47.00 | 346.1 | 232.7 | 1.104 | 0.142 | 10.73 | 319.9 |
| 55.77 | 345.4 | 231.4 | 1.104 | 0.147 | 10.78 | 321.9 |
| Run 5206 | | | $G = 0.1046$ kg/sec, $P_{in} = 0.569$ MPa, $T_{fin} = 278.7$ K | | | |
| 1.20 | 517.3 | 243.7 | 1.138 | 0.149 | 19.86 | 317.9 |
| 2.19 | 471.8 | 243.4 | 1.151 | 0.167 | 19.86 | 322.0 |
| 3.78 | 437.3 | 243.0 | 1.163 | 0.185 | 19.88 | 325.9 |
| 6.37 | 407.6 | 242.2 | 1.173 | 0.206 | 19.90 | 330.2 |
| 11.15 | 383.3 | 240.8 | 1.182 | 0.230 | 19.93 | 335.1 |
| 19.12 | 352.8 | 238.5 | 1.194 | 0.268 | 19.98 | 342.5 |
| 27.88 | 342.5 | 236.1 | 1.196 | 0.290 | 20.04 | 347.4 |
| 38.23 | 333.6 | 233.3 | 1.196 | 0.312 | 20.10 | 352.7 |
| 46.99 | 326.2 | 231.0 | 1.197 | 0.332 | 20.16 | 357.2 |
| 55.76 | 322.1 | 228.7 | 1.195 | 0.348 | 20.21 | 361.2 |
| Run 5209 | | | $G = 0.1037$ kg/sec, $P_{in} = 0.562$ MPa, $T_{fin} = 279.6$ K | | | |
| 1.20 | 513.8 | 240.8 | 1.234 | 0.296 | 33.66 | 346.3 |
| 2.19 | 467.3 | 240.3 | 1.256 | 0.337 | 33.67 | 353.4 |
| 3.78 | 431.7 | 239.5 | 1.276 | 0.379 | 33.70 | 360.3 |
| 6.37 | 401.1 | 238.2 | 1.294 | 0.428 | 33.76 | 367.7 |
| 11.15 | 372.6 | 235.9 | 1.310 | 0.492 | 33.85 | 376.9 |
| 19.11 | 339.2 | 232.1 | 1.330 | 0.591 | 34.00 | 398.1 |
| 27.87 | 326.3 | 228.2 | 1.331 | 0.657 | 34.18 | 399.0 |
| 38.22 | 316.9 | 223.6 | 1.328 | 0.723 | 34.39 | 407.9 |
| 46.98 | 307.4 | 219.9 | 1.327 | 0.788 | 34.56 | 416.0 |
| 55.74 | 302.7 | 216.3 | 1.321 | 0.842 | 34.73 | 422.9 |

**Table 21** Annulus with $d_1/d_2 = 0.108$ (*Continued*)

| $x_2/d$ | $Nu_{21}$ | $Re \cdot 10^{-3}$ | $\Psi_2'$ | $q_1 \cdot 10^{-3}$ W/m² | $q_2 \cdot 10^{-3}$ W/m² | $T_{w2}$ K |
|---|---|---|---|---|---|---|
| Run 5214 | | $G = 0.1031$ kg/sec, $P_{in} = 0.555$ MPa, $T_{fin} = 278.7$ K | | | | |
| 1.20 | 502.7 | 239.3 | 1.438 | 0.753 | 61.58 | 403.5 |
| 2.19 | 450.7 | 238.4 | 1.485 | 0.890 | 61.58 | 418.4 |
| 3.78 | 410.3 | 237.0 | 1.526 | 1.041 | 61.66 | 433.2 |
| 6.37 | 375.1 | 234.7 | 1.565 | 1.228 | 61.81 | 449.4 |
| 11.14 | 340.6 | 230.6 | 1.600 | 1.494 | 62.08 | 469.6 |
| 19.10 | 303.3 | 224.0 | 1.635 | 1.921 | 62.52 | 497.4 |
| 27.85 | 287.2 | 217.0 | 1.631 | 2.230 | 63.13 | 515.5 |
| 38.20 | 273.8 | 209.8 | 1.617 | 2.572 | 63.75 | 533.8 |
| 46.94 | 262.9 | 204.2 | 1.607 | 2.857 | 64.26 | 549.7 |
| 55.69 | 256.3 | 199.0 | 1.589 | 3.186 | 64.80 | 562.8 |
| Run 5218 | | $G = 0.1060$ kg/sec, $P_{in} = 0.567$ MPa, $T_{fin} = 278.8$ K | | | | |
| 1.20 | 515.2 | 245.3 | 1.612 | 1.385 | 88.51 | 453.5 |
| 2.19 | 457.1 | 244.0 | 1.682 | 1.682 | 88.50 | 476.1 |
| 3.78 | 411.9 | 242.0 | 1.744 | 2.020 | 88.62 | 498.7 |
| 6.37 | 373.6 | 238.8 | 1.799 | 2.451 | 88.93 | 522.8 |
| 11.14 | 333.8 | 233.0 | 1.852 | 3.121 | 89.46 | 554.3 |
| 19.09 | 290.0 | 223.7 | 1.904 | 4.286 | 90.30 | 598.7 |
| 27.83 | 269.9 | 214.8 | 1.895 | 5.179 | 91.58 | 627.8 |
| 38.17 | 254.0 | 205.5 | 1.866 | 6.131 | 92.82 | 655.5 |
| 46.90 | 241.6 | 198.2 | 1.843 | 7.087 | 93.76 | 679.3 |
| 55.64 | 234.6 | 191.6 | 1.808 | 7.913 | 94.81 | 697.5 |
| Run 5221 | | $G = 0.1062$ kg/sec, $P_{in} = 0.576$ MPa, $T_{fin} = 280.1$ K | | | | |
| 1.20 | 500.6 | 243.9 | 1.809 | 2.534 | 11.49 | 512.1 |
| 2.19 | 439.4 | 242.2 | 1.907 | 3.146 | 11.47 | 544.2 |
| 3.78 | 392.0 | 239.6 | 1.992 | 3.870 | 11.48 | 576.3 |
| 6.36 | 340.8 | 235.6 | 2.073 | 4.882 | 11.50 | 612.2 |
| 11.13 | 306.7 | 228.2 | 2.148 | 6.504 | 11.56 | 658.4 |
| 19.07 | 263.4 | 217.1 | 2.201 | 9.183 | 11.63 | 718.0 |
| 27.81 | 246.1 | 206.7 | 2.160 | 11.02 | 11.83 | 752.0 |
| 38.14 | 220.7 | 195.8 | 2.103 | 13.25 | 11.97 | 787.1 |
| 46.87 | 216.9 | 187.3 | 2.059 | 15.37 | 12.06 | 816.6 |
| 55.60 | 209.4 | 179.9 | 2.006 | 17.17 | 12.19 | 839.5 |

**Table 22** Annulus with $d_1/d_2 = 0.205$

| $x_2/d$ | $Nu_{2\perp}$ | $Re \cdot 10^{-3}$ | $\Psi'_2$ | $q_1 \cdot 10^{-3}$ W/m² | $q_2 \cdot 10^{-3}$ W/m² | $T_{w2}$ K |
|---|---|---|---|---|---|---|
| Run 3206 | | $G = 0.0972$ kg/sec, $P_{in} = 0.495$ MPa, $T_{fin} = 282.0$ K | | | | |
| 1.34 | 418.2 | 206.6 | 1.134 | 0.144 | 17.85 | 320.5 |
| 2.45 | 382.5 | 206.4 | 1.146 | 0.161 | 17.85 | 324.4 |
| 4.24 | 357.3 | 206.0 | 1.156 | 0.177 | 17.86 | 327.9 |
| 7.15 | 334.7 | 205.4 | 1.166 | 0.196 | 17.88 | 331.8 |
| 12.51 | 316.2 | 204.3 | 1.173 | 0.217 | 17.90 | 336.3 |
| 21.45 | 293.9 | 202.4 | 1.183 | 0.250 | 17.93 | 343.0 |
| 31.28 | 284.4 | 200.4 | 1.186 | 0.271 | 18.00 | 347.9 |
| 42.90 | 278.6 | 198.2 | 1.185 | 0.292 | 18.06 | 352.7 |
| 52.72 | 272.5 | 196.3 | 1.186 | 0.292 | 18.11 | 357.1 |
| 62.55 | 272.2 | 194.4 | 1.182 | 0.316 | 18.16 | 360.4 |
| Run 3210 | | $G = 0.0984$ kg/sec, $P_{in} = 0.525$ MPa, $T_{fin} = 282.3$ K | | | | |
| 1.34 | 418.2 | 208.7 | 1.239 | 0.304 | 31.91 | 350.9 |
| 2.45 | 381.1 | 208.3 | 1.261 | 0.345 | 31.90 | 358.1 |
| 4.24 | 353.8 | 207.6 | 1.279 | 0.385 | 31.93 | 364.1 |
| 7.14 | 329.9 | 206.5 | 1.312 | 0.434 | 31.98 | 372.0 |
| 12.51 | 307.3 | 204.5 | 1.330 | 0.496 | 32.06 | 381.0 |
| 21.44 | 281.1 | 201.4 | 1.333 | 0.593 | 32.19 | 393.9 |
| 31.27 | 260.6 | 197.9 | 1.330 | 0.661 | 32.36 | 403.0 |
| 42.88 | 261.6 | 193.9 | 1.329 | 0.731 | 32.55 | 412.1 |
| 52.70 | 254.0 | 190.8 | 1.296 | 0.732 | 32.70 | 420.2 |
| 62.52 | 252.1 | 187.7 | 1.321 | 0.831 | 32.87 | 426.2 |
| Run 3214 | | $G = 0.0999$ kg/sec, $P_{in} = 0.533$ MPa, $T_{fin} = 282.4$ K | | | | |
| 1.34 | 423.8 | 211.3 | 1.369 | 0.580 | 50.19 | 388.8 |
| 2.45 | 383.1 | 210.7 | 1.406 | 0.673 | 50.18 | 400.7 |
| 4.24 | 352.3 | 209.6 | 1.437 | 0.770 | 50.22 | 412.0 |
| 7.14 | 325.3 | 207.9 | 1.465 | 0.890 | 50.32 | 424.3 |
| 12.50 | 297.9 | 204.9 | 1.493 | 1.058 | 50.48 | 440.3 |
| 21.43 | 268.6 | 199.9 | 1.520 | 1.318 | 50.74 | 461.8 |
| 31.25 | 254.5 | 194.8 | 1.521 | 1.517 | 51.11 | 477.1 |
| 42.85 | 244.2 | 189.2 | 1.511 | 1.727 | 51.51 | 492.0 |
| 52.66 | 235.3 | 184.8 | 1.505 | 1.731 | 51.82 | 505.0 |
| 62.48 | 231.7 | 180.8 | 1.489 | 2.054 | 52.18 | 514.6 |
| Run 3218 | | $G = 0.101$ kg/sec, $P_{in} = 0.646$ MPa, $T_{fin} = 282.5$ K | | | | |
| 1.34 | 425.7 | 212.1 | 1.531 | 1.081 | 72.78 | 435.9 |
| 1.45 | 380.7 | 211.1 | 1.588 | 1.288 | 72.75 | 454.5 |
| 4.24 | 346.8 | 209.6 | 1.636 | 1.512 | 72.82 | 472.4 |
| 7.14 | 317.1 | 207.2 | 1.679 | 1.800 | 73.02 | 491.8 |

**Table 22** Annulus with $d_1/d_2 = 0.205$ (*Continued*)

| $x_2/d$ | $Nu_{21}$ | $Re \cdot 10^{-3}$ | $\Psi_2^*$ | $q_1 \cdot 10^{-3}$ W/m² | $q_2 \cdot 10^{-3}$ W/m² | $T_{w2}$ K |
|---|---|---|---|---|---|---|
| 12.50 | 286.9 | 202.9 | 1.719 | 2.218 | 73.37 | 516.6 |
| 21.41 | 255.0 | 195.9 | 1.754 | 2.895 | 73.95 | 550.1 |
| 31.22 | 239.0 | 189.1 | 1.749 | 3.441 | 74.81 | 573.8 |
| 42.81 | 226.5 | 181.9 | 1.729 | 4.049 | 76.71 | 596.8 |
| 52.62 | 222.3 | 176.3 | 1.696 | 4.058 | 76.55 | 610.5 |
| 62.41 | 212.9 | 171.0 | 1.681 | 5.018 | 77.22 | 630.2 |

Run 3222   $G = 0.1011$ kg/sec, $P_{in} = 0.553$ MPa, $T_{fin} = 282.6$ K

| $x_2/d$ | $Nu_{21}$ | $Re \cdot 10^{-3}$ | $\Psi_2^*$ | $q_1 \cdot 10^{-3}$ W/m² | $q_2 \cdot 10^{-3}$ W/m² | $T_{w2}$ K |
|---|---|---|---|---|---|---|
| 1.34 | 422.8 | 212.5 | 1.714 | 1.937 | 97.55 | 489.3 |
| 2.45 | 375.2 | 211.2 | 1.793 | 2.357 | 97.42 | 515.5 |
| 4.23 | 337.7 | 209.2 | 1.863 | 2.842 | 97.43 | 541.9 |
| 7.13 | 305.3 | 206.1 | 1.925 | 3.484 | 97.62 | 570.1 |
| 12.49 | 271.5 | 200.4 | 1.981 | 4.478 | 97.97 | 607.0 |
| 21.39 | 236.7 | 191.6 | 2.022 | 6.103 | 98.59 | 651.1 |
| 31.19 | 219.0 | 183.3 | 2.006 | 7.458 | 99.81 | 688.7 |
| 42.76 | 204.4 | 174.6 | 1.967 | 9.004 | 100.9 | 721.0 |
| 52.55 | 193.3 | 167.8 | 1.937 | 9.031 | 101.6 | 748.0 |
| 62.35 | 188.9 | 161.4 | 1.884 | 11.40 | 102.7 | 765.4 |

Run 3224   $G = 0.1001$ kg/sec, $P_{in} = 0.547$ MPa, $T_{fin} = 283.5$ K

| $x_2/d$ | $Nu_{21}$ | $Re \cdot 10^{-3}$ | $\Psi_2^*$ | $q_1 \cdot 10^{-3}$ W/m² | $q_2 \cdot 10^{-3}$ W/m² | $T_{w2}$ K |
|---|---|---|---|---|---|---|
| 1.34 | 420.1 | 209.6 | 1.848 | 2.882 | 116.2 | 530.4 |
| 2.45 | 360.8 | 208.1 | 1.947 | 3.560 | 115.9 | 563.5 |
| 4.23 | 330.6 | 205.1 | 2.033 | 4.354 | 115.8 | 596.3 |
| 7.13 | 295.7 | 202.1 | 2.111 | 5.472 | 115.9 | 632.7 |
| 12.48 | 260.2 | 195.4 | 2.175 | 7.190 | 116.1 | 678.7 |
| 21.38 | 223.2 | 185.3 | 2.219 | 10.11 | 116.5 | 739.3 |
| 31.17 | 204.9 | 176.3 | 2.189 | 12.52 | 117.9 | 780.2 |
| 42.73 | 189.6 | 166.6 | 2.129 | 15.24 | 118.9 | 818.5 |
| 52.50 | 178.7 | 159.1 | 2.080 | 15.28 | 119.4 | 849.0 |
| 62.30 | 174.3 | 152.8 | 2.010 | 19.23 | 120.7 | 868.1 |

**Table 23** Annulus with $d_1/d_2 = 0.373$

| $x_2/d$ | $Nu_{21}$ | $Re \cdot 10^{-3}$ | $\Psi_2^*$ | $q_1 \cdot 10^{-3}$ W/m² | $q_2 \cdot 10^{-3}$ W/m² | $T_{w2}$, K |
|---|---|---|---|---|---|---|
| Run 1203 | | $G = 0.1820$ kg/sec, $P_{in} = 0.537$ MPa, $T_{fin} = 298.7$ K | | | | |
| 1.697 | 596.0 | 325.1 | 1.027 | 0.029 | 7.096 | 308.8 |
| 3.111 | 544.0 | 325.0 | 1.029 | 0.032 | 7.093 | 307.7 |
| 5.373 | 506.0 | 324.9 | 1.031 | 0.035 | 7.089 | 308.4 |
| 9.049 | 478.8 | 324.7 | 1.033 | 0.038 | 7.085 | 308.2 |

**Table 23** Annulus with $d_1/d_2 = 0.373$ (*Continued*)

| $x_2/d$ | $Nu_{21}$ | $Re_2 \cdot 10^{-3}$ | $\Psi_2$ | $q_1 \cdot 10^{-3}$ W/m² | $q_2 \cdot 10^{-3}$ W/m² | $T_{w2}$ K |
|---|---|---|---|---|---|---|
| 15.84 | 468.8 | 324.3 | 1.034 | 0.040 | 7.077 | 309.7 |
| 27.15 | 443.6 | 323.7 | 1.035 | 0.044 | 7.061 | 311.0 |
| 39.59 | 425.0 | 323.1 | 1.037 | 0.048 | 7.044 | 312.1 |
| 54.29 | 427.7 | 322.3 | 1.036 | 0.050 | 7.026 | 312.8 |
| 60.73 | 416.5 | 321.7 | 1.037 | 0.047 | 7.009 | 313.8 |
| 79.17 | 421.8 | 321.0 | 1.036 | 0.050 | 6.994 | 314.4 |

Run 1207      $G = 0.1820$ kg/sec, $P_{in} = 0.527$ MPa, $T_{fin} = 299.2$ K

| $x_2/d$ | $Nu_{21}$ | $Re_2 \cdot 10^{-3}$ | $\Psi_2$ | $q_1 \cdot 10^{-3}$ W/m² | $q_2 \cdot 10^{-3}$ W/m² | $T_{w2}$ K |
|---|---|---|---|---|---|---|
| 1.696 | 381.9 | 324.3 | 1.075 | 0.091 | 19.68 | 322.1 |
| 3.110 | 537.2 | 324.1 | 1.081 | 0.100 | 19.67 | 324.2 |
| 5.372 | 500.8 | 323.8 | 1.087 | 0.109 | 19.67 | 326.3 |
| 9.047 | 472.4 | 323.2 | 1.092 | 0.118 | 19.68 | 328.5 |
| 15.83 | 448.3 | 322.2 | 1.096 | 0.129 | 19.68 | 331.0 |
| 27.14 | 423.9 | 320.6 | 1.101 | 0.144 | 19.69 | 334.3 |
| 39.58 | 413.8 | 318.8 | 1.102 | 0.153 | 19.71 | 337.0 |
| 54.27 | 413.1 | 316.7 | 1.101 | 0.159 | 19.72 | 339.4 |
| 66.71 | 401.7 | 315.0 | 1.103 | 0.166 | 19.74 | 342.1 |
| 79.15 | 400.0 | 313.3 | 1.102 | 0.171 | 19.75 | 344.2 |

Run 1211      $G = 0.1827$ kg/sec, $P_{in} = 0.529$ MPa, $T_{fin} = 293.4$ K

| $x_2/d$ | $Nu_{21}$ | $Re_2 \cdot 10^{-3}$ | $\Psi_2$ | $q_1 \cdot 10^{-3}$ W/m² | $q_2 \cdot 10^{-3}$ W/m² | $T_{w2}$ K |
|---|---|---|---|---|---|---|
| 1.696 | 583.0 | 320.2 | 1.130 | 0.167 | 34.83 | 334.6 |
| 3.109 | 530.4 | 329.9 | 1.149 | 0.184 | 34.83 | 338.2 |
| 5.370 | 501.7 | 329.3 | 1.159 | 0.213 | 34.83 | 342.1 |
| 9.044 | 471.1 | 328.4 | 1.168 | 0.224 | 34.85 | 346.0 |
| 15.83 | 444.2 | 326.6 | 1.177 | 0.249 | 34.92 | 350.8 |
| 27.13 | 417.3 | 323.6 | 1.184 | 0.281 | 34.98 | 357.0 |
| 39.56 | 407.2 | 320.5 | 1.185 | 0.302 | 35.06 | 361.6 |
| 54.25 | 403.1 | 316.7 | 1.183 | 0.319 | 35.10 | 365.9 |
| 66.68 | 391.7 | 313.7 | 1.184 | 0.340 | 35.11 | 370.7 |
| 79.11 | 387.5 | 310.8 | 1.182 | 0.355 |  | 374.5 |

Run 1216      $G = 0.1832$ kg/sec, $P_{in} = 0.533$ MPa, $T_{fin} = 295.1$ K

| $x_2/d$ | $Nu_{21}$ | $Re_2 \cdot 10^{-3}$ | $\Psi_2$ | $q_1 \cdot 10^{-3}$ W/m² | $q_2 \cdot 10^{-3}$ W/m² | $T_{w2}$ K |
|---|---|---|---|---|---|---|
| 1.695 | 578.1 | 329.2 | 1.275 | 0.423 | 69.77 | 377.7 |
| 3.108 | 533.2 | 328.5 | 1.297 | 0.475 | 69.77 | 385.2 |
| 5.367 | 493.3 | 327.2 | 1.318 | 0.536 | 69.82 | 393.3 |
| 9.038 | 460.9 | 325.4 | 1.337 | 0.602 | 69.93 | 401.7 |
| 15.82 | 429.0 | 321.9 | 1.354 | 0.693 | 70.13 | 412.3 |
| 27.11 | 396.2 | 316.0 | 1.370 | 0.820 | 70.48 | 426.4 |
| 39.52 | 379.3 | 310.2 | 1.372 | 0.921 | 70.90 | 437.2 |
| 54.20 | 369.4 | 303.7 | 1.366 | 1.015 | 71.38 | 447.3 |
| 66.61 | 355.3 | 298.5 | 1.367 | 1.123 | 71.77 | 457.8 |
| 79.02 | 348.2 | 293.5 | 1.362 | 1.208 | 72.18 | 466.3 |

**Table 23** Annulus with $d_1/d_2 = 0.373$ (*Continued*)

| $x_2/d$ | $Nu_{21}$ | $Re_2 \cdot 10^{-3}$ | $\Psi_2^*$ | $q_1 \cdot 10^{-3}$ W/m² | $q_2 \cdot 10^{-3}$ W/m² | $T_{w2}$ K |
|---|---|---|---|---|---|---|
| Run 1220 | | $G = 0.1830$ kg/sec, $P_{in} = 0.530$ MPa, $T_{fin} = 295.9$ K | | | | |
| 1.695 | 571.4 | 327.4 | 1.407 | 0.783 | 102.8 | 418.7 |
| 3.106 | 523.5 | 326.3 | 1.441 | 0.895 | 102.8 | 430.6 |
| 5.364 | 482.1 | 324.5 | 1.474 | 1.027 | 102.9 | 443.2 |
| 9.032 | 446.4 | 321.8 | 1.503 | 1.185 | 103.2 | 456.8 |
| 15.80 | 410.4 | 316.7 | 1.531 | 1.409 | 103.7 | 474.2 |
| 27.08 | 373.6 | 308.6 | 1.554 | 1.738 | 104.5 | 497.1 |
| 39.48 | 356.2 | 300.6 | 1.551 | 1.995 | 105.5 | 513.4 |
| 54.14 | 344.0 | 201.9 | 1.537 | 2.256 | 106.6 | 528.9 |
| 66.53 | 320.8 | 285.0 | 1.533 | 2.551 | 107.6 | 544.8 |
| 78.92 | 322.3 | 278.4 | 1.520 | 2.791 | 108.5 | 557.4 |
| Run 1225 | | $G = 0.1850$ kg/sec, $P_{in} = 0.527$ MPa, $T_{fin} = 296.8$ K | | | | |
| 1.693 | 559.3 | 328.7 | 1.726 | 2.338 | 181.5 | 517.4 |
| 3.102 | 502.9 | 326.8 | 1.798 | 2.785 | 181.4 | 542.5 |
| 5.355 | 458.9 | 323.7 | 1.858 | 3.291 | 181.6 | 566.9 |
| 9.014 | 416.9 | 318.9 | 1.917 | 4.010 | 182.1 | 395.4 |
| 15.77 | 375.6 | 310.5 | 1.964 | 5.057 | 183.3 | 630.5 |
| 27.01 | 330.8 | 297.9 | 2.003 | 6.807 | 185.1 | 678.4 |
| 39.37 | 308.3 | 285.5 | 1.985 | 8.258 | 187.8 | 711.7 |
| 52.97 | 290.9 | 272.2 | 1.943 | 9.820 | 190.5 | 742.8 |
| 66.31 | 276.1 | 261.7 | 1.915 | 11.39 | 192.6 | 770.5 |
| 78.66 | 268.4 | 252.3 | 1.872 | 12.69 | 195.1 | 791.6 |

**Table 24** Annulus with $d_1/d_2 = 0.585$

| $x_2/d$ | $Nu_{21}^*$ | $Re_2 \cdot 10^{-3}$ | $\Psi_2^*$ | $q_1 \cdot 10^{-3}$ W/m² | $q_2 \cdot 10^{-3}$ W/m² | $T_{w2}$, K |
|---|---|---|---|---|---|---|
| Run 2321 | | $G = 0.1473$ kg/sec, $P_{in} = 0.471$ MPa, $T_{fin} = 302.1$ K | | | | |
| 2.57 | 412.7 | 225.3 | 1.099 | 0.118 | 28.26 | 332.7 |
| 4.71 | 382.7 | 225.1 | 1.107 | 0.130 | 28.27 | 335.4 |
| 8.13 | 358.2 | 224.6 | 1.113 | 0.145 | 28.29 | 338.2 |
| 13.69 | 339.1 | 224.0 | 1.119 | 0.158 | 28.33 | 341.2 |
| 23.95 | 321.0 | 222.8 | 1.125 | 0.175 | 28.41 | 345.1 |
| 41.05 | 304.2 | 220.8 | 1.130 | 0.197 | 28.54 | 350.4 |
| 59.86 | 291.7 | 218.7 | 1.133 | 0.215 | 28.68 | 355.7 |
| 82.08 | 289.1 | 216.3 | 1.132 | 0.226 | 28.86 | 360.1 |
| 100.89 | 283.0 | 214.3 | 1.132 | 0.240 | 29.00 | 364.6 |
| 119.69 | 282.4 | 212.4 | 1.130 | 0.259 | 29.15 | 368.3 |

**Table 24** Annulus with $d_1/d_2 = 0.585$ (*Continued*)

| $x_2/d$ | $Nu_{21}$ | $Re_2 \cdot 10^{-3}$ | $\Psi_2$ | $q_1 \cdot 10^{-3}$ W/m² | $q_2 \cdot 10^{-3}$ W/m² | $T_{w2}$ K |
|---|---|---|---|---|---|---|
| Run 2324 | | | $G = 0.1481$ kg/sec, $P_{in} = 0.477$ MPa, $T_{fin} = 308.2$ K | | | |
| 2.57 | 411.9 | 222.4 | 1.187 | 0.282 | 55,51 | 367.4 |
| 4.70 | 381.5 | 221.9 | 1.201 | 0.315 | 55,50 | 372.6 |
| 8.12 | 354.3 | 221.1 | 1.215 | 0.355 | 55,52 | 378.5 |
| 13.67 | 331.4 | 219.9 | 1.227 | 0.397 | 55,58 | 384.9 |
| 23.92 | 308.3 | 217.8 | 1.239 | 0.454 | 55,68 | 393.5 |
| 41.0 | 286.5 | 214.3 | 1.249 | 0.529 | 55,86 | 404.8 |
| 59.77 | 273.2 | 210.7 | 1.252 | 0.594 | 56,08 | 414.7 |
| 81.96 | 266.6 | 206.6 | 1.247 | 0.648 | 56,35 | 424.0 |
| 100.73 | 257.2 | 203.4 | 1.247 | 0.707 | 56,55 | 433.1 |
| 119.49 | 253.4 | 200.3 | 1.243 | 0.775 | 56,77 | 440.5 |
| Run 2327 | | | $G = 0.1472$ kg/sec, $P_{in} = 0.485$ MPa, $T_{fin} = 309.9$ K | | | |
| 2.56 | 403.4 | 219.6 | 1.277 | 0.495 | 81,49 | 398.1 |
| 4.70 | 371.9 | 219.0 | 1.299 | 0.559 | 81,50 | 406.3 |
| 8.11 | 344.1 | 217.9 | 1.320 | 0.640 | 81,56 | 415.4 |
| 13.66 | 319.6 | 216.2 | 1.339 | 0.730 | 81,71 | 425.5 |
| 23.89 | 294.1 | 213.1 | 1.357 | 0.858 | 81,99 | 439.3 |
| 40.94 | 270.2 | 208.3 | 1.371 | 1.036 | 82,48 | 457.0 |
| 59.68 | 256,6 | 203.3 | 1.371 | 1.192 | 83,09 | 471.9 |
| 81.83 | 248.4 | 197.9 | 1.361 | 1.338 | 83,79 | 486.1 |
| 100.56 | 239.3 | 193.5 | 1.358 | 1.493 | 84,35 | 499.6 |
| 119.28 | 235.4 | 189.2 | 1.347 | 1.654 | 84.94 | 510.6 |
| Run 2330 | | | $G = 0.1390$ kg/sec, $P_{in} = 0.495$ MPa, $T_{fin} = 302.4$ K | | | |
| 2.56 | 383.7 | 210.8 | 1.478 | 1.063 | 128.3 | 451.2 |
| 4.69 | 349.8 | 209.6 | 1.519 | 1.242 | 128.3 | 466.5 |
| 8.10 | 320.1 | 207.9 | 1.557 | 1.466 | 128.4 | 483.2 |
| 13.63 | 292.9 | 205.1 | 1.592 | 1.743 | 128.7 | 502.3 |
| 23.82 | 263.6 | 200.3 | 1.625 | 2.169 | 129.2 | 528.5 |
| 40.80 | 235.7 | 192.8 | 1.644 | 2.804 | 130.1 | 561.6 |
| 59.48 | 221.2 | 185.4 | 1.630 | 3.365 | 131,4 | 587.0 |
| 81.54 | 211,2 | 177.4 | 1,600 | 3.933 | 132.9 | 611.0 |
| 100.19 | 201.8 | 171.1 | 1.580 | 4.521 | 134.0 | 632.8 |
| 118.86 | 197.9 | 165.4 | 1.550 | 4.984 | 135,3 | 649.8 |
| Run 2333 | | | $G = 0.1347$ kg/sec, $P_{in} = 0.494$ MPa, $T_{fin} = 306.7$ K | | | |
| 2.56 | 368.0 | 201.3 | 1.637 | 1.935 | 168.7 | 508.2 |
| 4.68 | 332.1 | 199.9 | 1.695 | 2.313 | 168.6 | 530.6 |
| 8.08 | 302.3 | 197.7 | 1.745 | 2.776 | 168.6 | 553.9 |
| 13.59 | 273.3 | 194.3 | 1.794 | 3.407 | 168.9 | 581.8 |

**Table 24** Annulus with $d_1/d_2 = 0.585$ (*Continued*)

| $x_2/d$ | $Nu_{21}$ | $Re_2 \cdot 10^{-3}$ | $\Psi_2^*$ | $q_1 \cdot 10^{-3}$ W/m² | $q_2 \cdot 10^{-3}$ W/m² | $T_{w2}$ K |
|---|---|---|---|---|---|---|
| 23.76 | 241.7 | 188.4 | 1.838 | 4.421 | 169.4 | 620.4 |
| 40.68 | 212.4 | 179.5 | 1.853 | 5.962 | 170.4 | 667.7 |
| 59.29 | 197.4 | 171.0 | 1.821 | 7.299 | 172.2 | 701.8 |
| 81.28 | 186.7 | 161.2 | 1.766 | 8.675 | 174.0 | 733.4 |
| 99.88 | 178.5 | 154.5 | 1.726 | 9.988 | 175.3 | 760.2 |
| 118.51 | 176.1 | 148.4 | 1.671 | 10.94 | 177.1 | 779.1 |
| Run 2338 | | $G = 0.1404$ kg/sec, $P_{in} = 0.478$ MPa, $T_{fin} = 312.1$ K | | | | |
| 2.57 | 407.1 | 209.2 | 1.044 | 0.056 | 13.17 | 326.2 |
| 4.71 | 376.8 | 209.1 | 1.048 | 0.061 | 13.17 | 327.5 |
| 8.13 | 350.6 | 208.9 | 1.051 | 0.068 | 13.17 | 328.9 |
| 13.69 | 333.6 | 208.6 | 1.053 | 0.074 | 13.17 | 330.2 |
| 23.96 | 319.4 | 208.1 | 1.056 | 0.080 | 13.17 | 331.9 |
| 41.07 | 301.4 | 207.2 | 1.058 | 0.089 | 13.17 | 334.5 |
| 59.87 | 286.6 | 206.3 | 1.061 | 0.096 | 13.17 | 337.2 |
| 82.14 | 284.7 | 205.2 | 1.060 | 0.099 | 13.18 | 339.3 |
| 100.96 | 277.4 | 204.3 | 1.061 | 0.102 | 13.18 | 341.5 |
| 119.78 | 277.1 | 203.4 | 1.061 | 0.105 | 13.19 | 343.3 |

## 8. EXPERIMENTAL DATA ON HEAT TRANSFER IN ANNULI HEATED FROM BOTH SIDES

**Table 25**

| $x_1/d$ | $x_2/d$ | $Nu_{12}$ | $Nu_{21}$ | $Re_1 \cdot 10^{-3}$ | $Re_2 \cdot 10^{-3}$ | $\Psi_1$ | $\Psi_2$ | $q_1 \cdot 10^{-3}$ W/m² | $q_2 \cdot 10^{-3}$ W/m² | $T_{w1}$ K | $T_{w2}$ K |
|---|---|---|---|---|---|---|---|---|---|---|---|
| \multicolumn{12}{l}{Run 4002    $G = 0.098$ kg/sec, $P_{in} = 0.5276$ MPa, $T_{fin} = 277.7$ K} |

Run 4002    $G = 0.098$ kg/sec, $P_{in} = 0.5276$ MPa, $T_{fin} = 277.7$ K

| $x_1/d$ | $x_2/d$ | $Nu_{12}$ | $Nu_{21}$ | $Re_1 \cdot 10^{-3}$ | $Re_2 \cdot 10^{-3}$ | $\Psi_1$ | $\Psi_2$ | $q_1 \cdot 10^{-3}$ W/m² | $q_2 \cdot 10^{-3}$ W/m² | $T_{w1}$ K | $T_{w2}$ K |
|---|---|---|---|---|---|---|---|---|---|---|---|
| 1.34 | 1.34 | 459.8 | 454.5 | 213.0 | 213.0 | 1.062 | 1.058 | 8.752 | 7.466 | 295.2 | 292.8 |
| 2.46 | 2.46 | 436.0 | 417.6 | 212.8 | 212.8 | 1.065 | 1.058 | 8.771 | 7.463 | 296.3 | 294.3 |
| 4.25 | 4.25 | 413.4 | 389.5 | 212.6 | 212.6 | 1.069 | 1.062 | 8.802 | 7.461 | 297.7 | 295.8 |
| 7.15 | 7.15 | 396.1 | 364.0 | 212.3 | 212.3 | 1.072 | 1.066 | 8.852 | 7.458 | 299.1 | 297.5 |
| 12.52 | 12.52 | 396.9 | 346.9 | 211.7 | 211.7 | 1.072 | 1.069 | 8.946 | 7.450 | 300.2 | 299.3 |
| 21.46 | 21.46 | 402.3 | 328.1 | 210.7 | 210.7 | 1.072 | 1.072 | 9.101 | 7.435 | 301.9 | 301.9 |
| 31.29 | 31.29 | 409.5 | 323.5 | 209.6 | 209.6 | 1.071 | 1.072 | 9.269 | 7.421 | 303.6 | 303.8 |
| 42.91 | 42.91 | 403.2 | 325.1 | 208.4 | 208.4 | 1.072 | 1.070 | 9.462 | 7.403 | 306.3 | 305.7 |
| 52.75 | 52.75 | 406.4 | 320.6 | 207.3 | 207.3 | 1.072 | 1.070 | 9.629 | 7.387 | 308.2 | 307.6 |
| 62.58 | 62.58 | 432.6 | 322.3 | 206.2 | 206.2 | 1.068 | 1.069 | 9.798 | 7.371 | 309.0 | 309.2 |

Run 4005    $G = 0.103$ kg/sec, $P_{in} = 0.558$ MPa, $T_{fin} = 277.7$ K

| $x_1/d$ | $x_2/d$ | $Nu_{12}$ | $Nu_{21}$ | $Re_1 \cdot 10^{-3}$ | $Re_2 \cdot 10^{-3}$ | $\Psi_1$ | $\Psi_2$ | $q_1 \cdot 10^{-3}$ W/m² | $q_2 \cdot 10^{-3}$ W/m² | $T_{w1}$ K | $T_{w2}$ K |
|---|---|---|---|---|---|---|---|---|---|---|---|
| 1.34 | 1.35 | 492.4 | 455.3 | 223.0 | 223.0 | 1.137 | 1.110 | 20.94 | 15.55 | 316.6 | 309.1 |
| 2.46 | 2.46 | 464.1 | 418.2 | 222.8 | 222.8 | 1.145 | 1.120 | 20.96 | 15.55 | 319.3 | 312.2 |
| 4.25 | 4.25 | 440.0 | 389.2 | 222.3 | 222.3 | 1.153 | 1.128 | 20.99 | 15.56 | 322.2 | 315.3 |
| 7.15 | 7.15 | 416.1 | 365.6 | 221.7 | 221.7 | 1.161 | 1.136 | 21.03 | 15.57 | 325.7 | 318.5 |
| 12.52 | 12.52 | 409.4 | 344.3 | 220.4 | 220.4 | 1.162 | 1.142 | 21.13 | 15.59 | 328.3 | 322.7 |
| 21.46 | 21.46 | 403.2 | 325.7 | 218.3 | 218.3 | 1.162 | 1.147 | 21.28 | 15.61 | 332.3 | 328.0 |
| 31.29 | 31.29 | 406.5 | 321.1 | 216.0 | 216.0 | 1.158 | 1.146 | 21.45 | 15.64 | 335.5 | 331.9 |
| 42.91 | 42.91 | 394.4 | 320.2 | 213.4 | 213.4 | 1.160 | 1.143 | 21.63 | 15.68 | 341.0 | 336.0 |
| 52.75 | 52.75 | 389.2 | 316.0 | 211.3 | 211.3 | 1.160 | 1.142 | 21.80 | 15.71 | 345.2 | 339.9 |
| 62.58 | 62.58 | 406.8 | 315.1 | 209.2 | 209.2 | 1.150 | 1.139 | 21.97 | 15.74 | 346.7 | 343.3 |

Run 4009    $G = 0.105$ kg/sec, $P_{in} = 0.573$ MPa, $T_{fin} = 277.8$ K

| $x_1/d$ | $x_2/d$ | $Nu_{12}$ | $Nu_{21}$ | $Re_1 \cdot 10^{-3}$ | $Re_2 \cdot 10^{-3}$ | $\Psi_1$ | $\Psi_2$ | $q_1 \cdot 10^{-3}$ W/m² | $q_2 \cdot 10^{-3}$ W/m² | $T_{w1}$ K | $T_{w2}$ K |
|---|---|---|---|---|---|---|---|---|---|---|---|
| 1.34 | 1.35 | 489.3 | 452.3 | 224.9 | 224.9 | 1.266 | 1.217 | 40.48 | 30.46 | 353.8 | 339.5 |
| 2.46 | 2.47 | 459.2 | 414.3 | 224.4 | 224.4 | 1.282 | 1.235 | 40.54 | 30.46 | 358.3 | 345.6 |
| 4.25 | 4.26 | 434.3 | 383.3 | 223.6 | 223.6 | 1.297 | 1.252 | 40.63 | 30.49 | 364.5 | 352.0 |
| 7.15 | 7.16 | 409.1 | 357.6 | 222.3 | 222.2 | 1.312 | 1.267 | 40.78 | 30.53 | 371.5 | 358.8 |
| 12.52 | 12.52 | 398.2 | 334.1 | 219.8 | 219.8 | 1.315 | 1.279 | 41.08 | 30.60 | 377.4 | 367.3 |
| 21.46 | 21.46 | 390.1 | 311.5 | 215.8 | 215.8 | 1.312 | 1.288 | 41.57 | 30.71 | 385.1 | 378.3 |
| 31.29 | 31.29 | 391.5 | 303.3 | 211.6 | 211.7 | 1.301 | 1.284 | 42.11 | 30.84 | 391.2 | 386.3 |
| 42.91 | 42.91 | 377.1 | 299.5 | 206.8 | 206.8 | 1.301 | 1.275 | 42.70 | 31.00 | 402.3 | 394.3 |
| 52.74 | 52.74 | 371.2 | 292.8 | 202.9 | 203.0 | 1.296 | 1.270 | 43.21 | 31.12 | 410.4 | 402.2 |
| 62.57 | 62.57 | 384.7 | 290.4 | 199.4 | 199.4 | 1.277 | 1.262 | 43.77 | 31.24 | 413.8 | 408.9 |

## Table 25 (*Continued*)

| $x_1/d$ | $x_2/d$ | $\mathrm{Nu}_{12}$ | $\mathrm{Nu}_{21}$ | $\mathrm{Re}_1 \cdot 10^{-3}$ | $\mathrm{Re}_2 \cdot 10^{-3}$ | $\Psi_1$ | $\Psi_2$ | $q_1 \cdot 10^{-3}$ W/m² | $q_2 \cdot 10^{-3}$ W/m² | $T_{w1}$ K | $T_{w2}$ K |
|---|---|---|---|---|---|---|---|---|---|---|---|
| \multicolumn{12}{l}{Run 4013    $G = 0.103$ kg/sec, $P_{in} = 0.548$ MPa, $T_{fin} = 278.0$ K} |
| 1.34 | 1.36 | 478.6 | 443.0 | 220.2 | 220.2 | 1.417 | 1.335 | 62.30 | 46.27 | 396.4 | 373.4 |
| 2.46 | 2.48 | 445.7 | 402.0 | 219.4 | 219.4 | 1.445 | 1.366 | 62.41 | 46.29 | 406.0 | 383.8 |
| 4.25 | 4.26 | 419.5 | 368.8 | 218.2 | 218.2 | 1.468 | 1.394 | 62.60 | 46.34 | 415.4 | 394.5 |
| 7.15 | 7.17 | 391.5 | 341.8 | 216.1 | 216.1 | 1.493 | 1.417 | 62.89 | 46.46 | 427.4 | 405.7 |
| 12.52 | 12.53 | 377.2 | 327.0 | 212.5 | 212.5 | 1.497 | 1.421 | 63.46 | 46.68 | 437.5 | 415.5 |
| 21.46 | 21.46 | 367.3 | 289.5 | 206.7 | 206.6 | 1.487 | 1.449 | 64.55 | 46.93 | 449.2 | 438.4 |
| 31.29 | 31.29 | 366.9 | 279.4 | 200.5 | 200.5 | 1.463 | 1.438 | 65.67 | 47.30 | 459.2 | 451.4 |
| 42.90 | 42.91 | 350.3 | 272.8 | 194.0 | 194.0 | 1.457 | 1.419 | 66.87 | 47.74 | 476.9 | 464.5 |
| 52.73 | 52.73 | 343.6 | 265.0 | 188.9 | 188.9 | 1.444 | 1.408 | 67.94 | 48.07 | 489.5 | 477.1 |
| 62.53 | 62.55 | 355.3 | 261.6 | 184.3 | 184.3 | 1.411 | 1.391 | 69.11 | 48.41 | 494.5 | 487.5 |
| \multicolumn{12}{l}{Run 4017    $G = 0.102$ kg/sec, $P_{in} = 0.539$ MPa, $T_{fin} = 277.9$ K} |
| 1.34 | 1.37 | 470.2 | 435.8 | 216.9 | 216.9 | 1.619 | 1.494 | 91.20 | 67.48 | 454.1 | 419.4 |
| 2.46 | 2.48 | 433.8 | 392.6 | 215.7 | 215.7 | 1.664 | 1.542 | 91.36 | 67.51 | 469.8 | 435.1 |
| 4.25 | 4.27 | 405.5 | 357.6 | 213.9 | 213.9 | 1.700 | 1.585 | 91.68 | 67.63 | 484.7 | 452.1 |
| 7.16 | 7.17 | 373.8 | 327.9 | 210.9 | 210.9 | 1.740 | 1.621 | 92.15 | 67.88 | 504.5 | 470.2 |
| 12.52 | 12.54 | 356.5 | 298.0 | 205.8 | 205.8 | 1.742 | 1.650 | 93.26 | 68.26 | 520.7 | 493.5 |
| 21.45 | 21.47 | 341.4 | 270.2 | 197.5 | 197.5 | 1.721 | 1.660 | 95.13 | 68.92 | 540.6 | 521.0 |
| 31.28 | 31.29 | 339.2 | 257.3 | 189.6 | 189.6 | 1.673 | 1.637 | 97.21 | 69.76 | 554.0 | 542.7 |
| 42.90 | 42.90 | 321.8 | 248.9 | 181.3 | 181.3 | 1.650 | 1.598 | 99.29 | 70.75 | 579.8 | 561.1 |
| 52.72 | 52.72 | 315.7 | 240.8 | 174.9 | 174.9 | 1.619 | 1.573 | 101.2 | 71.50 | 597.0 | 580.1 |
| 62.53 | 62.54 | 328.0 | 237.2 | 168.9 | 168.9 | 1.559 | 1.539 | 103.4 | 72.25 | 602.5 | 595.1 |
| \multicolumn{12}{l}{Run 4023    $G = 0.107$ kg/sec, $P_{in} = 0.590$ MPa, $T_{fin} = 278.7$ K} |
| 1.34 | 1.37 | 479.8 | 452.2 | 227.2 | 227.2 | 1.994 | 1.777 | 151.6 | 111.8 | 563.6 | 502.5 |
| 2.46 | 2.49 | 436.1 | 400.8 | 225.4 | 225.3 | 2.075 | 1.862 | 151.7 | 111.8 | 592.3 | 531.4 |
| 4.25 | 4.28 | 403.7 | 359.6 | 222.3 | 222.3 | 2.131 | 1.934 | 152.1 | 111.9 | 618.1 | 560.9 |
| 7.16 | 7.18 | 365.5 | 322.8 | 217.8 | 217.8 | 2.197 | 1.997 | 152.7 | 112.3 | 653.4 | 594.0 |
| 12.52 | 12.54 | 341.6 | 286.4 | 209.6 | 209.6 | 2.191 | 2.037 | 154.6 | 112.9 | 682.4 | 634.6 |
| 21.45 | 21.46 | 317.8 | 251.2 | 197.6 | 197.6 | 2.140 | 2.040 | 157.9 | 113.9 | 718.3 | 684.9 |
| 31.27 | 31.28 | 312.0 | 235.2 | 186.6 | 186.6 | 2.036 | 1.980 | 161.9 | 115.6 | 737.8 | 717.8 |
| 42.88 | 42.88 | 294.5 | 222.4 | 175.1 | 175.1 | 1.960 | 1.901 | 165.4 | 117.2 | 773.4 | 750.1 |
| 52.69 | 52.69 | 291.6 | 211.8 | 166.5 | 166.5 | 1.879 | 1.846 | 169.3 | 118.3 | 793.0 | 778.8 |
| 62.49 | 62.49 | 308.4 | 206.9 | 159.1 | 159.1 | 1.764 | 1.779 | 174.3 | 119.4 | 792.7 | 799.6 |

# REFERENCES

1. Monin, A. S., and Yaglom, A. M. 1967. *Statisticheskaya Gidromekhanika* (*Statistical Fluid Mechanics*), parts 1 and 2. Moscow: Nauka Press [Engl. transl. MIT Press, part 1, 1970, part 2, 1975].
2. Hinze, J. O. 1959. *Turbulence*. New York: McGraw-Hill [Russ. transl. 1963].
3. Schlichting, H. 1968. *The Boundary Layer Theory*. New York: McGraw-Hill [Russ. transl. 1974].
4. Ievlev, V. M. 1975. *Turbulentnoye Dvizheniye Vysokotemperaturnykh Sploshnykh Sred* (*Turbulent Motion of High-Temperature Continuous Media*). Moscow: Nauka Press.
5. Hall, W. B. 1962. Heat transfer in channels having rough and smooth surfaces. *Mech. Eng. Sci.* 4(3):287–291.
6. Kjellström, B., and Hedberg, S. 1966. *On Sheer Stress Distribution for Flow in Smooth or Partially Rough Annuli*.
7. Maubach, K., and Rehme, K. 1972. Negative eddy diffusivities for asymmetric turbulent velocity profiles. *Int. J. Heat Mass Transfer* 15:425.
8. Dalle Donne, M., and Meerwald, E. 1973. Heat transfer and friction coefficients for turbulent flow of air in smooth annuli at high temperatures. *Int. J. Heat Mass Transfer* 16(4):787–809.
9. Quirrenbach, F. I. 1960. Wärmeübergang bei turbulenten strömung in ringspalten. *Allg. Wärmetech.* 9(13):271–276.
10. Doroshchuk, V. Y., and Frid, F. P. 1959. Investigation of heat transfer in annuli. In *Teploobmen pri vysokikh temperaturakh i drugikh spetsial'nykh usloviyakh* (*Heat Transfer at High Temperatures and Other Special Conditions*), pp. 101–110. Moscow-Leningrad.
11. Puchkov, P. I., and Vinogradov, O. S. 1963. Heat transfer in annuli. *Energomashinostroyeniye* 11:22–24.
12. Puchkov, P. I., and Vinogradov, O. S. 1965. Heat transfer and hydraulic resistance in annular channels with smooth and rough heat transfer surfaces. In *Teplo- i Massoperenos* (*Heat and Mass Transfer, Proc. 2d All-Union [Minsk] Heat and Mass Transfer Conference*), vol. 1, pp. 76–91. Minsk [Engl. transl. RAND Report R-451-PR, 1967, pp. 80–96].
13. Kays, W. M., and Leung, E. Y. 1963. Heat transfer in annular passages—Hydrodynamically

developed turbulent flow with arbitrarily prescribed heat flux. *Int. J. Heat Mass Transfer* 6(7):537–557.
14. Petukhov, B. S., and Royzen, L. I. 1963. Heat transfer in annuli. *Inzh. Fiz. Zhurn.* 6(3): 3–11.
15. Petukhov, B. S., and Royzen, L. I. 1963. Experimental study of heat transfer in turbulent gas flows in annuli. *Teplofizika Vysokikh Temperatur* 1(3):416–424.
16. Petukhov, B. S., and Royzen, L. I. 1965. Heat transfer for gas flows in annular tubes. In *Teplo- i Massoperenos (Heat and Mass Transfer, Proc. 2d All-Union [Minsk] Heat and Mass Transfer Conference)*, vol. 1, pp. 66–75, Minsk [Engl. transl. RAND Report R-451-PR, 1967, pp. 67–79].
17. Petukhov, B. S., and Royzen, L. I. 1974. Correlations for heat transfer in annuli. *Teplofizika Vysokikh Temperatur* 12(3):565–569.
18. Bobkov, V. P., Ibragimov, M. Kh., and Savanin, N. K. 1975. Heat transfer in turbulent flows of various coolants in annuli. *Teplofizika Vysokikh Temperatur* 13(4):779–786.
19. Buleyev, N. I., Mosolova, V. A., and Yel'tsova, L. D. 1967. Transfer of heat in turbulent flows of fluids in annular and plane spaces. 1967. In *Zhidkiye metally (Liquid Metals)*, p. 123. Moscow: Atomizdat Press [Engl. transl. NASA TT F-522, Washington, D.C., 1969, p. 120].
20. Dwyer, O. E., Tu, P. S. 1961. Bilateral heat transfer to liquid metals flowing turbulently through annuli. *Nucl. Sci. Eng.* 21(1):90–105.
21. Petrikevich, B. B. 1974. Convective heat transfer in channels with various cross sections. *Inzh. Fiz. Zhurn.* 27(2):215–222.
22. Wilson, N. W., and Medwell, J. O. 1968. Analysis of heat transfer for fully developed turbulent flow in concentric annulus. *J. Heat Transfer* 90(1):43.
23. Wilson, N. W., and Medwell, J. O. 1971. An analysis of the developing turbulent hydrodynamic and thermal boundary layers in an internally-heated annulus. *J. Heat Transfer* 93(1):25.
24. Dalle Donne, M., and Meerwald, E. 1966. Experimental local heat transfer and average friction coefficients for subsonic flow of air in an annulus at high temperatures. *Int. J. Heat Mass Transfer* 9(12):1361–1378.
25. Presser, K. H., Pietralla, G., and Hart, R. 1967. Wärmeübergang und druckrerlust an innenbeheizten ringspalten bei hochdruckgaskühlung. *Atomkernenergie* 12, Ig., H. ½.
26. Nemira, M. A., and Vilemas, J. V. 1975. Heat transfer from cylindrical and helically twisted rods in an annulus to turbulent flow of air with variable physical properties. *Trudy Akad. Nauk LitSSR, Ser. B.* 3(88):127–136 [Engl. transl. *Int. Chem. Eng.* 16(3):412–417, 1976].
27. Vilemas, J. V., and Nemira, M. A. 1976. Effect of variability of physical properties on heat transfer to turbulent air flows in annuli. In *Teplo- i Massoperenos (Heat and Mass Transfer, Proc. 5th All-Union Conference on Heat and Mass Transfer*, vol. 1, part 1, pp. 285–289. Minsk [Engl. transl. *Heat Transfer—Soviet Research* 9(4):151–155, 1977].
28. Galin, N. M., and Yesin, V. M. 1977. Calculation of heat transfer to turbulent flow in circular and annular tubes with correction for the temperature dependence of thermophysical properties of the coolant. *Teplofizika Vysokikh Temperatur* 15(5).
29. Petukhov, B. S., and Popov, V. N. 1963. Analytic calculation of heat transfer and friction drag in turbulent pipe flows of incompressible fluids with variable physical properties. *Teplofizika Vysokikh Temperatur* 1(1):85–101.
30. Petukhov, B. S. 1970. Heat transfer and friction coefficients for turbulent pipe flow with variable physical properties. *Adv. Heat Transfer* 6:503–564.
31. Lel'chuk, V. L., and Dyadyakin, B. V. 1959. Heat transfer from a wall to turbulent air flows in a pipe and the hydraulic drag at high temperature differences. In *Voprosy Teploobmena (Problems of Heat Transfer)*, pp. 123–192. U.S.S.R. Academy of Sciences Press.
32. Lel'chuk, V. L., and Elfimov, G. I. 1968. Results of studies of local heat transfer from the wall of a tube heated to 1000°C to gases with different atomicity in turbulent flow. In *Teplo- i Massoperenos (Heat and Mass Transfer, Proc. 3d All-Union Heat and Mass Transfer Conference)*, vol. 1, pp. 465–474. Moscow: Energiya Press.
33. Lel'chuk, V. L., Shuyskaya, K. F., and Sorokin, A. G. 1968. Investigation of heat transfer to

a turbulent gas flow in a tube heated to very high temperatures ($t_w = 1000°C$). In *Teplo- i Massoperenos (Heat and Mass Transfer, Proc. 3d All-Union Heat and Mass Transfer Conference)*, vol. 1, pp. 485–493. Moscow: Energiya Press.
34. Dalle Donne, M., and Browdich, F. H. Local heat transfer and average friction coefficients for subsonic laminar and turbulent flow of air in a tube at high temperature. *Nucl. Sci. Eng.* 8(80):20–29.
35. Taylor, M. F. 1968. Correlation of local heat transfer coefficients for single phase turbulent flow of hydrogen in tubes with temperature ratios to 23. NASA TND, 4332.
36. Perkins, H. C., and Vorsoe-Schmidt, P. 1965. Turbulent heat and momentum transfer for gases in a circular tube at wall to bulk temperature ratios to seven. *Int. J. Heat Mass Transfer* 8(7):1011–1031.
37. Petukhov, B. S., Kirillow, V. V., and Maidanik, V. N. 1966. Heat transfer experimental research for turbulent gas flow in pipes at high temperature difference between wall and bulk fluid temperature. *AIChE J.* 1:285–292.
38. Petukhov, B. S., Kurganov, V. A., and Gladuntsov, A. I. 1970. Experimental study of heat transfer to turbulent flow of many atomic gases at high temperature differences. *Teplofizika Vysokikh Temperatur* 8(5):1311–1315.
39. Petukhov, B. S., Kirillov, B. B., Tszu-Tszyi-Sian, and Maidanik, B. N. 1965. Experimental study of the effect of the temperature factor on heat transfer to turbulent pipe flows of gases. *Teplofizika Vysokikh Temperatur* 3(1):102–108.
40. Petukhov, B. S., Kurganov, V. A., and Gladuntsov, A. I. 1972. Heat transfer in turbulent pipe flow of gases with variable properties. In *Teplo- i Massoperenos (Heat and Mass Transfer, Proc. 4th All-Union Heat and Mass Transfer Conference)*, vol. 1, pp. 117–127. Minsk [Engl. transl. *Heat Transfer—Soviet Research*, 5(4):109–116. 1973].
41. Kurganov, V. A., and Petukhov, B. S. 1974. Analysis and correlation of experimental data on heat transfer in turbulent pipe flows of gases with variable physical properties. *Teplofizika Vysokikh Temperatur* 12(2):304–315.
42. Kurganov, V. A., and Petukhov, B. S. 1974. On the calculation of heat transfer in pipes, nonuniformly heated along their length in turbulent flows of gases with variable physical properties. *Teplofizika Vysokikh Temperatur* 12(5):1038–1044.
43. Deissler, R. G. 1954. Heat transfer and fluid friction for fully developed turbulent flow of air and supercritical water with variable fluid properties. *Trans. ASME* 76(1):73–86.
44. Deissler, R. G. 1955. Analysis of turbulent heat transfer, mass transfer and friction in smooth tubes at high Prandtl and Schmidt numbers, pp. 1–14. NASA, 1210.
45. Deissler, R. G., and Presler, A. F. 1961. Computed reference temperatures for turbulent variable-property heat transfer in a tube for several common gases. In *Int. Heat Transfer Conf.*, part III, pp. 579–584. Colorado.
46. Kutateladze, S. S., and Leont'yev, A. N. 1962. *Turbulentnyy Pogranichnyy Sloy Szhimayemogo Gaza (The Turbulent Boundary Layer in Compressible Gases)*. Siberian Division of the USSR Academy of Sciences Press [Engl. transl. Academic Press, 1964].
47. Bankston, C. A., and McEligot, D. M. 1970. Turbulent and laminar heat transfer to gases with varying properties in the entry region of circular ducts. *Int. J. Heat Mass Transfer* 13(2):319–344.
48. McEligot, D. M., Smith, S. B., and Bankston, C. A. 1970. Quasideveloped turbulent pipe flow with heat transfer. *J. Heat Transfer* 92(4):641.
49. Patankar, S. V., and Spalding, B. B. 1970. *Heat and Mass Transfer in Boundary Layers*. Intertext [Russ. transl. 1971].
50. Popov, V. N. 1970. Heat transfer and drag in turbulent axial flow of air over a plate. *Teplofizika Vysokikh Temperatur* 8(5):1032–1042.
51. Van Driest, E. R. 1959. Convective heat transfer in gases. In *Turbulent Flows and Heat Transfer*, ed. C. C. Lin, pp. 339–427. Princeton, N.J.: Princeton University Press [Russ. transl. 1963].
52. Žukauskas, A., and Slančiauskas, A. 1973. *Teploperedacha v Turbulentnom Potoke Zhidkosti (Heat Transfer in Turbulent Flows)* (*Teplofizika*, 5 [Thermophysics, 5]). Vilnius: Mintis Press.

53. Jackob, M., and Dow, W. M. 1946. Heat transfer from a cylindrical surface to air in parallel flow with and without unheated starting sections. *Trans. ASME* 68(2):123–134.
54. Eckert, H. U. 1952. Simplified treatment at the turbulent boundary layer along a cylinder in compressible flow. *J. Aeron. Sci.* 19:23–29.
55. Sparrow, E. M., Eckert, E. R. G., and Minkowycz, W. J. 1963. Heat transfer and skin friction for turbulent boundary-layer flows longitudinal to a circular cylinder. *J. Appl. Mech.* 30(1):37.
56. Ginevskiy, A. S., and Solodkin, Ye. Ye. 1958. Effect of transverse curvature on the parameters of an axisymmetric turbulent boundary layer. *Prikladnaya Matematika Mekhanika* 22(6):819–825.
57. Hughes, G. 1954. Friction and form resistance in turbulent flow, and a proposed formulation for use in model and ship correlation. *Trans. Inst. Navel Architects* 96(4):314–376.
58. Kemph, G. 1924. Über den reibungswiderstand von flächen verschiedener form. In *Proc. First Int. Cong. Appl. Mech.*, pp. 439–449. Delft, Netherlands.
59. Vasserman, A. A., Kazavchinskiy, Ya. Z., and Rabinovich, V. A. 1966. *Teplofizicheskiye Svoystva Vozdukha i Yego Komponentov* (*Thermophysical Properties of Air and Air Components*). Moscow: Nauka Press. [Engl. transl. Israel Program for Scientific Translations, Jerusalem, 1971].
60. Vargaftik, N. B. 1963. *Spravochnik po Teplofizicheskim Svoystvam Gazov i Zhidkostey* (*Handbook on the Thermophysical Properties of Gases and Liquids*). Moscow: GIFML Press.
61. Preobrazhenskiy, V. P. 1953. *Teplofizicheskiye izmereniya i pribory* (*Thermophysical Measurements and Instruments*). Moscow-Leningrad: GEI Press.
62. Makarov, A. N., and Sherman, M. Ya. 1958. *Raschet Drossel'nykh Ustroystv* (*Design of Throttling Devices*). Moscow: Metallurgizdat Press.
63. Povkh, I. L. 1974. *Aerodinamicheskiy Eksperiment v Mashinostroyeniyii* (*Aerodynamic Experimentation in Machine Building*). Leningrad: Mashinostroyeniye Press.
64. Shaw, R. 1960. Influence of hole dimensions on static pressure measurements. *J. Fluid Mech.* 7(4):550–564.
65. Ray, A. K. 1956. Influence of the pressure tap diameter upon the accuracy of statical pressure measurement at different Reynolds numbers. *Ing. Arch* 24(3):171–181.
66. Rekstin, F. S. 1959. Micrometer-screw device for investigating flows in stationary components of flow passages of a centrifugal compressor. *Nauchno-tekhn. Inform. Byull. LPI im. M. I. Kalinina* 6, pp. 48–53.
67. Anisimov S. A., Rekstin, F. S., and Shkarbul, S. N. 1959. Comparison of small-size static-pressure tubes. *Nauchno-tekhn. Inform. Byull. LPI im. M. I. Kalinina* 12, pp. 80–91.
68. Anisimov, S. A., Rekstin, F. S., and Shkarbul', S. N. 1959. Comparison of small-size total-pressure tubes. *Nauchno-tekhn. Inform. Byull. LPI im. M. I. Kalinina* 6, pp. 54–62.
69. Sakiadis, B. C. 1961. Boundary layer behavior on continuous solid surfaces. III. The boundary layer on continuous cylindrical surface. *AIChE J.* 7(3):467–472.
70. Liepmann, H. W., and Skinner, G. T. 1954. Shearing stress measurements by use of a heated element. NASA TN, 3268.
71. Konstantinov, N. I. 1955. *Comparative Study of Friction Stress on the Surfaces of Bodies. Trudy LPI* (*Leningrad Polytechnic Institute*), *176*, pp. 201–214. Moscow–Leningrad: Mashgiz Press.
72. Pankhurst, R. C., and Holder, D. W. 1952. *Wind Tunnel Technique.* Pitman.
73. Gorlin, S. M., and Slezinger, I. I. 1964. *Aerodinamicheskiye Izmereniya* (*Aerodynamic Measurements*). Moscow: Nauka Press [Engl. transl. Israel Program for Scientific Translations, Jerusalem, 1966].
74. Hool, J. N. 1956. Measurements of skin friction using surface tubes. *Aircraft Eng.* 28(324):52–54.
75. Preston, J. H. 1954. The determination of turbulent skin friction by means of Pitot tubes. *J. R. Aero. Soc.* 58:109–121.
76. Staff of the Aerodynamics Division, N. P. L. 1958. On the measurement of local surface friction on a flat plate by means of Preston tubes pp. 1–23. ARC Rep. Mem., 3185.

77. Dean, W. R. 1936. Note on the slow motion of fluid. *Cambridge Phil. Soc. Proc.* 32:598–613.
78. Dean, W. R. 1950. Slow motion of viscous liquid in a semi-liquid channel. *Cambridge Phil. Soc. Proc.* 47:127–141.
80. Rechenberg, I. 1963. Messung der turulenten Wandschubspannung. *Z. Flugwiss.* 11:429–438.
81. Clauser, F. 1959. Problemy mekhaniki (*Problems of Mechanics* [collection of translations]), 2.
82. Česna B. A., Stasiulevučius, J. K., and Šlančiauskas, A. A. 1972. Effect of the temperature factor on skin friction of a cylinder in axial flow. *Tr. Akad. Nauk LitSSR, Ser. B.*, 2(69):129–139.
83. Kjellström, B., and Hedberg, St. 1968. Calibration experiments with a DISA hot-wire anemometry. Swedish Committee on Atomic Energy.
84. Andreyev, P. A., Gremilov, D. I., and Fedorovich, Ye. D. 1965. Swedish Committee on Atomic Energy. *Teploomennyye Apparaty Yadernykh Energeticheskikh Ustanovok (Heat Exchangers for Nuclear Power Plants).* Leningrad: Sudostroyeniye Press.
85. Petukhov, B. S., and Mukhin, V. A. 1966. Experimental study of heat transfer to supersonic gas flows in a circular tube (inlet length). *Teplofizika Vysokikh Temperatur* 2.
86. Isachenko, V. P., Osipova, V. A., and Sukomel, A. S. 1975. *Teploperedacha (Heat Transfer).* Moscow: Energiya Press.
87. Blokh, A. T. 1962. *Osnovy Teploobmena Izlucheniyem (Fundamentals of Radiative Heat Transfer).* Moscow: Gosenergoizdat Press.
88. Seban, R. A., and Bond, R. 1951. Skin friction and heat transfer characteristics of a laminar boundary layer on a cylinder in axial incompressible flow. *J. Aeron. Sci.* 18:671–675.
89. Kelly, H. R. 1954. A note on the laminar boundary layer on a circular cylinder in axial incompressible flow. *J. Aeron. Sci.* 21:634.
90. Van Driest, E. R. 1956. On turbulent flow near a wall. *J. Aeron. Sci* 23(11):1007–1011.
91. Clauser, F. 1956. The turbulent boundary layer. In *Advances in Applied Mechanics*, eds. H. L. Dryden and Th. von Karman, New York: Academic Press, [Russ. transl. 1956].
92. Rotta, I. K. 1962. Turbulent boundary layers in incompressible flow. In *Prog. Aeronaut. Sci.* 2:1–219 [Russ. transl. 1967].
93. Šlančiauskas, A. A., Ulinskas, R. V., and Žukauskas, A. A. 1959. Turbulent heat transfer on a plate in flows of various fluids with variable viscosity. *Tr. Akad. Nauk LitSSR, Ser. B.* 4(59):163–177.
94. Romanenko, P. N. 1971. *Teplomassoobmen i Treniye pri Gradientnom Techenii Zhidkostey (Heat and Mass Transfer and Friction in Gradient Flows of Liquids).* Moscow: Energiya Press.
95. Česna, B. A., Survila, V. J., and Adomaitis, E. J. 1976. Effect of surface-type turbulence promoters on local heat transfer of a cylinder in axial flow of air. *Tr. Akad. Nauk LitSSR, Ser. B.* 6(97):49–55 [Engl. transl. *Int. Chem. Engn.* 19(2):276–280, 1979].
96. Deissler, R. G. 1955. Turbulent heat transfer and friction in the entrance region of smoother passages. *Trans. ASME* 77(8):1221–1233.
97. Chi, S. W., and Spalding, D. B. 1966. Influence of temperature ratio on heat transfer to a flat plate through a turbulent boundary layer in air. In *Proc. Third Int. Heat Transfer Conf.*, part III, vol. 2, pp. 41–49. Chicago.
98. Solodkin, Ye. Ye., and Ginevskiy, A. S. 1956. The turbulent boundary layer and frictional drag of a cylinder with correction for the effect of the transverse curvature of the surface. *Trudy TsAGI* 690:1–22.
99. Česna, B. A., Stasiulevičius, J. K., Šlančiauskas, A. A., and Survila, V. J. 1971. Skin friction and velocity profiles in the turbulent boundary layer on a cylinder in axial flow of air. *Tr. Akad. Nauk LitSSR, Ser. B.* 4(67):101–113.
100. Popov, V. N. 1970. Heat transfer and friction drag in longitudinal turbulent flow of a gas with variable physical properties over a plate. *Teplofizika Vysokikh Temperatur* 8(2):333–345.
101. Survila, V. J., and Stasiulevičius, J. K. 1968. Effect of the temperature factor on heat transfer in turbulent axial air flow over a cylinder. *Tr. Akad. Nauk LitSSR, Ser. B.* 4(55):145–157.

102. Survila, V. J., and Stasiulevičius, J. K. 1969. Effect of Transverse Curvature of a Cylinder in Axial Flow on Its Heat Transfer, *Tr. Akad. Nank LitSSR, Ser. B* 4(59):179–189, [Engl. transl. *Heat Transfer—Soviet Research* 2(3) 1970, p. 142].
103. Survila, V. J., and Stasiulevicius, J. K. 1969. Heat transfer from a cylinder in axial flow with induced turbulization of the incoming boundary layer. *Tr. Akad. Nauk LitSSR, Ser. B* 2(57):113–121, [Engl. transl. *Heat Transfer-Soviet Research* 2(3) p. 151, 1970.
104. McDonald, H., and Kreskovsky, J. P. 1974. Effect of free stream turbulence on the turbulent boundary layer. *Int. J. Heat Mass Transfer* 17(17):705–716.
105. Moretti, P. M. 1965. Heat transfer to a turbulent boundary layer with varying free-stream velocity and varying surface temperature—An experimental study. *Int. J. Heat Mass Transfer* 8(9):1187–1202.
106. Coon, C. W., and Perkins, H. C., Jr. 1970. Transition from the turbulent to the laminar regime for internal convective flows with large property variations. *J. Heat Transfer* 92(3):506.
107. Bankston, C. A. 1970. The transition from turbulent to laminar gas flow in a heated pipe. *J. Heat Transfer* 92(4):569.
108. Higgins, R. W., and McDonald, T. J. 1962. Axial midpoint spacers improve EGCR heat transfer. *Nucl. Sci. Eng.* (1):22–26.
109. Seban, R. A. 1965. Heat transfer to the separated flow of air downstream of a step in the surface of a plate. *J. Heat Transfer* 86(2):259–264.
110. Ktalkherman, M. G. 1966. Heat transfer to a plate downstream of an obstacle. *Zh. Prikl. Mekh. Tekhn. Fiz.* (5):130–133.
111. Humble, L. V., Lowdermilk, W. H., and Desmon, L. G. 1951. Measurements of average heat transfer and friction coefficient for subsonic flow of air in smooth tubes at high surface and fluid temperatures, pp. 1–15. NASA 1020.
112. McEligot, D. M. 1959. Ph.D. dissertation, Department of Mechanical Engineering, Stanford University, Stanford, Calif.
113. Lowdermilk, W. H., Weinland, W. F., and Livingood, J. N. 1954. Measurement of heat transfer and friction coefficients for flow of air in noncircular ducts at high surface and fluid temperatures. NASA research mem. L53J07.
114. Filonenko, G. K. 1954. Hydraulic drag of pipelines. *Teploenergetika* (4):40–44.
115. Quarmby, A. 1967. An experimental study at turbulent flow through concentric annuli. *Int. J. Mech. Sci.* 9:205–221.
116. Paegle, K. K., Savel'yev, P. A., and Chaksi, S. A. 1967. Investigation of heat transfer from tubes with helical cross section. *Izv. LatSSR, Ser. Fiz. Tekhn. Nauk* (2):62–65.
117. Kidd., G. I., Jr. 1970. The heat transfer and pressure drop characteristics of gas flow inside spirally corrugated tubes. *J. Heat Transfer* 92(3):513.
118. Lawson, C. G., Keld, R. Y., and McDonald, R. E. 1966. Enhanced heat transfer tubes for horizontal condensers with possible applications in nuclear power plant design. *Trans. ANS* 9(2):565–566.
119. Shurig, Ch. 1965. Eine brennstoff element variante für WWR-S reaktoren mit erhöter leistung. Int. Konf. Phys. Tech. Forschungsreaktoren, Budapest, 15–20 November, 1965.
120. Shchukin, V. K., 1970. *Teploobmen i Gidrodinamika Vnutrennikh Potokov v Polyakh Massovykh Sil* (*Heat Transfer and Fluid Dynamics of Internal Flows in Mass-Force Fields*). Moscow: Mashinostroyeniye Press.
121. Kalinin, E. K., Dreytser, G. A., and Yarkho, S. A. 1972. *Intensifikatsiya Teploobmena v Kanalakh* (*Augmentation of Heat Transfer in Channels*). Moscow: Mashinostroyeniye Press.

# INDEX

Acceleration parameter, 132, 160
Anemometer techniques, 36, 39
Argon, 8
Asbestos insulation, 50
Augmentation methods, 135
Axial flows, 55, 133
    bundles in, 124, 134
    cylinders, 57–58, 80, 91, 106, 151
    skin friction and, 77–90

Baffle systems, 137
Balance method, 35
Blasius equations, 57
Blowers, 16
Blunt edge effects, 179
Bobkov-Ibragimov-Savanin formulas, 118, 121, 160
Boundary conditions, 56, 129
Boundary layers:
    acceleration parameter and, 133
    approximations for, 7
    curvature and, 152
    over cylinders, 74, 90, 151
    dimensions of, 16
    enhancement in, 95
    on flat plate, 11, 70, 152
    formation of, 11, 91
    of heated air, 104
    hydrodynamic parameters, 32–43
    instability of, 155
    local disturbances in, 15
    outer part, 72
    Prandtl number and, 152
    pressure gradients in, 57, 91
    self-similar behavior, 62, 78
    shear stress in, 12
    skin friction and, 90
    steady-state, 56
    surface curvature and, 89, 91, 156
    temperature profiles in, 55–76, 155, 157
    thermophysical properties in, 10
    thickness of, 81, 102, 129, 152, 154–155
    thin, 152
    transition region and, 152
    variable physical properties, 84, 87
    velocity profiles in, 55–76, 152, 155, 163
    virtual origin of, 78, 81
    viscosity over, 65
Bundles:
    in axial flow, 124, 134
    of circular bars, 150

# INDEX

Calorimeter, 25
Carbon dioxide, 8
Clauser method, 36–37
Compressibility effects, 16
Concave perimeter, 149
Continuity equations, 56
Control volume, 78
Convection effects, 63
    components of, 120, 133
    transfer coefficient, 93
    turbulence and, 10
Curvature, of surfaces, 72
    correction factor for, 87
    for cylinders, 87, 97, 102, 191
    friction and, 77, 87
    power exponent and, 104
    of surfaces, 72
    variable properties and, 99, 103
Cylinders:
    in axial flow, 57–58, 80, 91, 106, 151–155, 163
    boundary layer on, 90, 151
    curvature of, 87, 97, 102, 191
    flat plates and, 10, 87
    infinite radius, 151
    leading edge, 78
    power law for, 97, 99
    skin friction of, 77–90, 156
    small-diameter, 101
    surface friction, 155
    surface geometry, 87, 91, 191
    temperature profiles for, 68, 99, 101
    transfer to air, 104
    transfer coefficients for, 91, 93–94
    turbulence promoters and, 171–172
    velocity profiles on, 11, 83

Damping function, 58, 66, 154
Dean formula, 38
Density effects, 71
Diffusion method, 36
Dimensionless variables, 9, 58, 65
Displacement thickness, 34–35, 46
Drag (see Hydraulic drag)
Ducts, heat transfer in, 109

Eddy diffusivity, 3, 72
Eddy viscosity, 10
Emissivity, 48
Energy equation, 3

Entrance regions, 11, 31, 77, 117, 146
    of annulus, 19, 91–108, 151
    boundary layer at, 15–16
    equation for, 119
    flow conditions in, 91
    friction in, 12
    heat transfer in, 43
    inner tube transfer and, 157
    length of, 5
    nonheated, 8
    Nu variation, 9
    separation zone in, 138
    variable properties and, 12, 123
Excess-velocity law, 37
Experimental techniques, 15–54
Externally heated annulus, 119–122, 205

Filonenko formula, 140
Fins, separation and, 135
Flat plates, 152
    annulus and, 121
    boundary-layer profiles, 11, 152
    curvature effects, 103
    drag over, 11
    gas flow over, 157
    parallel, 160
    skin friction coefficient for, 10, 80, 82
    turbulence and, 103, 152
    velocity profiles for, 152
Flat slot, 131
Flow-metering orifices, 19
Flow patterns:
    fully developed flow, 109–140
    models of, 10
    (See also specific effects, types)
Fourier law, 49
Freestream turbulence, 91, 94, 107, 159
Friction, 12, 15, 79
    axial air flow, 173–178
    boundary layers and, 90
    Clauser method for, 37
    coefficients of, 11, 79, 139–140
    curvature and, 156
    cylinders and, 39, 77–90, 156
    determination of, 35–36
    drag force and, 90, 139
    equation for, 157
    flat plates and, 80, 103
    Ludwieg-Tillman equation, 83
    nondimensional equation, 85
    physical properties variation, 86
    protruding plank method, 38

Friction (*Cont.*):
  self-similarity and, 78
  skin friction, 35
  Sparrow method, 58, 82, 89
  surface curvature and, 77, 157
  temperature factor and, 83
  turbulence and, 151
Fuel-element assemblies, 135
Fully developed flow, 24, 109–140

Geometries, complex, 134, 141
  (*See also* Curvature, of surfaces)
Grids, 13, 16
  spacing, 135
  turbulence-generating, 39, 41

Heat exchangers, design of, 109
Heat transfer coefficients, 1, 91, 94, 111, 148
Heating, of annulus, 8, 120–129
Helical tubes, 28, 141
  wall temperature of, 142
Hot wall region, 148
Hot-wire anemometer, 39, 43
Hydraulic drag, 1, 139
  annulus and, 161
  coefficients of, 15, 37, 50
  cylinders and, 155
  determination of, 47
  equation for, 90, 139
  fully developed flow and, 158
  in heated gases, 151
  oval-shaped tube and, 150
  temperature factor and, 140
  turbulence and, 158
  twisting and, 149–150
  variable properties and, 139
  (*See also* Friction)

Ideal gas equation, 5
Incompressible flows, 105
Inlets, temperature at, 133
Inner tubes, 6, 117, 130
  drag and, 140
  flow separation and, 135
  helically twisted, 149–150
  oval-shaped, 144, 149
  rotating, 148
  three-lobe shaped, 143
  twisting pitches, 145

variable properties and, 118
wall temperatures of, 134
Insulation:
  asbestos, 50
  vacuum, 111, 120
Integral method, 7
Integral momentum equation, 34
Internal heating, 132, 197–204

Kinematic viscosity, 82
Kurganov-Petukhov equations, 131, 160

Laminarization, 132, 154
Leading edge, shape of, 179–181
Least squares method of, 113
Logarithmic distribution, 152
Loop oscillograph, 41
Ludwieg-Tillman formula, 82
Lyon-type integrals, 3

Maclaurin series, 90
Manometer, water-type, 17
Meshes, 16
Mixing:
  centrifugal forces and, 148
  equation for, 58
  turbulent flow and, 70
  velocity gradient and, 58
Molecular viscosity, 152
Momentum equation, 34, 37, 78

Navier-Stokes equations, 56
Newton-Richman law, 43
Nonisothermal flow, 10, 148
Nonlinear equations, 114
Nonseparated flow, 138
Nuclear reactors, 1–2, 141
Nusselt numbers, 2
  adiabatic temperatures and, 4
  for constant property flow, 120, 133, 160
  formula for, 2, 144
  heated wall, 5
  for inner tubes, 160
  Lyon integrals and, 3
  obstacles and, 136, 138
  one-sided heating and, 2
  physical properties variation and, 111
  power law for, 99
  Reynolds number and, 143

Nusselt numbers (*Cont.*):
   thermal entrance length and, 7
   wall temperatures and, 142

Obstacles, in annulus, 134–138
One-sided heating, 2, 110, 126, 131, 161
Outer tube, 110, 130
   insulation shield, 120
   temperatures of, 134–135
Oval geometry, 144

Petukhov correlation, 50
Petukhov-Popov equation, 7
Petukhov-Royzen equation, 118, 121–122, 128, 160
Physical properties, variation of (*see* Variable properties)
Pipe flows (*see* Cylinders; Tubes)
Pitot-Prandtl tubes, 33–34
Plates (*see* Flat plates)
Power-law equations, 62, 84, 96, 99, 112
   exponents in, 88, 139
   for friction, 87, 140
Prandtl number, 70
   boundary layer and, 152
   entrance region, 160
Prandtl resistance law, 140
Prandtl theory, 58
Pressure drop, 35, 45
   determination of, 46
   friction-induced, 133
   at high Re, 18
   on protruding plank, 36
   sensors for, 19
   transfer measurement and, 30
   water-type, 17
Pressure gradient, 57
Pressure-tap hole, 21
Preston tube, 36
Projecting plank method, 24, 36–38, 155
Projection method, 79

Quasi-stabilization flow, 7

Reattachment region, 136
Reliability, geometry and, 141
Reynolds number, 97
   curvature and, 103

diameter ratio, 2
enhancement methods and, 95
incompressible flows and, 105
physical properties variation and, 111
power exponent of, 104, 137–138
self-similarity and, 78
skin friction coefficients and, 90
stabilization and, 146
surface curvature and, 102
turbulence intensity and, 157
virtual origin and, 78
viscosity and, 16
Ring, on inner tube, 109
Rod bundles, 124
Rotation, of tubes, 148
Roughness elements, 106
Royzen correlations, 50

Scaling factors, 58
Scaling temperature, 70
Schultz-Grunow law, 80
Secondary flows, 148
Self-similar behavior, 55, 62, 78
Separated flow, 135
Shape factors, 82
Shear stresses, 12, 57, 59
   Blasius law for, 87
   distribution of, 70
   skin friction and, 90
   turbulence and, 71
   at wall, 60
Skin friction (*see* Friction)
Slot leakage, 35
Spacing grids, 135, 138
Sparrow method, 58, 82, 89
Stabilization length, 117
Stagnation enthalpy, 8
Stanton tube, 36
Static-pressure measurements, 139
Stresses, 59
Superposition, 3, 10
Surface curvature (*see* Curvature, of surfaces)
Surface drag, 24, 33
   pressure drop and, 35

Teflon packing, 26
Temperature:
   axial distribution, 92
   hydraulic drag and, 139–140

Temperature (*Cont.*):
  power law for, 84, 99
  skin friction and, 84
  (*See also* Temperature factor; Thermocouples; Wall, law of)
Temperature factor, 115, 122, 126, 137
  cylindrical geometry and, 154
  effect of, 113
  friction and, 86
  heat transfer and, 7
  obstacles and, 138
  power exponent of, 162
  surface curvature and, 95
Tensometric method, 39, 42
Thermal entrance region (*see* Entrance region)
Thermocouples:
  in annulus walls, 115
  calibration of, 51
  chromel-alumel, 19, 24
  copper-constantan, 21, 51
  emf values of, 31
  micrometer-screw mechanism, 24
  platinum-rhodium, 51
  shielded, 33
  traversing, 24
  in tube, 23
Thermodynamic critical temperature, 8
Thin flow region, 57
Thin laminar sublayer, 152
Transition region, 152–155
  velocity field, 66
Translational motion, 152
Triangular flow, 150
Tubes, 7, 130
  centrifugal forces in, 148
  circular, 2
  electrical resistance of, 111
  fully developed region, 146
  hydraulic drag in, 141
  oval-shaped, 145–147
  power law for, 111
  smooth, 140
  temperature along, 133
  thermocouples in, 115
  three-lobed, 142
  twisting pitch of, 144–147
  (*See also* Cylinders; Inner tubes; Outer tube; specific shapes, types)
Turbulence:
  boundary layer and, 37, 151
  enhanced flow and, 71
  entrance region and, 91
  equations for, 1
  freestream, 91–94, 107, 155–159
  fully developed flows and, 91
  grids for, 39
  intensity of, 39, 42
  momentum transfer in, 77
  shear stress and, 57, 71
  skin friction and, 38
  transfer coefficients and, 59, 66
  virtual origins and, 78, 158
  (*See also* Turbulence promoters)
Turbulence promoters, 55, 72, 74, 171
  auger-shaped, 141
  boundary layer and, 73
  decay of effects, 74
  effectiveness of, 74
  physical properties variation, 106
  power exponent of, 10
  roughness of, 106
  surface-type, 23, 74, 106
  temperature variation and, 149
Twisting, effects of, 146–149
Two-sided heating, 5, 15, 132–133, 158, 213–214

Universal coordinates, 61, 65, 69

Vacuum, insulation and, 111, 120
Vacuum chamber tests, 30
Variable properties, 47, 73, 119, 145
  boundary layer and, 129
  circular tubes and, 10, 131
  curvature and, 9
  drag and, 161
  effects, 6–8, 113
  friction and, 162
  heated inlet and, 8
  high-rate heating, 132
  high-temperature turbulence, 157
  Nusselt number and, 112
  one-sided heating and, 123, 131
  power laws, 85
  stabilization and, 132, 158
  surface geometry and, 124, 149
  temperature factor, 6, 9, 129, 155, 169
  turbulence and, 74, 157
  velocity profiles and, 65, 71, 94, 155
Velocity profiles, 7, 21, 33
  distortion of, 73
  free stream turbulence and, 155

Velocity profiles (*Cont.*):
   logarithmic, 152
   1/7 law, 87
   probe for, 34
   turbulence promoter and, 75
   for variable properties, 65, 71, 94, 155
   velocity-defect law, 152
Virtual origins, 78
Viscosity:
   boundary layer and, 55, 65
   curvature and, 87
   laminar, 59
   skin friction, 87
   sublayer and, 152, 154
   turbulence transport coefficient and, 77
   variable properties and, 155

Wall:
   flow at, 57
   heat flux ratios, 123
   heat transfer through, 93
   law of, 60, 62, 66, 69, 83, 152–153
   Nusselt numbers for, 2
   obstacles near, 136
   physical properties variation and, 6, 127
   shear stresses in, 38, 57
   superposition principle, 3
   temperature values near, 69, 113, 154
   thin flow region, 57
   thin viscous sublayer, 152
   variable physical properties, 6, 66, 127
   velocity profile near, 58
Water cooling loop, 16
Wind tunnels, 18

RAYMOND H. FOGLER LIBRARY
DATE DUE

BOOKS ARE SUBJECT TO
RECALL AFTER TWO WEEKS

7 1987